Printing of Textile Substrates Machineries and Methods

Printing of Textile Substrates Machineries and Methods

Mathews Kolanjikombil

Senior Textile Technologist

CRC Press
Taylor & Francis Group
Boca Raton London New York

CRC Press is an imprint of the
Taylor & Francis Group, an **informa** business

WOODHEAD PUBLISHING INDIA PVT LTD

New Delhi

First published 2025
by CRC Press
4 Park Square, Milton Park, Abingdon, Oxon, OX14 4RN

and CRC Press
2385 NW Executive Center Drive, Suite 320, Boca Raton FL 33431

© 2025 Woodhead Publishing India Pvt. Ltd.

CRC Press is an imprint of Informa UK Limited

The right of Mathews Kolanjikombil to be identified as author of this work has been asserted in accordance with sections 77 and 78 of the Copyright, Designs and Patents Act 1988.

British Library Cataloguing-in-Publication Data
A catalogue record for this book is available from the British Library

Print edition not for sale in South Asia (India, Sri Lanka, Nepal, Bangladesh, Pakistan or Bhutan)

ISBN13: 9781032840642 (hbk)
ISBN13: 9781032840659 (pbk)
ISBN13: 9781003511038 (ebk)

DOI: 10.1201/9781003511038

Typeset in Times New Roman
by Bhumi Graphics, New Delhi

This book is dedicated to my brothers M/s K. Thomas and K. Johnson and my Sister Mrs. Mercy Alex who have been great support and strength in all ups and downs in my life.

Contents

Preface

This is the 6th Book in the series. Printing is an area where lots of secrets exist. When I started my career way back in the 1970s, Printing Managers, especially those in famous print houses, never used to reveal the recipes to the subordinates. Either they themselves weigh the recipes secretly, or they keep some confidant to weigh the materials for or often some chemicals used to be labelled in codes so that the secrets won't be exposed. Today things are different, thanks to the various dyes and chemical manufacturers and their experts, there have been a lot of technical knowledge been transferred to the shop floor technicians and are not secret anymore.

In the first half of 20th century (1912), Dr. Edmund Knecht and James Best Fothergill has written a book gathering most of the information of printing at that time in 'Principles and Practice of Textile Printing'. After this, I have not seen such an effort made in printing area. Printing Technology has marched forward so much that those recipes are not much relevant today and new type of printing machines and many new fibres have been introduced further. But I have not seen many books written by printers giving the recipes, helping the novice and experienced technicians in their works. By this book I have tried to collect and put down all the recipes and techniques of printing and methods necessary for the printer for managing his department.

In the volume I of Printing of Textile Substrates I have included the historical details and development of printing technique and machinery, design making details, engraving let it be flat or rotary screen, different styles of printing, ingredients of print pastes, including detailed notes on the machines and equipment and technology used in printing. Some details given may not be much in practice now, but are included for academic interest of students or printing enthusiasts.

Any suggestions or corrections are welcome and can be included in the next edition. I request the readers to please inform me of such omissions and corrections.

Hoping this book will be accepted by all related technicians and students as in the case of my other books.

Mathews Kolanjikombil
Bangalore

1
Introduction

1.1   Historical background

Printing is technically defined as localised dyeing. Incidentally, in the first century, the historian Pliny has written about Indian textile printing art, unaware of this definition as, 'They dye cloth in a wonderful way. First, they place colourless substances on the cloth then dip in dye liquids, after which it is seen that the dye has taken only where these substances have been applied. Also, without changing the dye liquid, different colours are dyed where different (basic) material has been applied.' He would have been explaining about a resist style of printing, if the explanation is that dye has been taken only where these substances are not applied. It can be a mistake of observation. Or it can be that the colourless substances applied on the material can be mordant.

The weaving of fabric would have developed from Egyptian plait technique (Sprang system). But in this system only one type of threads were used (warp). In Polynesia, plaited articles for general use such as floor mats, baskets, sails and fans were made using coconut fibres. They employed the simplest form of plaiting in which each crossing weft over and under one warp element, resulting in a chess board design. The natives worked over the body with black or red strands to form patterns in the front only. Geometrical designs weremade on boarders or spread all over made by taniko technique. In basketry, the Bankans knew twill weave in stone age.

To make the fabrics more attractive ornamentation was done personally in ancient times and slowly it was developed as an art form and exchanged by barter system by people who were specialists in this area. Often it was presented to kings and officials to please them and to get gifts. These techniques of ornamentation by plaiting or weaving were known in many parts of the world. But ornamentations by colouration of textiles were a technique known to Indians and Chinese from immemorial. The colours used were of vegetable origin and Indians were more into cotton colouration and Chinese into silk and cotton.

Coming to printing, one of the earliest methods of decorating fabrics with colourful designs must have been a painting or printing process. The art of printing by stamping or actual printing supposed to have originated in the eastern world in India and China mainly. Hand printing with wooden blocks was known to Indians even before 500 B.C. But discoveries in the Pyramids, and other Egyptian tombs, of fragments of cloth undoubtedly decorated by some process other than dyeing alone prove conclusively that the ancient Egyptians also were acquainted with some form of the art. But these can be traded from other parts of the world also, but confirms that the ornamentation by printing was known to human time immemorial.

Several researches and investigations have shown that the Indians and Chinese have to be given the credit for the invention of block printing. Many travellers like Pliny the Elder (A.D. 23 to A.D. 79) points out that the Egyptians were producing designs on the dyed cloth by some unknown process. But it is on record that letterpress printing from engraved wood blockwas extensivelypracticed in China nearly two thousand years ago, and, further, that valuable

Indian fabrics printed in colour were known to, and esteemed by, the Romansin the days of the Empire. From this latter information, itshows that the Indians had already gained a worldwide reputation for the excellence oftheir printed cottons. Thus, it may be concluded that block printing, even at this early date, was an old-established industry in India and that any contemporary work produced in Egypt was either not executed by the process used by the Indians, or was still in the experimental stage, from which it has emerged as an industry. The idea may have been first used by Assyrians who employed carved wooden stamps for impressing cuneiform symbols on wet clay tablets and bricks. But in fact, the rival claims of China and India to the claim of the inventor of block printing, which was the primitive form of printing, are based upon pure circumstantial evidence alone. Oddly, after so much of development in this area, block printing is still used in in India for special handmade printed materials.

In the course of years, certain methods have been developed which permit mass production. This is due to the great variety in design which the trade expects of textile fabrics. Wherever the demand permits the production of fabrics with individual character and artistic originality, printing methods are used which make such an individual production economically possible. This is the reason why block printing is still in use at the present time. They used wood blocks for carving the design which was then transferred to the cloth with the help of a suitable colour pad. Very fine artistic work was produced by block printing, which was highly appreciated by Imperial Rome. Enormous trade, in attractive printed calicos, was carried on between India and Europe by land through Persia, Asia Minor to Germany and France. Later the art of printing also followed the same route and France was the first country in Europe to start calico printing in the second half of the seventeenth century. At the close of the same Century, it was introduced into England.

Where textile printing originated whether in India, Egypt, and China or elsewhere-is not clear. Printing with wooden blocks were practiced by the Hindus and the Chineseeven earlier than 500 B.C. Ancient Egyptians also has the knowledge of decorating the fine linen cloth with small motifs and borders in blue, amber and brown which was evident from the cloths obtained from mummies. It is presumed that the first patterncould be produced by applying beeswax as a resist to the dye liquoror by tying threads tightly aroundthe areas to be resisted. In earlier days, printing effects were made on fabric by dyeing methods like batik, tie and dye methods or by hand painting. Then printing has developed through machine method to most productive method like Rotary Printing machines today.

We can look at printing history through the following developments:

1. Hand Painting
2. Resist Printing - Stencilling
3. Resist Printing - Batik, Tie and Dye, Plangi
5. Hand Block Printing
6. Engraved Roller
7. Silk Screen Printing
8. Rotary Screen Printing
9. Ink-Jet Printing

As early as the first century AD hand painted fabric has been produced in India and the use of madder was by then long established. By the 14thcentury the use of wooden blocks in printing was established in France, Italy and Germany, but mineral pigments were used rather than dyes. In the early 17thcentury hand-painted Indian cotton reaching Europe in big quantities. They were colourful and colour fast, and introduced a richness of novel and stimulated design styles. Paisley designs for example, were derived directly from one of these styles.

The English term "printing "was established in the 18thcentury and comes from a Latin word meaning pressure. Around that time printing was normally done using wood blocks, which had raised printing surfaces, which needed pressure to obtain good contact between the fabric and the colour on the block.

1.1.1   Hand painting

First trial of making designs on fabric would have been probably hand painting. But this needed a lot of skill and look lot of time to finish one design on fabric. This made people venture on different methods to print a design on fabric faster. Thus, different methods of printing evolved.

1.1.2   Stencilling

Even though stencilling is used for different purposes from time immemorial, and is being used even today, may not be for textiles. But the credit goes to Japanese who had developed it as art of printing textiles. The technique and beauty of their designs have not been equalled by any other craftsmen. For centuries these people have been making imprints on fabrics. Whenever a baby is born, its mother gives a design or even a self-prepared stencil to the printer with instructions to use a particular combination of colours for printing the baby's first garment. Even today, though it has been adapted as a method of cloth printing by almlost all the nations of the world, yet all of them are mostly using it for printing a coarse cloth, like hangings, curtains, table-covers, etc., but the Japanese are using it for printing fine cotton muslin or fine silk cloth. They excel all other nations in this art.

Stencilling is the most exacting master of simplicity in printing. First part of the process is cutting of the stencil. Stencil plates are merely sheets of paper or thin metal, with perforations arranged to form a pattern. They are laid on the work, and colour is brushed through their cut-out parts on to whatever material shows beneath them, thus producing a coloured pattern corresponding to the shape of the perforations. Ties are not only used for outlines, but must be freely introduced, for strength, wherever long lines, extended spaces, or intricate patterns are cut, otherwise the plate will be too fragile, and apt to lie unevenly on the work, and thereby give rise to faulty execution. Although an unbroken outline cannot be stencilled, it may be reserved or left uncut in cases where both ground and pattern are cut out on one stencil, or where two or more plates are employed for the realisation of a multicoloured design—in which cases it (the outline) appears in the original colour of the material to be ornamented.

A design can be made in two ways. After placing the stencil on the fabric to be printed spraying liquid dye with an atomizer permits not only the usual direct colouring of the design areas, but also of resist stencilling, in which a light design is produced on a dark background. A small dark design on a large light background is stringy and thin. The light seems to eat into the edge of the design and minimise its importance. On the other hand, a small light design on a large dark background is magnified in importance. While for proper control in either case we must adjust the space relations of the design and background areas, yet it is equally important that one be able to adjust the colour relations of the design and background areas.

The process for "resist" stencilling is as follows: The stencil is laid upon the surface to be decorated, and the open pattern is carefully covered with a thin layer of library paste or paste made from flour and salt water. Flour added to a solution of salt in water is the best preparation we have found. A palette knife, or a case knife may be used to spread this paste. The stencil is at once lifted and the colour desired for the background is sprayed in a flat tone over the entire surface. The paste acts as a resist, preventing the penetration of the colour. The entire surface is immediately wet with cold water and the resist washed off. The stencil is carefully and thoroughly cleaned in the same way and then pressed and dried. The stencilling prints were done in olden days especially for book covers and end paper designs, or mats, where the pattern does not partake of the nature of a repeat, as it is better to remove the resist while wet. It is interesting to stencil the open pattern in one or more colours, then apply the resist and give another colour to the background.

Stencil designs

Most interesting stencilling is done with two or more stencils. To make these, a stencil should be cut for each colour. Use the original design sheet for one colour. Transfer other colour areas to new sheets. Make all sheets, including the tracing paper, the same size. Before tracing, lay sheets and tracing paper together and punch coinciding holes in the upper corners. Keep these holes coincident during the process of tracing. By means of these holes the respective stencils are easily applied so that the colour scheme is accurately reproduced.

1.1.2.1   Stencil making

Stencils are made of paper or metal sheet (copper or Zinc). For paper stencil the paper is made transparent or a transparent paper can be used. The paper is placed over the full design and the required colour is only traced onto this paper and it is then placed on a smooth surface e.g., a thick glass plate and the parts are cut out with a sharp knife. A second colour is traced on another paper and cut and so on with all the colours. In case of metallic plates (Zinc sheet is more serviceable than a copper sheet). The design can be transferred onto the metals sheet either white paper is pasted on the metal plate and the design is traced over it by putting a carbon-paper in between the plate and the design paper or a paper stencil of the design is cut out and the cut stencil is directly pasted on the plate or the design is brushed through the cut parts after placing the stencil on the plate carrying gummed white paper on its surface. Cutting of the plate is carried out in two ways - (a) With a hand fret-saw or (b) with a machine fret-saw.

Colour can be applied after placing the stencil on the material to be printed by brushing, sponging, dabbing, spraying by (a) hand or (b) machine.

1.1.2.2   Features of stencil prints

A stencil is much easier to handle and adjust than a block, since it is lighter than the block and the colour is applied directly after placing the stencil in the right position on the cloth to be printed. Because of lightness the designer is not restricted in the size of his pattern. The printer can, by proper handling of the colour-brush, show the skill in the grading of shades and in the production of mixed shades from one stencil. An artist printer can show his art or skill of shading effects in stencil printing which is not possible in block printing. Fine or broad blotch effects can be produced by or included in, one and the same stencil which is not possible in block printing. In stencilling, colour is applied directly to the cloth through the open parts of the design plate. In block printing it is applied indirectly; the wooden' block first receives the colour which is then transferred to the cloth. But there are disadvantages like an uninterrupted circle, or an outline cannot be produced by stencilling, as it is necessary to keep some ties. But a skilled artist can, however, skilfully arrange places which are to be covered up by succeeding stencils of the design or design in such a way that does not need ties, etc. (See designs above).

Paper or metal stencils are placed on the cloth to be printed and the colour is brushed or sprayed over them. To avoid the ties marks Japanese introduced a new method in which a stencil in a twofold paper so that two identical stencils of the same design are simultaneously produced. The double paper is then unfolded, and a solution of glue is applied to the inner sides of the folded surfaces. Raw silk threads are then laid over one of the glued surfaces sufficiently close to form a net-work pattern. The other part of the stencil is folded over again as before and pressed and ironed so that whole arrangement becomes one stencil. When the colour pastes are brushed, the loose structured silk does not obstruct the passage of the colour underneath them as they are above the cloth resting on the bottom stencil. Thus, making the design object print on the fabric without any tie marks. Either one colour is applied with one stencil or blending of two or more colours is also possible in one stencil-an advantage not available in any other method of printing. A special feature of using both sides of the stencil gives some advantages. For e.g., if a circle has to be printed the stencil can be cut as a semicircle or even as quadrant and printed flipping the stencil two times (in case of semicircle) or four times (in case of quadrant) The feature of using both sides of a printing device is not available in any other method of printing.

1.1.2.3   Advantages of stencil printing

1. Any broad pattern can be easily printed.
2. A continuous line design can be produced with greater accuracy than by block printing.
3. Unlike block printing, repeats can be joined up accurately producing no misfits.
4. The cutting of stencil does not require the delicate skill necessary for block cutting. Carelessness in cutting the latter can result in the wastage of an expensive wooden block. Mistakes made in the cutting of a stencil can be easily remedied by soldering up the spoiled metal area or by repairing a paper stencil with glue.
5. A detached pattern can be more successfully printed with a stencil than with a block.
6. A small order of goods requiring a small pattern can be economically executed by using a paper stencil.
7. The stencil allows the printer great latitude to exhibit his artistic talent and individuality in gradation of colours and tints.

1.1.3   Spray printing (Aerograph printing)

Spray printing has found a limited use for pattern printing of natural silk fabrics. The pattern is applied on the fabric by spraying a dye solution through a stencil with holes cut in the form of the pattern required. An aerograph, which is the device used for spraying the dye onto the fabric, is shown in fig. The dye liquor is poured into tank 1 and branch pipe 2 is connected by a rubber.

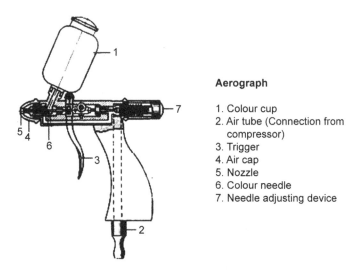

Aerograph

1. Colour cup
2. Air tube (Connection from compressor)
3. Trigger
4. Air cap
5. Nozzle
6. Colour needle
7. Needle adjusting device

Pipe with the compressed air bulb. By depressing lever 3, compressed air is let into the aerograph and at the same time needle 4 closing the hole for ink passage is shifted aside. The air exhausted through annular aperture 5 takes with it the dye liquor coming from tank 1 and sprays it on the fabric. By lowering or raising the aerograph above the fabric, it is possible to obtain colours of different intensity with a gradual transition from one tint to another. In this way, it is possible to obtain different shades of colour using one stencil. By employing several stencils, it is possible to achieve such effects which cannot be attained by other printing techniques. The aerograph makes possible gradual transitions of colour from dark to light with blending of various colours. The operator should not only be skilled in the art, but should also have a good artistic taste. Printing with an aerograph is carried out on special rotating tables provided with glass hoods and exhaust fans for exhausting the excess of sprayed dye. Printing textiles by spraying is a technique of low efficiency and is not employed for fabrics in large assortments. It gives a unique effect in the production of separate articles and for the execution of individual art works.

The colour used in aerograph printing is much thinner than .in any other method of printing. If any thickening is necessary at all it is advisable to use a soluble starch in place of gum or ordinary starch, as a thickening in preparation of aerograph paste. The nozzle of the spray is so fine that it gets easily choked by thickening agents, causing a lot of trouble to the printer.

1.1.4   Hand block printing

A design is carved in raised form into a wooden block and while using the block is placed on a colour pad which contains the necessary printing colour and fixing agents. Once the raised pattern on the block becomes uniformly charged with printing colour it is placed on the cloth to be printed and is smartly tapped on the back of the block either with the fist or with a wooden mallet. Thus, the coloured impressions are transferred to the fabric. Only one colour, at a time, may be applied from a single block (there is method of getting a multicolour effect with one block at a time, which is described later) and it is necessary, while producing: multi-coloured pattern, to employ a separate block prepared with the requisite alternative colour so that the complete pattern is built up by stages. An intelligent designer can minimise the number of blocks required for a given pattern, by arranging for some of the colours to overlap e.g., allowing a yellow to fall on a blue in some parts of the design, to give a green. It is usual for the first block to print the major outlines of the design and for the following blocks to fillin other colours to build up the desired design.

1.1.4.1   The block

Wood employed in block manufacture must be of a very good quality: freedom from knots is essential and fineness and regularity of grain as well as adequate seasoning are important factors. The most suitable woods for this purpose are Box-wood, Holly-wood, Lime-wood, Sycamore, Shisham and Teakwood. These possess a fine grain and have a good and durable cutting surface.

The wood is usually cut across the grains in the form of irregular circular pieces 4 to 5 inches thick. One face of this cross section is smoothened by filing or rubbing on a fine sand spread over a uniform stone surface or cement slab. The entire surface of the piece must be in one plane. This is verified or checked by sprinkling fine chalk powder uniformly over the surface of a thick glass-sheet. The smooth end surface of the block is placed over the chalk coated glass-sheet. It is gently pressed to bring in closer contact with the chalk surface. The block is then lifted and the face examined. If it has picked up a uniform coating of chalk all over its surface, then the surface is in one plane. If there are some parts left unchalked, they are below the general level of the surface. The block is rubbed again on the sand surface and tested. These operations are repeated several times till the whole surface comes in one level.

After a smooth uniform surface is obtained in one plane, the next step is to transfer the design on to it. The design to' be transferred is first drawn on a piece of paper with all the colours duly filled in. The paper is then placed on the smoothened surface of the block, the design-side touching the wood surface i.e., face downward. The paper is then secured in position by driving pins or wire nails at the four corners of the paper.

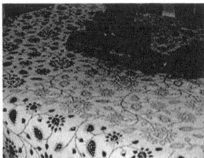

Print block making   Print with a block

The outline, which is generally in black, is first punched through with light strokes on a sharp steel pointer. This transfers the design on the wooden surface in the form of fine pin holes. After the whole outline gets punched the four pins or nails at the four corners are pulled out and the design paper is lifted away from the block. This procedure is followed for each colour in the design till all the coloured parts are transferred on different blocks-one colour per block. Four coloured designs will require four blocks. One and the same design paper is employed for all the blocks and for all the colours in it. This procedure avoids the possibility of a misfit of different coloured parts of the design in printing the prepared blocks. Moreover, the perforations in the paper clearly indicate the parts which have been already transferred, this avoids duplication.

Printing blocks   Paisley designs on printing blocks

Size of Block:Printing blocks may be of different sizes and may vary from, for example, 3 in. x 4 in. upto 9 in. x 12 in. in width and length respectively. Blocks larger than about 12 in. square, are not easily manageable by hand. Blocks of large size, e.g., 36 in. square, are often employed in the printing of table-covers, kerchiefs, and similar articles, requiring no repeat of the design. Such blocks cannot be hand operated and these are slung from the ceiling of the printing room (or from a convenient frame) on elastic or spring suspenders which allow the blocks to be lowered for printing or for charging with colour from an adjacent colour pad of the required size. During use, the block is lowered on,to the colour pad and then lifted and lowered into position on the cloth to be printed; it is then gently tapped all over its back with a mallet so that a uniform coloured impression of the design is obtained. Where, more than one colour are to be printed, the table carrying the cloth is moved on wheels to a position beneath the secondblock and adjacent to a second colour-pad.

1.1.4.2   Features of block printing

It is very difficult to produce very fine lines or small dots with sharp outlines except perhaps with a new block. The cutting of very fine lines or dots, on even closely grained wood, is a tedious and difficult operation and even if the craftsman succeeds in producing a design in this way, the block will very quickly get damaged in use by constant knocking. and washing. To strengthen the fine lines or dots the craftsman resorts to the method of cutting the design out of the wood, with an obtuse slant and avoids any formation of perpendicular walls. In this way, the design in relief, is narrower at the face than at the base and this enables the pattern to withstand the strain and stress of the printing, cleaning, and washing operations for longer periods. The process is very laborious. Mainly direct prints are only done by Block printing. Hence, there are limitations for the colours, print designs, etc. which can be done by block printing.

1.1.4.3   Developments of the wooden block

As usual, for any prints more and more intricate demands have come from customers, the manufacturers started reviewing the ways to print fine lines and objects. Thus, the restrictions of wooden blocks involving loss of fineness of lines were overcome by preparing fine and more delicate forms from copper or brass strips and wires. The thicker parts of the design are cut out of the wood in the normal way, but finer detail is produced by cutting into the wood a seat for metal strips while wires are directly driven into the wood for obtaining a fine dot design, such blocks are styled as "Combined Blocks."

The most important part of the block manufacturer was to make various parts of the raised part of the design same level whether it is metal or wood so that all parts of the design can become uniformly charged with colour and give an even transfer to the cloth. In a effort to reduce the cost and weight of the blocks, especially big were made by making the design on good quality wood and using a cheaper and lighter wood like deal or pine, for the handle and back support ot the block and screwed together.

Since the wood required for making blocks were expensive not freely available in Europe, there evolved a new method of making blocks to make the block cheaper. Although a block may be about 40 mm (1–5 in) thick the cutting necessary to raise the design in relief seldom penetrates deeper than 5 mm (0–2 in). By bonding together, a layer of carefully selected, fairly hard, close-grained wood (e.g., sycamore, pear or lime) with cheaper woods, economies were possible. The layers in the bonded wood are arranged so that the grain runs in directions at right angles to each other. The size of block may vary slightly to allow for a design to be fitted in and have an irregular outline to encompass the pattern area. Because the block must be lifted and transferred to the cloth repeatedly during the printing operation, its weight, and therefore its average size, must be restricted. Dimensions greatly in excess of 500 mm × 500 mm × 40 mm (18 in × 18 in × 1–5 in) are unusual but may be required for special designs.

In India where block printing is still in practice, the method of making blocks are different. Instead of multi-layer blocks in India blocks are made from a single piece of wood and are smaller in size. The type of design and fabric to be printed determines the method of block making, particularly the area of a given colour required. It is difficult to apply printing paste evenly over large areas from a wooden surface; therefore, any large, plain areas of the design are usually recessed, leaving a wooden outline wall, the cut-out portions then being filled in with a hard wool felt pad. Again, very fine lines in the design are fragile if left in wood, and these are therefore built-up with copper strip. Double copper strip outlines interleaved with felt are also widely employed.

1.1.4.4 Stuffed blocks

When broad and solid portions are required to be produced in a block design, e.g., large leaves, squares as those of diamonds in playing cards, it is difficult to obtain a uniform pick-up of colour from the pad. This difficulty is overcome by a device known as "Stuffing" and the resulting blocks are called "Stuffed Blocks". These are prepared by cutting out the outline or alternatively, the outline may be built up with metal strips, driven into a wooden block. The hollow area within the outline is then packed with felt or loose wool. The operations of packing with felt and with loose wool are known as "Felting" and "Fluffing" respectively.

Typical hand black printind designs

1.1.4.5   Printing tables

The tables are usually prepared from flat planks of even and uniform surface. They are supported either on thick wooden or iron frame or at times made from flat slabs of stone slate, concrete and occasionally of iron too. The tables are covered with thick pieces of canvas or hessian cloth or printer's blanket, faced with cotton pieces. The printing table surface must be uniformly resilient so that it yields easily when the block is hammered on to the fabric. The width and length (usually 5 m) of the tables are naturally governed by the local conditions and styles. An iron axle is passed through the centre and the roller is placed at one end of the printing table on a suitable stand. The top end of the cloth is then drawn and placed over the table. After printing, one table length with all colours of the design, the piece is drawn at the other end by some mechanical device and exposed to air for drying while the adjacent portion is spread for printing on the table and printed.

1.1.4.6   Block printing

First part of the printing operation is the gumming down of the fabric to be printed.

The cloth is wound round a wooden tube, called a shell (Roll). A suitable gum paste is then placed at the end of the table and spread along it in as thin a film as possible, using a rubber squeegee. The roll of cloth is unrolled along the table so gummed, and lightly pressed into place either by hand or by careful ironing. As an alternative to gumming, plasticised resin preparations are available. Their use avoids the need to wash the table each time a fresh length of cloth is to be printed. For gumming, a gumming roller also can be used. Sometimes, for shorter lengths pinning on the sides are also practised.

For applying the printing paste on the block a shallow box of dimensions 0-6 m × 0–9 m (2 ft × 3 ft) fitted on a table that can be wheeled up and down alongside the printing table is used. It is called a swimming tub, is placed some old semi-fluid starch paste or other thickening paste, and upon this floats a wooden-sided tray, the bottom of which consists of a sheet of waterproof fabric. On this is placed a piece of blanket or a sieve impregnated with the colour paste to be printed (the sieve is simply the blanket stretched in a light wooden frame).

1.1.4.7 Printing operation

First, the block is placed in the above mentioned colour box over the blanket, so that it picks up an even layer of paste. The block is placed on the first position to be printed and struck by the handle of a small but heavy mallet or maul to affect even transfer of print paste and thorough penetration. The sieve is resupplied evenly with colour bymeans of a small whitewash brush called a tiering brush and the next print is done. The process is repeated and one colour in the print is completed if it is a multicolour design. The sieve is then removed, replaced by a clean one, and the second colour is applied in the same way, the process being repeated until all the different colours have been separately applied. Care has to be taken to print each joint or repeat accurately. To enable the printer to fit his block lays in correct register on the cloth,'pitch pins' are often fixed round the block sides. These pins are arrangeto coincide with certain well defined points in the pattern printed during the previous lay. The first row of impressions made must be precisely at a right angle to the cloth selvedge, otherwise as printing along the cloth progresses the pattern would be found to gradually run off to one side. On completion of printing, the goods are dried and given different treatments as per style of printing.

A 4 colour wooden block

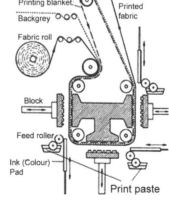

Perrotine

Hand block printing has been used in Europe from the 18th and 19thcentury until the mid-20th. (Switzerland for wool and silk scarves). It still can be seen in India and Indonesia in some traditional style printing. These methods of printing like Batik, tie and dye, was too laborious and less productive. They were produced only by specially skilled personnel and was not universally practiced.

1.1.4.8   Kalam Kari prints

When we are talking about oldtime fabric printing, we cannot miss the ancient Kalamkari prints which used to be exceptionally attractive and mage completely with dyes of plant origin. This method of printing was more popular in Srikalahasti in the Chitoor district Machlipatnam in Krishna district Andhra Pradesh,and probably some villages in Tamil Nadu (Sickinaickanpettai) by migrants from Andhra Pradesh and over the generations has constituted their livelihood. The preparation and printing of the fabric used to follow special routs which made the print attractive to most kings and members of the royal family and gave many artisans their patronage and later was promoted during the 18th century, as a decorative design on clothing by Britishers in India.

Kalamkari is an ancient style of hand painting done on cotton or silk fabric with a tamarind pen, using natural dyes. The word Kalamkari is derived from a Persian word where 'kalam' means pen and 'kari' refers to craftsmanship. This art involves 23 tedious steps of dyeing, bleaching, hand painting, block printing, starching, cleaning and more. Motifs drawn in Kalamkari spans from flowers, peacock, paisleys to divine characters of Hindu epics like Mahabharata and Ramayana.

Centuries ago, folk singers and painters used to wander from one village to other, narrating stories of Hindu mythology to the village people. But with course of time, the process of telling tales transformed into canvas painting and that's when Kalamkari art first saw the light of day. This colourful art dates back to more than 3000 B.C. According to historians, fabric samples depicting Kalamkari art was found at the archaeological sites of Mohenjo-Daro.

In the Middle Ages, this artwas promoted by the wealthy Golconda sultanate, Hyderabad. The Mughals who patronised this craft in the Coromandel and Golconda province called the practitioners of this craft "qualamkars", from which the term "kalamkari" evolved.The Pedana Kalamkari craft made at Pedana nearby Machilipatnam in Krishna district, Andhra Pradesh, evolved with the patronage of the Mughals and the Golconda sultanate. Owing to the said patronage, this school was influenced by Persian art under Islamic rule.

Kalamkari art has been practiced by many families in Andhra Pradesh, some villages in TamilNadu (Sickinaickanpettai) by migrants from Telugu speaking families and over the generations has constituted their livelihood. Kalamkari had a certain decline, then it was revived in India and abroad for its craftsmanship. Since the 18th century, the British have enjoyed the decorative element for clothing.

The process of making Kalamkari involves 23 steps. From natural process of bleaching the fabric, softening it, sun drying, preparing natural dyes, hand painting, to the processes of air drying and washing, the entire procedure is a process which requires precision and an eye for detailing.As mentioned earlier, the old method of production of Kalamkari prints, the preparation of the fabric also used to be a long process. Cotton fabric used for Kalamkari is first treated with a solution of cow dung and bleach. After keeping the fabric in this solution for hours, the fabric gets a uniform off-white colour. After this, the cotton fabric is immersed in a mixture of buffalo milk and Myrobalans (the dried astringent fruit of several East Indian trees of the genus Terminalia (as T. Chebula and T. Bellerica used chiefly in tanning). This avoids smudging of dyes in the fabric when it is painted with natural dyes. Later, the fabric is washed under running water to get rid of the odor of buffalo milk. The fabric, likewise, is washed twenty times and dried under the sun. Once the fabric is ready for painting, artist's sketch motifs and designs on the fabric. Post this, the Kalamkari artists prepare dyes using natural sources to fill colours within the drawings.

Process flow chart for kalam kari prints
Dyes for the cloth are obtained by extracting colours from various roots, leaves, and mineral salts of iron, tin, copper, and alum. Various effects are obtained by using cow dung, seeds, plants and crushed flowers to obtain natural dye. Along with buffalo milk, myrobalan is used in kalamkari. Myrobalan is also able to remove the odd smell of buffalo milk. The fixing agents available in the myrobolan can easily fix the dye or colour of the textile while treating the fabric. Alum is used in making natural dyes and also while treating the fabric. Alum ensures the stability of the colour in kalamkari fabric.

Incorporating minute details, the Kalamkars use tamarind twig, bamboo or date palm stick pointed at one end with a bundle of fine hair attached to this pointed end to serve as the brush or penas pen, to sketch beautiful motifs of Krishna Raas-Leela, Indian god and goddesses like Parvati, Vishnu, Shri Jaganath; designs of peacock, lotus; and scenes from the Hindu epics like Mahabharata and Ramayana. This pen is soaked in a mixture of fermented jaggery and water; one by one these are applied, then the vegetable dyes.

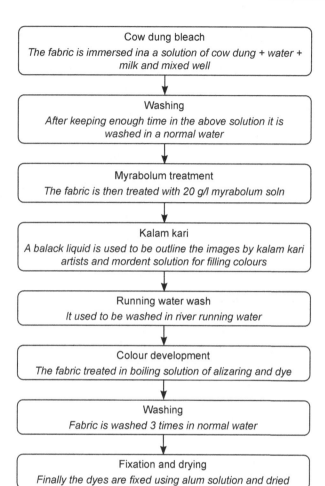

Cow dung bleach
The fabric is immersed ina a solution of cow dung + water + milk and mixed well

↓

Washing
After keeping enough time in the above solution it is washed in a normal water

↓

Myrabolum treatment
The fabric is then treated with 20 g/l myrabolum soln

↓

Kalam kari
A balack liquid is used to be outline the images by kalam kari artists and mordent solution for filling colours

↓

Running water wash
It used to be washed in river running water

↓

Colour development
The fabric treated in boiling solution of alizaring and dye

↓

Washing
Fabric is washed 3 times in normal water

↓

Fixation and drying
Finally the dyes are fixed using alum solution and dried

Dyes for the cloth are obtained by extracting colours from various roots, leaves, and mineral salts of iron, tin, copper, and alum. Various effects are obtained by using cow dung, seeds, plants and crushed flowers to obtain natural dye. Along with buffalo milk, myrobalan is used in kalamkari. Myrobalan is also able to remove the odd smell of buffalo milk. The fixing agents available in the myrobolan can easily fix the dye or colour of the textile while treating the fabric. Alum is used in making natural dyes and while treating the fabric. Alum ensures the stability of the colour in kalamkari fabric.

Different colours are applied almost like Indonesian Batik technique. The red, black, brown, and violet portions of the designs are outlined with a mordant and cloth is then placed in a bath of alizarin. The next step is to cover the cloth, except for the parts to be dyed blue, in wax, and immerse the cloth in indigo dye. The wax is then scraped off and the remaining areas are painted by hand.

1.1.5   Perrotine Machine

As the demand went higher, they manufacturers were not able to supply as per demand and newer methods to apply print on fabric was tried. At the advent of Industrial Revolution there was a tendency to replicate printing by machines so that production can be increased. Thus Perrotine was come into use in many print-works in Europe(~1850 –1950). As you can see from the fig, it is an automatic block printing machine involving mechanical movement of the charging of the sieve with colour, the loading of the block with colour, the transfer of the colour from block to cloth and the precise forward movement of the cloth into the next exact position.

Each block is worked independently of one another. The block is made with many repeats (or as one design) equal to the width of the cloth so that the full width of the cloth is printed in one Operation. Each colour position will have one block, a colour pad, a brush and a colour box. The colour pad has to and fro motion. A roller in contact with another roller partly dipped in colour paste colour to the surface of the pad. The pad, once it takes the colour from the roller in the to and fro motion applies colour on the block which at that position is disengaged from printing the fabric. In the next operation when the block fully saturated with the colour paste comes forward and gives its impression on the cloth to be printed which is resting on a three faced table. After receiving the impression, the block recedes, the cloth moves forward equal to the length of the repeat and the block is applied with colour again as explained earlier. Thus, the whole cycle of operations of colour supply, smoothing, colour application to the block and impressing the design on the cloth is repeated each time for each impression of the repeat of the design.

The machine was developed in Switzerland and was popular till the roller printing was developed. It is not very popular because of the complexity of operation and often gives defective prints even with an experienced operator.

1.1.6   Peg printing (Surface printing)

The development of printing techniques were moving towards the final roller printing machine which was the first continuous printing machine which probably was used for more than a thousand years. The peg printing machine was a more sophisticated perrotine machine and more resembling roller printing. Instead of using wooden blocks, in this machine the designs are carved on a round wooded roller giving the look of a roller with pegs attached to it. Thus, the machine got the name Peg Printing. The peg roller gets the colour applied on when it comes into contact with a cloth covered wooden roller which is kept partly dipped in colour paste. In another improved model a continuous colour supply blanket, which in turn gets the colour supply from a wooden roller partly dipped in colour paste. On the upper side of the design

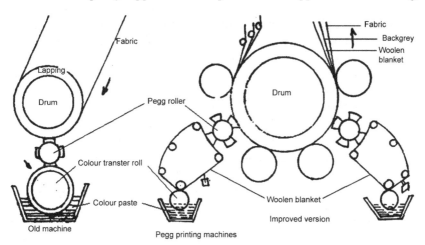

Pegg printing machines

roller there is a big drum carrying a cushion of cotton lapping round which the cloth to be printed travels. This drum is pressed at will against the design roller to get the impression transferred on to the cloth. A similar woollen blanket as is used in the colour paste transfer is arranged to run continuously round the central pressure drum to give additional cushion surface to the lapped drum. A back grey cloth is run in between this woollen blanket and the fabric to be printed to take up the impression at the back side of the printed cloth which may otherwise stain the costly woollen blanket. The back grey can be washed and reused after many times usage. Many pegs and colour paste transfer arrangement can be placed around the drum (those days may be 3 -4 colours max.) and print a multicolour design in one go.

1.1.7   Tie and dye and plangi

Technically these are all resist dyeing processes giving printing look which are unique. The origin of tie and dye or plangi technique cannot be traced to a single place. It is true that bunched and raised points of material can only be tied from thin, pliable fabrics, which, in their turn, presuppose a measure of skill in textile production. On the other hand, there is so close a resemblance between the technical basis of the plangi process and that of the wrap and tie resist methods as to suggest that plangi is not an entirely new invention either but the result of adapting an old idea to new materials. Mostly the techniques, which may be little different to one another, has been developed in different areas independently. The areas this type of dyeing/printing has been developed is spread all over the world – Polynesia and North America. Mexico and Central America, Europe. South America, Africa, Melanesia and Asia, including Japan and Indonesia. In each of the areas, there are many places where it was introduced and no two places the techniques were same. Probably the self-developed techniques were kept a secret in each area to get better value for their product. In Asia, the plangi tie and dye technique is reliably recorded in very early times. From Japan have come cloths going back to the seventh and eighth centuries A.D. They may have been imports from China. Japanese accounts refer to the technique as "yuhata" ("yu", tying together; "hata", fabric) as early as the year 667. In Central Asia (East Turkistan) scraps of Chinese cloth have been excavated which are patterned according to the plangi process and derive from the same period. From the paintings of Ajanta caves in India, it can be seen that the this type dyeing was available before the sixth and seventh centuries. Indonesian plangi cloths from Java and Bali were described in 1680 by the Dutch botanist Georg Eberhard Rumphius. According to this report, the cloths were tied up and then dyed with yellow and red kasumba (safflower) in such a manner that the entire fabric was covered with figures and stripes in red, yellow, and white. By tying portions of fabric into elevated points and subsequent red-dyeing, cloths were produced strewn with a pattern of "white marble flowers", a description which may be safely taken to refer to genuine plangi fabrics. The so-called endalan cloths of Java, which are characterised by large-dimensioned resists and by stitch-resist technique, occur in still earlier records, viz., in 1580, as goods shipped from the port of Tjilatjap, and in 1654 at the court of Mataram in Central Java. In this island, the two techniques last mentioned may therefore be even older than the plangi process proper, with which they are so often combined to-day. The inference seems even more justified as endalan cloths play an important part in the religious customs of Javanese society.

The plangi technique is one of the several patterning methods known as resist processes.

They are technically mechanical resists. The designations of this reserving method either point to the technique or describe the type of pattern obtained. "Plangi", the Malay term, is increasingly accepted by European specialists and variously explained as meaning "many-coloured" or "reserved dot". On the Indian subcontinent, the familiar terms are "bandhana" (from "band", "to bind"), and" chundri". "Bandan(n)a" has been adopted into English for certain forms of the process. In Japan, both the processes and its product are called "shibori", that is, "tied" or "knotted". The technical equipment required for plangi tie and dye work is of the simplekind. Selected portions of thefabric are protected from the action of the dye by covering and tying them with leaves, bast, string, or yarn. After dyeing, these parts emerge as reserved patterns -scattered over the material or arranged in decorative figures and groups.

They fall into six distinct groups. The first of these comprises methods by which folded portions of the material to be decorated serve to protect other parts against absorption of the dye. The second group to be distinguished also dispenses with special resist materials, the reserve patterns being obtained by knotting or plaiting the article. Like folding, however, knotting or plaiting alone gives little scope to the designer and all three devices are suited to hardly more than yarn or narrow strips of fabric. Much wider possibilities are offered by the third group of resist processes. This employs resist shields or stencils which are sewn or pasted to one side of the fabric or clipped on to it, thus producing the reserve patterns on the covered portions.

The fourth reserving process is based on the principle of wrapping cord, bast, or ribbon round folded or furled portions of a fabric or round hanks of yarn. Cloths woven from warps or weft yarns patterned in this way are known as "ikats". The methods of plangi patterning must be regarded as special forms of the wrap or tie techniques for producing reserves. Their distinctive feature consists in the open fabric being resist-tied in places only. The device of stitching thread, yarn, or bast into the material and drawing this together is known as "tritik". By their nature, both plangi and tritik always involve reserves arising from folds in the material. The same may be said of certain other types of tie resists.

The plangi-tritik patterns show up white, yellow, and red on a green ground.

The fifth group includes processes based on the application of resist materials such as mud, paste, wax, etc. After dyeing, these coatings are readily broken up and removed by tapping the fabric or dissolved in hot water. The "batik" technique, which has attained its highest development in Java, belongs to this group.

Negative reserves make up the sixth and last group. Selected parts of the fabric are pre-treated in such a way that they alone retain the dye. Some of the most notable examples of the technique are furnished by the patterning methods developed on the Indian sub-continent in conjunction with mordant colours. First, the fabric is painted or printed with the mordants, then placed in the dye bath, and finally washed, when the dye disappears from all parts not impregnated with the mordant.

As explained, the tie and dye design is produced by tying the cloth with threads at different areas and with different shapes usually geometrical. Yarn and bast are not well suited for tyingup large areas of fabric, hence Indonesiansgenerally apply pieces of dried banana leaf, and it was a method specially followed in Indonesia only. Several other methods were practiced at different areas like the Baule natives of theIvory Coastfold oblong portions of cloth zigzag-fashion, tying them up inplaces so that the colouring matter takesgreater effect on the outside of the unprotectedparts than on the inside.In this way,radiating patterns having no sharply definedcontours are produced. The effect is typicalof much plangi work in other areas as well,but its conscious use as a decorative elementseems to be restricted to Senegal, ChineseTurkistan, and Japan. Circular and diamondshapedsections of fabric are gathered radiallyso that the folds converge in a point. Thewhole of the folded portion is wound roundwith yarn or bast in such a way as to leaveoccasional gaps between successive coils. Thedye liquor is enabled to penetrate the resistedportions here and there and to dye the outerparts of the folds. Accordingly, in triangularsections of cloth with a common apex, thereserves exhibit dyed transverse stripes.In some cases, to achieve large, resisted designs of a particular shapes many articles like seeds or fruits, pebbles, wooden pegs or glass beads,etc., were tied up tightly at the required place on the fabrics and resist dyed.

In Japanese method large reserves were obtained by pushing up the parts of the fabric by pointed wooden pegs. The resulting conical forms protruded about half an inch and were well protected with waxed thread. There was a similar method used in India (Gujarat) in which a wooden block which is provided with raised iron tacks arranged according to the pattern desired is pressed on a steamed material to be dyed when it is still damp when the material is pressed upwards in the shape of the motif. The raised portion is seized and tied up with bast or waxed cotton thread. In most cases a continuous thread is often used to tie up many nearby motifs successively. The material is dyed to get the reserved design undyed.

In some places the yarn used for tying has been dyed (or probably dipped) and used while it is moist so that the tied portion is dyed by the time it is dried. This can give a different colour to the boundary of the reserved area at the same time acting as a mechanical resist for the reserved area. Nowadays after resist dyeing the areas reserved as white is again painted to different colours for ornamentation rather than dyeing again to get different coloured motifs. Another method of getting multicoloured designs by tie and dye is to start with large designs such as triangles and palmettes are painted on the fabric and their outlines stitched. Before drawing the thread tight and applying resists to the enclosed area of fabric, smaller reserves are prepared within the large patterns, stitched round, and drawn together. The unprotected portions within the primary design are dabbed with paint. Only then is the main design reserved by drawing the material together on the main thread and tying in the usual manner. This principal resist accordingly protects both the painted areas and the secondary resists it encloses.

In the earlier days all the dyes used were vegetable dyes. The ground of the plangi cloths, however, was dyed fast with lac dye, cochineal, or morinda (red), or with indigo and fustic (green). After about 1900, natural dyes were used less and less. In Indonesia they were replaced by synthetic dyes both for plangi and kembanganpatterns, whose ordinarily florid effects now often assumed an unpleasant gaudiness.

1.1.8   Batik printing

Batik is an Indonesian or basically Java native method of producing printed effect by wax resist. Classical batiks were originally prepared by the daughters of Javanese nobility who worked diligently for many months on a single sarong using only the 'tjanting'. Batiks produced in this way were outstanding examples of technique, comparable in pride of workmanship with the hand embroideries of their European sisters of former times. Java batiks are normally marketed either in large oblong pieces which form the 'sarong', or in smaller square pieces to form a head cloth, 'kain kapella'. Their designs, especially the use of border areas, have this end use predominantly in mind. Attractive European garments such as dresses, beach shirts, dressing-gowns and even ties can be made from sarong lengths.

In every true batik, wax is painstakingly applied to the cloth to *resist* successive dyes so that wherever the cloth is waxed, dyes cannot penetrate. For example, if the desired design is a red flower on a blue background, wax is first applied to the area that will become the flower. The white cloth is then immersed in blue dye and dried. After drying, the wax which covered the flower pattern is scraped from the cloth. Because the wax resisted the blue dye, there is now a white flower on a blue background. To make the flower red, the blue background is then covered with wax and the entire cloth is immersed in red dye. When the wax is scraped from the cloth for a second time, a red flower emerges on a blue background.

This process is repeated as more colours are used. The finest batik is reversible. Motifs are drawn, waxed, and dyed, first on one, then the other side of the fabric. Since the greatest Javanese batik is multicoloured, it is not surprising that designers, waxers, dyers, and finishers take twelve months or more to complete a single piece of a yard or two.

Both silk and cotton are used for batik, and in certain areas, such as Juana on Java's north coast, silk is particularly popular. Unlike cotton, silk requires little preparation; its fibers are quite receptive to wax and dye without the elaborate series of treatments needed by cotton. Nevertheless, among Javanese batik makers the overwhelming preference is for cotton.

The wax is mixed with resins: *gandarokan* (resin of the eucalyptus tree), *matakucing* (the Javanese word for "cat's eye," another resin), and *endal*(the fat from cows). Because the composition of the wax mixture affects the appearance of the finished product, the recipe varies according to the type of design, and the proportions are always a well-guarded secret.

A javanese batik unit

The basic tool for drawing designs on the fabric is the *canting* (also spelled *tjanting)* with which liquid wax is drawn on cloth. The *canting* works much like a fountain pen. It has a bamboo or reed handle, about six inches long, with a small, thin copper cup from which a tiny pipe protrudes. (Copper is used for both cup and pipe because it conducts heat and keeps the wax warm and fluid.) The diameter of the spout is varied to accord with the width of the line it is desired to trace and the very small bore 'tjantings' are naturally the most difficult to make and use. A woman holds the *canting* by its bamboo handle, scooping up the heated wax and blowing through the tip of the pipe to keep the wax fluid. Then, using the *canting's* pipe as a pen, she draws the design on the fabric, outlining with wax instead of ink.

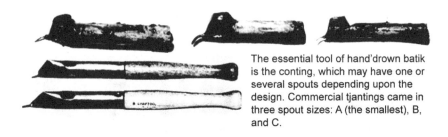

The essential tool of hand'drown batik is the conting, which may have one or several spouts depending upon the design. Commercial tjantings came in three spout sizes: A (the smallest), B, and C.

Before wax is applied, the cloth is draped over a bamboo frame called *gawungan* and weighted on one side to keep it from blowing in the wind while the waxing takes place. The batiker sits between the cloth and the pan of wax and begins her work with the *canting*. Her free hand supports the underside of the fabric, and she covers her lap with a napkin or *taplak* to protect herself from dripping hot wax.

Around 1840 the invention of the *cap* (or *tjap)*—a copper block that applies an entire design onto the cloth with a single imprint—revolutionised the batik industry. The *cap,* made by soldering copper shapes into the desired pattern, resembles a flat iron and is held by a metal handle attached to the back. In the *cap* process, the cloth is spread on a padded table, and the design is applied by dipping the copper block in wax (the all-metal construction allows its use in hot wax) and stamping it on the fabric. Small metal pins attached to the corners of the *cap* are used to align one *cap* impression with the next. The method of construction allows a completed block to be carefully sawn in two, thus forming a pair of mirror image 'tjaps'. Printing on both sides of the fabric, essential in high-class work, is thus not only possible, but also achieves a high degree of precision even in intricate designs. It may take a skilled craftsman up to a month to make a pair of 'tjaps'. These blocks are seldom seen outside Java.

As with *tulis* batik, in fine *cap* work wax is applied to both sides of the cloth. Often two or three different caps are used for one batik, one for each successive colour or design. Quite commonly, both *tulis* and *cap* techniques are combined to produce a piece of batik. Another tool, used in both *tulis* and *cap* batiks, is the *cemplogen* (also *tjemplogen).* Especially common to the Indramayu region on Java's north coast, the *cemplogen* is a block of short gold, silver, or steel needles attached to a wooden handle—rather like a wire brush. Using this tool to puncture wax on solid background areas, and the dye then penetrates the small holes, producing hundreds of tiny dots.

Originally, the colours found in batik reflected the place of origin, as well as the cultural attitudes of the people who produced them. That is usually not the case today. Though synthetic dyes have now largely replaced natural dyes, it was from nature that batik received its original colors.

The cemplogen, often used in tulis batik, punctures wax that has been poured onto a solid bakground. Dye penetrates these holes, producing hundreds of tiny dots on the finished batik.

Printing block or cap

Printing block, or cap, is the work of men. Each man stands at a padded table with his own wax and caps, imprinting the white cotton.

Among the many colors found in Javanese batik, four are by far the most popular. The most common one, also believed to be the oldest, is indigo, derived from the plant of the same name and called *torn* by the Javanese. Batik using this blue is called *biron* (from *biru,* the Javanese word for "blue"). A second common dye is *mengkudu,* a deep red from the bark and roots of the *Morinda citrifolia* plant; batik using this colour is called *bangbangan* from *abang,* meaning red. *Tegerang,* from the *Cudriana javanesis* plant, is yellow. And *soga* is a rich, uniquely Javanese brown characteristic of batik from the central Javanese towns of Yogyakarta and Surakarta; it comes from the bark of the *Pelthophorum ferrugineum* tree.

After each waxing cycle in the batik process, the cloth is ready to be dyed. In batches of twelve *kodi–* one *kodi* equals twenty pieces of batik— the cloths are placed in appropriate dye baths three times a day for ten days. They are then put into a bath of lime and water, which sets the dye.The cloth is immersed as often as necessary in a colouring vat to achieve the desired shade. It is then soaked in another solution of lye and water, to fix the dye. After each colour has been set the wax is scraped off and reapplied; sometimes additional designs are drawn on the cloth between dyeings. Overdyeing is used to produce certain colours. Green, for example, starts off as a light blue (from indigo), which is then overdyed in a yellow bath; black is produced similarly, by overdyeing indigo with red or brown.

After dyeing is completed and the last of the wax is either scraped or boiled from the cloth, the finished batik is draped over bamboo racks or laid on the ground to dry. It is then folded and put under a press for "ironing."

There are several native finishing processes given to the fabric based on which area to which it is supplied.

Batik dyed fabric placed on floor for final drying

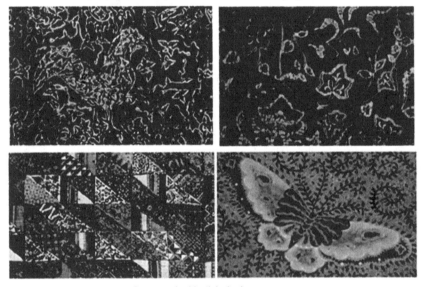
Some typical batick designs

The commercial production of batiks still occupies a considerable number of people in modern Java. The basic design is normally applied by 'tjap' and the 'tjanting' is used to complete the later stages and dyeing operation. Batiks produced in this way normally have rich but sober colourings with brown and oranges predominating. Indigo is used as a basis for deep shades of brown or black.

To save time and labour, sometimes an alternative type of batik is characterised by striking designs of birds, animals or floral motifs executed in very bright colourings. These are produced in central Java (especially

Pekalongan) by a slightly different technique. First the basic design is either stamped on by 'tjap' or hand drawn by 'tjanting'. Afterwards, dyestuff solution is painted over suitable areas using a small piece of bamboo fanned-out at one end to form a type of brush. In this way a design whose outlines were defined in wax, may receive many different colours without a separate dyeing operation for each. When painting is completed the whole design area is covered with wax on both sides of the fabric using a coarse-spouted 'tjanting'. A background colouring to the whole design is then applied by dipping the fabric in a dye solution which penetrates all unwaxed areas.

1.1.8.1   Practical Batik printing

The method of batik printing mainly based on 3 operations:
- Application of Wax resist
- Application of Dye (dyeing)
- Removal of Wax
- After treatments (soaping, etc.)

1. Application of the wax resist

A well prepared ready for dyeing fabric is selected for batik printing and treated with athin starch solution and ironed; this process helps to prevent undue penetration of the wax into the cloth. For a good batik work wax is printed on both the face and the back of the fabric. Care has to be taken to see that the wax does not penetrate the cloth but rather to form layers on the two surfaces, which will give sharper prints.

Pure wax is not used for batik, a mixture of wax and rosin is usually employed. The proportion of these depends on the batik effect desired. For example, if a batik showing a 'crackle effect' is required, a resist consisting of a mixture of paraffin wax and resin is used. A suitable mixture is one of 4 parts of rosin with 1 part of paraffin wax, or 3 parts of rosin with 1 part of paraffin wax should the rosin be a high-melting variety. The higher the proportion of paraffin wax in the mixture the more brittle becomes the print and the more pronounced the 'cracking'.

The wax-printed goods are allowed to hang for a few days and during this time the wax hardens and becomes more brittle. Should the hardening be incomplete the final 'crackle' effect will lack the characteristically fine veined effect which is the hallmark of a high-class batik of this type. To increase the crackle effect the printed and dried pattern is sometimes worked for a short time under cold water to intensify the 'crazing' of the wax. Batiks showing a crackle effect are not normally required in Java. They a'e mainly made for outside markets.

Application of wax is a very skilled work, and the proportions of the wax mixture also may not fully depend on the final effect needed. Other than the main ingredients of wax and rosin many other ingredients are used depending on the areas where the batik is produced. The Java batik is produced by block printing the designs on both sides of the fabric using two types of wax, each of which may have several components whose exact proportions differ from one batik printing establishment to another. The two main typesare a light easily removed type consisting essentially of paraffin wax, and a darker more adhesive type consisting essentially of beeswax with some paraffin wax and animal fat. Even the same fabric using different types of waxes having different adhesive properties can produce different effects of batiks. Sometimes after one dyeing, wax from one side is scraped off at areas where second overdyeing has to be done, keeping in mind the full appearance of the final design and how it is to be brought about. The two dyeing may be of different classes of dyes.

The fabric is stretched on a table covered with layers of cloth, to avoid the wax entering the tabletop while printing. On the tablecloth before the laying of the fabric an even layer china clay or sand or in some areas banana leaves previously soaked in caustic soda which will prevent the wax from adhering or penetrating. The object of the print is done using cap (see explanation above) or tjap and the outlines and fine details are done by canting (tjanting). In using the tjap (Printing block) the wax is kept molten in a simple waxing vessel with a piece of felt at the bottom which is a shallow dish of sufficient size to hold the block and kept heated by a kerosene lamp. Printing with wax is done like handblock printing.

2. Application of dye

All cold dyeing dyes of bright shades can be used for batik printing:

- Azoic colours
- Cold Brand Reactive dyes
- Solubilised vat dyes (for light shades)
- Indigo and cold dyeing vat dyes

Dyeing is normally done by dipping in shallow flat vessel of the size of the sarong or more approximately of the size 3 m × 1.3 m × 15 cm deep. The printed fabric is laid flat in the vessel and the dye solution is poured over it and gently rubbed to help in penetration. Once the fabric has absorbed the dye well it is taken out and hung over bamboo or metal rods. In a method called 'gasok' the fabric is turned from time to time. In some areas the dyeing is done in different ways, especially saving the dye solution much better. The dyeing is done in an inclined concrete trough (1.2 m × 0.9–1.3 m)with the bottom portion of which forms a V shaped canal. The slope is adjusted to allow for comfortable working from a standing position. The sarong is plaited into the liquor in folds. The folds are opened and laid on the concrete slope and the fabric is gently rubbed with liquor from the V canal. Extra dye solution is leached to the bottom V portion of the trough and the fabric is taken out and hung on rods over the trough so that the excess liquor flowing falls on the slope and collects at the base of the 'V'. In some other places a trough of the shape of small jig trough (1.3 m width and 22-23 cm depth) with one or two freely rotating wooden rods fixed at the bottom. Strings are attached at the corners of the fabric (sarong) and threaded through the wooden rods with the help of these strings. The dye solution is taken in the trough and fabric is passed through the liquor several times by pulling the string at each end alternatively. Here rubbing with the hand is not possible which may give some problems in heavily waxed batiks.

Dyeing Indigo on batiks is tedious and needs standing bath of high volume, which is taken in a bath as deep as the length of the fabric (sarong – 3 m) so that it can hang straight downwards in the liquor hence it is not widely used. The nearby shades are adjusted by using azoic or sulphur colours.

Examples of dye solutions used in batik printing (dyeing)

Indigo

Quantity	Unit	Additions
3	g	Caustic soda 38 0Be and
1	g	Sodium formaldehyde sulphoxylate dissolved in
950	g	Water
7	g	Indigo vat 60% grains avoiding as far as possible aeration of the solution*
1000	g	

*Note the deep trough used for dyeing indigo dyes

Vat dyestuffs (cold dyeing)

Normally IN and IW classes are mainly used. The required dye is taken in the dyeing vessel, mixed with water and heated to vatting temperature 40—60°C. The requisite amounts of caustic soda and sodium hydrosulphite are added, and the dyestuff is allowed to vat for 5—10 min with gentle stirring. After the completion of vatting the liquor is diluted as required with the addition quantity of chemicals and used for dyeing.

Vatting Solutions	Method A (IK)							Method B (IW)						
Dyes (g)	2	5	10	15	20	30	40	2	5	10	15	20	30	40
Water (ml)	20	50	100	150	200	300	400	20	50	100	150	200	300	400
Caustic soda (380Be)(ml)	1.25	3	6.25	9.5	12.5	18.75	25	1.25	2.5	5	6.25	7.25	8.75	10
Hydros (g)	0.5	1.25	2.5	3.75	5	7.5	10	0.5	1.25	2.5	3.75	5	7.5	10
Additions														
Caustic soda (380Be)(ml)	5	5	5	5	5	5	5	5	5	2.5	2.5	2.5	2.5	2.5
Hydros (g)	2	2	2	2	2	2	2	2	2	2	2	2	2	2
Sod. Sulphate/ Salt	5	10	20	25	30	35	40	10	20	40	50	60	70	80

The liquor is applied by repeated dipping or brushing and oxidising in the air for 10-20 min each time till the required depth is achieved. The material is oxidised and washed until free of alkali. It is always better to select the dyes which can be vatted at 500C or below.

Azoic Dyes

The main combinations of naphthols and fast salts used in batik printing are Naphthol ASG and Fast Scarlet GG salt (Yellow) and Naphthol AS-TR and Fast Red TR (Turkey Red shade) , or Fast Red B (Bluish Red), or Fast Bordeaux GP salt (Bordeaux) , or Fast Garnet GBC Salt (Garnet), or Fast Orange GC Salt (Orange).

Naphthol ASG

Normally cold dissolution method is used.

Quantity	Unit	Additions
15	g	Naphthol AS-G are pasted with
15	g	Methylated spirit followed by the addition of
9	g	Caustic Soda 380Be
30	g	Cold water is carefully stirred in, stirring is continued till a clear solution is obtained

The above solution is added dye bath of 910 parts of soft water containing 20 parts of Turkey Red oil and 2 parts Caustic soda 380Be (All total 1000 parts).

Naphthol AS-TR

Quantity	Unit	Additions
15	g	Naphthol AS-TR are pasted with a mixture of
22	g	Methylated spirit
30	g	Cold water
6	g	Caustic Soda 380Be
30	g	Cold water is carefully stirred in, stirring is continued till a clear solution is obtained

The above solution is added dye bath of 900 parts of soft water containing 20 parts of Turkey red oil. (All total 1000 parts).

If the fabric has to be fully dyed, it is dipped in the naphthol solution for a short while, excess solution is drained and allowed to dry away from sunlight. The material is the developed in Fast salts (prepared as given below).

In case the naphthol has to be applied locally the naphthol solution is slightly thickened by adding Gum Tragacanth (part of the water quantity in the dye bath can be replaced by the thickener) and applied using a brush. After a while the fast salt solution is applied in a similar way.

Fast Salts

Quantity	Unit	Additions
40	g	Fast salt is pasted with
300	g	water at room temperature or colder, make into a smooth paste and add
725	g	Cold water and dissolve.
1000	g	

Notes:

1. Add 25 g/l common salt in case of Garnet GBC salt.
2. When Fast Scarlet GG Salt is used along with Naphthol AS-G, add 4 parts of 40% acetic acid per 1000 parts to the developing bath before use. After developing the material should be washed well in water containing 3-5 parts of 20°Be hydrochloric acid per 1000 parts, followed by a thorough rinsing in cold water.

Reactive dyes

Cold brand dyes with high reactivity can be used safely for batik dyeing (printing). The solution is prepared (see below) and the fabric is dipped in the dye solution or locally applied using brush and hung dried while the fixation takes place.

Quantity	Unit	Additions
25	g	Urea is dissolved in
300	g	Boiling water and allowed to cool tom60-700C. And the add portion wise
10-50	g	Cold brand reactive dye and finally make with
565-525	g	cold water to make up to
1000	g	

Or follow the dye manufacturer's direction for dissolving.

Just before dyeing add

Quantity	Unit	Additions
4	g	Anhydrous sodium carbonate and
8	g	Sodium bicarbonate dissolved in
88	g	Water

Into the above solution. Once the alkali has been added the solution is stable for 2-3 hours only.

Solubilised Vat dyes

These dyes are used only for light shades. The dye is directly dissolved in water and the fabric dipped in a cold solution of dyestuff containing 5 parts of sodium nitrite per 1000 parts, squeezed out and then developed by a short treatment in a solution containing 2 parts by volume of sulphuric acid at 66°Be per 100 parts of water in the cold, washed, neutralised and dried.

1.1.9   Copper plate printing (Gravure or intaglio)

Copper plate printing is a method that first appeared in Europe in the mid-eighteenth century. This process involves images being incised or engraved onto a metal plate (usually copper) and is known as an intaglio method. Dye is applied to the surface of the plate; the excess is then scraped away, leaving the remainder only within the incised lines. This dye is then transferred to cloth by applying pressure.

A copper plate printed fabric (1759)   Tools for making copper plates and printed design

This process allowed the use of much finer linear marks and cross-hatching in printed designs and enabled more varied tonal qualities than had previously been possible. The types of images associated with this process were romanticised rural or mythological scenes and were typically printed in one colour: red, blue or purple.Another characteristic of this printing process was the use of layout and organisation. The copper plates were often very large and this presented difficulties in repealing the imagery; as a result, 'island designs' emerged, whereby the imagery sits alone within its own space, disconnected from the other elements. The best designs ensure that the space between the islands is managed in a way that provides a visual cohesion the distinct visual language of this process is often replicated today in contemporary textile design, either directly utilising similar romanticised scenes or ironically subverting them by using more contemporary or challenging images. Their main disadvantages were initial engraving of the plate is time-consuming and expensive and limitation - could be done only in monochrome. Butrepeats up to a meter in length in contrast to block which was limited to 46 cm [18"] Fine engraving made it difficult to overprint with a second colour.

Typical colours used were purple, red, sepia, all of which were derived from madder and China blue. Copper plate printing was popular between Popular during the years between 1760 and 1800.

1.1.10   Roller printing

Next significant development was Roller Printing, which was used widely up to 10-15 years ago. It is not completely obsolete, even now it is being used in wallpaper printing jobs and rarely in fabric printing. The first roller printing machine for mass production was a "relief printing" machine.

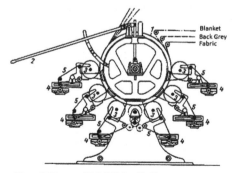

Schematic Diagram of Relief Printing Machine
1. Central Drum, 2. Lever, 3. Printing Roller,
4. Paste Tray, 5. Paste Transfer cloth

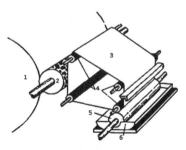

One printing position 1. Central Roller
2, Engraved roll 3. Colour Carrying cloth,
cloth, 4. Guide rolls, 5. Colour transfer roll
6.Colour tray

Copper plate printing on paper has been known since the 15th century. In the mid 18th century an entirely new type of printed fabric decorated with figures, landscapes and architecture made from engraved copper plates conquered the market - First in England then on the continent. In 1783 Thomas Bell patented the first "Copper Roller" printing machine. The inking cloth was later on replaced by colour transfer rolls and doctor blade. The doctor blade was a sharp steel 'doctor' blade to remove all the colour paste from the unengraved surface of the roller. The name given to the blade was derived from the word abductor, because it took away the unwanted ink.

The output of printed cotton in England alone increased from 30 million yards in 1796 to 210 million in 1821 and to 600 million in 1851. It peaked in 1911, when production from British print works amounted to 1400 million yards. Worldwide, roller printing production accounted for more than half of the total amount printed until 1976.

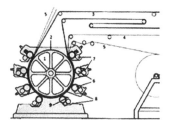

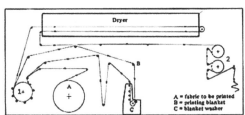

Roller Printing Machine: 1. Pressure bowl, 2. Lapping Roller, 3-Blanket, 4- Back grey, 5- Fabric, 6- Engraved roller, 7- Furnishing Roller, 8- Colour Box,

Schematic diagram of a 6 colour Roller Printing Machine with drier and blanket washer 1- Printing Machine, 2 - Drying Cans,

The next development in machine printing was the replacement of copper plates by copper rollers, which has really contributed to the increase in production. Copper roller machines have been ideal for high-volume printing of woven fabrics for almost 200 years. Although considerable skill was required to engrave and use copper rollers, the increase in productivity resulting from their use was so great that block printinginevitably declined. This was a revolution even more significant than those occurring in the spinning and weaving sectors, and there were inevitable disputes and strikes.

Since the early 1970s, the roller printer method has been more and more replaced by "Rotary Screen Printing" Some old Print works still keep 1 or 2 machines for special fine line and raster effects.

The copper rollers used are hollow cylinders of two main types – solid Roll, copper-shell.

Roller printing machine   Making of a copper roller

While copper roller machines proved ideal for high-volume, low-cost printing of woven cotton fabrics, there was always a market for small-scale production of individual designs, especially on silk, wool and, later, on man-made fabrics. For these, roller printing was not suitable at all. The costs of engraving and setting up the machine for each run were high, and long runs were therefore essential. Block printingsatisfied the demand for some time, but an alternative, fundamentally different approach emerged.

1.1.10.1   Manufacture of rollers

Shell rollers are made by the electrodepositing of copper onto a cast iron or steel tube. Whenever a design is turned off, more copper can be deposited to return the roller to its original circumference. Solid rollers are iron rollers covered with copper sheets.

1.1.10.2   Engraving the design

The manual method consists in transferring the design from plain to tracing paper and going over the contour with a paint containing sulphurous compounds (Na2S). The tracing paper is placed on the roller and kept on it for several hours. As a result, cupric sulphide is formed on the copper surface coming in contact with the paint, and a black pattern is obtained. Its outlines are incised by cutters, while the ground (evenly coloured part) is produced either by making a group of dots or parallel lines. The dots are called picots or stipples, and the lines, hatches. The depth of the hatches differs, but does not exceed 0.3 mm. The number of hatches per centimetre varies between 18-22. Half-tones are produced by means of picots and hatches of different sizes. In the process of printing, the colour remains in the incised hatches and stipples, while the roller surface between them serves as a support for the doctor blade.

The manual engraving method is costly and not very efficient; therefore, it is used in exceptional cases, for instance, for engraving portraits and other intricate designs with large repeats.

Engraved rollers   Engraved copper shel

1.1.10.3 Machine engraving

To engrave printing rollers by the milling method, a *die,* i.e., a small cylinder, is first made of soft steel, whose diameter and length must conform to the design repeat. If the repeat is small the die is made several times larger, but so that a whole number of repeats are arranged on its circumference. The die is hardened and then used for impressing a relief pattern on another, tempered steel roller mill *(mollette).* For this purpose, the die and the mill are placed into the bearings of a rotary press and pressed to each other. When the rollers rotate, a raised relief appears on the mill. After several revolutions in the press, the raised parts of the pattern are coated with special mastic and the mill is etched with nitric acid which affects only recessed parts not protected by the mastic. Etching and pressing are repeated several times after which the die is hardened and used for obtaining an engraving on the printing roller. Red copper is the most suitable metal for manufacturing printing rollers. Ordinary (calico) printing rollers have a circumference of 400 to 525 mm, while rollers for printing shawls (bigger repeats) are 780-815 mm in circumference. The roller length may be somewhat greater than the width of the cloth. First, the engraving is made in the form of a narrow strip corresponding to the length of the mill, and then the head of the machine and the support with the mill are shifted alone the axis of the printing roller and then the next portion of the design is engraved, and so on. After the engraving is completed, the printing roller is ground, and all protruding parts are retched. Grinding and etching are repeated several times to make the roller surface quite even.

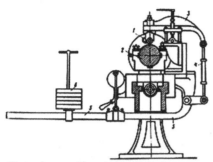

Engraving machine. 1. Mill, 2. Roller, 3,4,5. Levers, 6. Weight

Photoelectric pantograph

1.1.10.4 Pantograph

The operational principle of the machine is as follows. The design is photographed by a special camera (usually one repeat). When photographing, the size of the repeat is adjusted to suit the printing roller. The polished printing roller is fitted on a pin and coated with acid-proof mastic, after which it is placed into the bearings of the photoelectric pantograph. Then, electronic diamond cutters, in a number corresponding to the number of repeats, are mounted on the pantograph. The film with the design is placed in a special machine on a cylindrical aluminium roller of the required dimensions. One or several units with photoelectric cells are installed so that during machine operation they can scan the photograph of the design and transmit pulses to the electronic unit where they are amplified and directed to electronic diamond cutters. The diamond cutlers reproduce the design exactly on the roller of the machine. The points of the cutters pressed to the surface of the printing roller incise thin lines of the design on the surface of the printing roller, cutting through the mastic and leaving the metal bare when the design is engraved; the contours of the pattern are lines outlined and the background is hatched (16–32 hatches per 1 cm^2).

1.1.10.5 Photo engraving

Photoengraving is the best method of obtaining designs on printing rollers. It can be carried out in different ways, i.e.:

1. Engraving the design directly on copper rollers, use being made of a step-and-repeat machine (this method is used at most textile enterprises);

2. Engraving the design on a die with subsequent pressing of mills and engraving of copper rollers on a machine; and

3. Engraving the design directly on the copper roller without making use of the step-and-repeat machine.

In the photo engraving method, the design is transferred on to a roller coated with photographic solution. After drying the roller in a dark room, the design is transferred using light.After the design has been copied on the roller, the latter is removed from the machine, washed with water at a temperature of 30-35°C, then washed with a solution of methylene blue (10 *g/l)* for visual control, then again washed with water, dried, checked according to the layout, and heated in an electric oven at a temperature of 200-250°C. During heating, the violet colour disappears and the roller becomes golden-yellow, then brown, and finally chocolate brown. After cooling, the roller is mounted on the pin and all the defects and damaged places are filled with mastic. When the roller has dried, it is subjected to etching for 30-40 min. The roller is then taken out of the bath with the solution and washed with water, the mastic being removed with kerosene, and the points with the design are carefully wiped with turpentine, after which it is wiped to dryness and sent for grinding. To reduce the number of printing faults on the fabric and to increase the life of the engraving, the engraved copper rollers are covered with a thin layer of chromium by an electrolytic method.

Newer machines have given away with the central pressure bowl and introduced individual pressure bowls. (See the figure below)

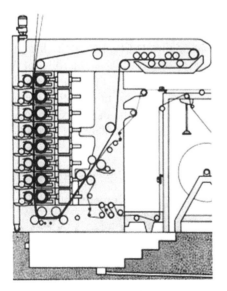

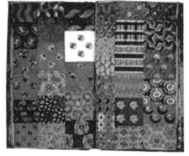

Roller printing pattern book
(1810-1820)

Modern roller printing Mc without central
pressure roll

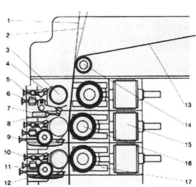

Newer Version of Roller Printing Machine 1. Fabric, 2. Back Grey, 3. Colour Doctor, 4. Engraved Roller, 5. Colour Doctor (released), 6. Lint Doctor (released), 7. Colour Tray, 8. Lever, 9. Furnishing Roller Pressing Device, 10. Colour Doctor Pressed, 11. Furnishing Roller Pressed, 12. Furnishing Roller, 13. Blanket, 14. Guide Roller, 15. Counter Roller Pressing Device, 16. Counter roller, 17. Counter roller Track.

1.1.11   Manual screen printing

Continuous efforts toward larger and cheaper production have led to the development of new processes which allow certain individuality but keep the cost of the finished goods within the reach of all. Screen process is such a new method. As a matter of fact, screen process should not be called a printing process because the colours are not printed but squeezed on the fabric. Screen printing is an extension of stencilling. In previous centuries, the Japanese developed the stencilling technique for textile printing and brought it to a fine art. Stencils made of tin-plate or paper is the predecessors of silk screen. Almost everyone has seen the metal stencils which are used in the shipping departments to mark parcels and cases. To prepare a block for block printing, a relief-like design has to be worked out of a wooden surface. The raised parts are then covered with an absorbent material, usually flock or felt. This surface absorbs the colour or dye and passes it on to the fabric during the printing operation. The preparation of an old-fashioned stencil for printing is entirely different. The design is cut out of paper or etched out of tin-plate. For stencil printing, as for block printing, separate stencils *for* each colour have to be prepared. However, these metal or paper stencils have a great disadvantage. The centers, or island parts of the stencils, such as the center of the letters D and 0, have to be held in place by bridges or ties. To eliminate any trace of the bridges in the finished print, it is necessary to have a second stencil for each colour in which the bridges are located in a different place. This is very impractical and results in a waste of time. Another disadvantage of these stencils is their limited use for relatively coarse designs only.

A B C D E F G

A B C D E F G

Top - Printed with metal stencils
Bottom - Printed with a silk stencil

The cost of such a stencil compared with the manufacturing cost of a block or a roller for machine printing is low. This is a decided advantage. The tin-plate is covered with an acid-resisting -lacquer. Then the design is copied onto the lacquered surface and its outlines are cut through the lacquer down to the metal surface with a sharp instrument. Afterwards, the tin-plate is put into a bath of nitric acid which eats through the metal along the outlines of the design. The stencil is ready for printing after a few finishing touches with a file. The Japanese gained extensive experience in the printing of highly artistic silk fabrics at an early date, and they probably were the first ones to use very fine silks threads, or even human hair, as ties to connthe center parts of the stencils with the outside. The fine silk thread did not obstruct the passage of the paint. Therefore, it became possible to print each colour of a design in one operation.

Japanese"Yuzen"screen

Gradually it was discovered that silk fabrics could be used for the making of stencils, provided that the meshes of the fabric were plugged with an impermeable substance wherever the passage of paint was not desired. It is quite evident that only fabrics with a relatively large open mesh area are suitable for this work, because close fabrics do not permit the necessary free passage of the printing paste. A screen is made of open and closed areas. The colour is forced by a blade through the open area onto the fabric.

Earlier days, silk bolting cloth has proved to be the most suitable stencil material. The stability of the screen is achieved by making the cloth in leno weave. The characteristic feature of this weave is the peculiar crossing of the warp threads with each other which locks the woof threads securely in place, thus preventing their shifting. Their meshes are so minute that the interlocking of the threads can only be observed with a magnifying glass. Nowadays the

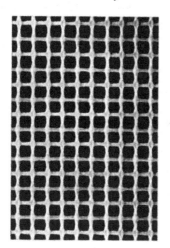

Silk Bolting Cloth (enlarged

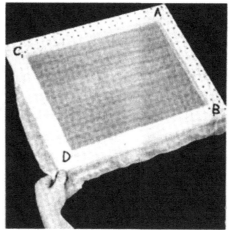

Making of a Hand Screen

bolting cloths are made using nylon, polyester or phosphor bronze gauze. Phosphor bronze, being metal a metal screen is more brittle compared to screens made of silk or synthetic fibres and hence can be dented or damaged very easily by any projections such as nails, pins, registration stops, etc. The dents on a metal screen cannot be easily repaired and hence handling has to be very careful.

1.1.11.1   Screen frames

Screen frames are usually made of either wood or metal. They should be light and at the same time sturdy. It should have dimensional stability and the smallest possible contact with the surface the fabric to be printed to avoid any marking off, as well as haloing of the designs and stains especially in wet-on-wet printing.

1. Metal frames

Metal frames ensure the above requirements more and hence are used widely. U-shaped, angle or closed profile tubes of iron or light metals like aluminium have been used for making screen frames but closed profile tubes are more common and gives better results and they are light. The tube profile can have different shapes with square, rectangular, trapezoidal, or triangular cross sections. Rounded rectangular or triangular shapes and other shapes have been developed for reducing contact surface while printing. Wooden frames reinforced by metal frames are also common. The disadvantages of the metal frames are that it can rust damaging the gauze and due to the tension there are chances of frame getting bend longitudinally. Prevention measures are lacquer coating of the frames and giving outward curvature on the longitudinal sides of the frame respectively(See fig. below).

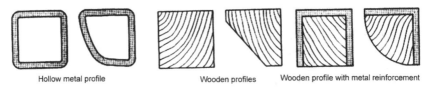

Hollow metal profile   Wooden profiles   Wooden profile with metal reinforcement

Some common profiles of screen frames

2. Wooden frames

Wooden frames are not very common due to the bending of the frame because of the tension of the gauze and warping. However, they are used for special requirements. The wood must be smooth and well-seasoned and only a soft wood with low water absorbency, resistant to warping and to varying degrees of humidity and heat, is suitable. The types of wood that satisfy these requirements are coniferous, and the most important are the following: Western Red Cedar, Yellow Pine, Kenya or other Cedars. Hard woods such as Teak are not recommended for the making of screen frames. Metal frames are made of rigid lightweight materials and will not warp or twist. They are costlier but can be used many times and can be used many times. Screen frames vary in size as per the design repeat, width of the fabric. But in special cases like printing of flag and bunting, etc., screens of sizes as large as 24 m × 1–8 m (9 ft. × 6 ft.) may be used. For printing longer length and avoid warping, etc., the frames must be made very rigid, and the corner joints can be strengthened with angle brackets. The side pieces of the frames are bevelled at a slight a ngle so that when the screen is laid flat on the table only the minimum area of the frame makes contact with the table. This helps to prevent marking-off on the following repeat. The outer edge of the wood is slightly rounded all-round the frame to prevent the gauze from being cut when stretched over the frame. (See fig. above)

3. Metal reinforced wooden frames
The likelihood of longitudinal distortion and warping can be avoided in this type of frames. The wood is exposed on the lower side of the frame so that the bolting cloth can be easily fastened on it. Normally it is inserted into U shaped or angular profiles and screwed. (See fig. above)

1.1.11.2   Bolting cloth

As mentioned above the oldest bolting cloth is made of silk. However, after the introduction of synthetic fibres, silk made bolting cloth is rarely used. Silk bolting cloth has limitation of using alkaline printing paste. Presently, following fibres are used for the manufacture of bolting cloths are:
  • Natural Silk fibres
  • Polyamide fibres (Perlon, nylon)
  • Polyester fibres
  Metal fibre or filaments (phosphor bronze, chromium nickel steel, glass filaments)

1. Silk Bolting cloth:
Silk fibre without the removal of sericin (not degummed is usually used for making the bolting cloth, usually made by gauze weave since it gives strong, anti-slippery and equal mesh size structure. Silk is resistant to acids and solvents but not stable to alkali. Silk bolting cloth gas been developed with higher resistance to higher pH and was used extensively. Mesh size of 6-9 was good for blotch prints, for fittings 10-14 and for finer outlines 14 – 16. It is advisable to clean and neutralise them with dilute acetic acid when alkaline pastes are used.

2. Polyamide bolting cloth
Polyamide bolting cloth using monofilament is widely used for screen making, with advantages of high breaking strength, stretchability, abrasion resistance and durability. It is resistant to solvents, alkalis but sensitive to oxidising agents, nitric acid, phenol, sulphuric acid, formic acid at higher concentrations. But they are resistant to most of the concentrations of acids at which print pastes are prepared.

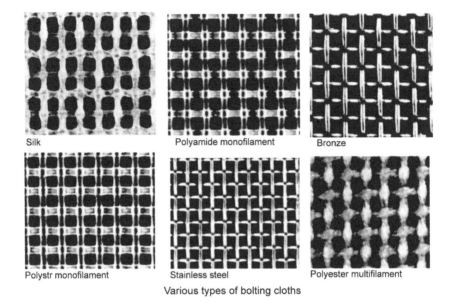

Silk   Polyamide monofilament   Bronze

Polystr monofilament   Stainless steel   Polyester multifilament

Various types of bolting cloths

3. Polyester bolting cloth

Polyester has some advantages like higher durability but lower abrasion resistance, and lower elasticity and hence it is used whenever it is difficult to cover the frames with polyamide fibre gauze. Since the fibre is hydrophobic, most of these properties are not affected in the moist state. It is resistant to alkalis, acids, oxidising agents and most of the solvents except some phenols. Most of the colours at room temperature do not stain the polyester. Thus, polyester gauze is also as popular as polyamide gauze.

4. Metal bolting cloth

Commonly used metal fibre for making gauze is phosphor bronze and stainless steel. They have excellent colour permeability, good resistance to chemicals. Being non-elastic and highly dimensionally stable it is well suited for bolting cloth withstands any distortion, but it seldom adapts itself to fabrics of coarse structure. Metal screens are very common in rotary printing.

5. Glass filament gauze

It is low priced, resistant to alkali, good colour permeability but low elasticity. Due to its high rigidity stretching on the screen frames is not easy. Glass filament gauze is not common except in special cases of small yardages.

The fastening of the silk to the frame should be done with great care.' A good and even screen and a clear print are only possible with a well stretched fabric. The following system of stretching can be recommended. Start at the corner marked A and tack this corner with five tacks as shown in the illustration. With hand at corner C, pull the fabric in the direction of corner C. Be careful to have the threads of the fabric run parallel to the bar A-C. Fasten the corner C with four or five tacks and then tack the entire side A-C. Place tacks about two centimeters apart and in two rows - a tack nearest the inside first, and then backed up with a second tack. Repeat this operation from corner A to corner B, always remembering that the bolting cloth should be taut (Stretched in all directions) and that the threads should run parallel to the frame. When two sides have been fastened the real, job of stretching begins. Stretch the fabric in the direction of D diagonally from A. At the same time equalise the pull on the fabric from C to D and from B to D. Tack this corner before letting go of the fabric. Now tack the two remaining sides, B-D and C-D starting from B and working towards D. Grip the fabric between thumb and index finger, stretch it taut, place two tacks, move on about two centimeters and repeat this operation until both sides are fastened.

Now there are stretching machines available which can stretch the bolting cloth equally to all sides and instead of tacking synthetic gums are used which sticks the stretched cloth firmly to the frame. Next, the fabric is coated with a light-sensitive substance after proper cleaning of the screen to remove all oily matters and any dirt and exposed to light in contact with a "positive" of the original design. It necessarily must be a "positive» and not a "negative", because the coloured parts of the design which have to be open in the finished stencil must be opaque on the "positive" to prevent exposure of the film. While the hardened photographicis not soluble in water but the unhardened part dissolves in water we get the printing image with the bolting cloth exposed. To be suitable for textile printing, the stencil first has to be protected with a special lacquer. This is done by completely coating one side of the screen with protective lacquer which is then removed in the open sections of the stencil from the other side. This operation is repeated for the other side also, if necessary, after the first side has hardened sufficiently.

Accurate printing of multi-colour designs requires stable screens. Screen fabric made from silk or other natural fiber is too hydrophilic to meet this requirement and water-based print pastes cause these fabrics to sag. The development of screen printing to its modern form ran parallel with improvements in the screens themselves.

1.1.11.3   Screen engraving – older methods

Even though these methods are explained as older methods, these can still be used. Today, professional chemicals for screen engraving from many manufacturers are easily available offthe shelf, one may not go for it, rather than making form the basic chemicals as explained here.

The gauze or bolting cloth, as it is generally called, is stretched on the frame manually or with the help of stretching machines.

1. Sensitising the screen

Most sensitising solutions are based on gelatine or polyvinyl alcohol, sensitised with either ammonium or potassium dichromate, or a mixture of these dichromates. Any one of the following sensitising solutions may be employed:

Gelatin based sensitising liquor (1)

	Quantity	Unit	Chemicals
A	125	Parts	Gelatine dissolved in
	552.5	Parts	Water at 50–60°C with minimum stirring

	Quantity	Unit	Chemicals
B	50	Parts	Zin oxide
	200	Parts	Water

	Quantity	Unit	Chemicals
C	13.5	Parts	Ammonium dichromate
	6.5	Parts	Potassum dichromate

	Quantity	Unit	Chemicals
D	50	Parts	Ammonia (Sp. Gr. 0.88)
	2.5	Parts	Wetting agent

Mix B to A, then C, and finally D, and bulk to 1000 parts. Filter and apply on screen using a rubber squeegee at 500C and dry at room temperature in the dark or orange safe light.

	Quantity	Unit	Chemicals
D	95	Parts	Gelatine dissolved in
	865	Parts	Water at 50–60°C with minimum stirring

	Quantity	Unit	Chemicals
B		Parts	Zinc oxide
		Parts	Water

	Quantity	Unit	Chemicals
C	13.5	Parts	Ammonium dichromate

	Quantity	Unit	Chemicals
D	20	Parts	Ammonia (Sp. Gr. 0.88)
	3.5	Parts	Vegetable oil

Mix same way as above bulk to 1000 parts and apply and dry as above. Polyvinyl alcohol based sensitising liquor.

	Quantity	Unit	Chemicals
A	475	Parts	Polyvinyl alcohol emulsion are stirred to
	375	Parts	Water at 60–70°C

	Quantity	Unit	Chemicals
B	30	Parts	Ammonium dichromate
	125	Parts	Water

B is added gradually to A while stirring. To make the exposed portion opaque and more visible 3% titanium dioxide or zinc oxide may be added to the mixture and mixed well and bulked to 100 parts before coating at 30-400C.

2. Process of transferring the design onto the screen

The design is examined by the artist for the repeat and the number of colours. Today, there are many easier ways to do this using software which will be explained later in detail, in earlier days this used to be done manually by the artist. Generally, the screens are made of a particular size and the repeat also is adjusted to a standard size by making small changes in the design, without upsetting the appearance of the design (and getting approval from the customer if necessary). Each colour is drawn on a transparent tracing paper and which is normally frosted on one side and smooth on the other side. There are specific papers available for this purpose which are stable to moisture and temperature so that no distortions happen in the accuracy of the design due to changes in temperature and moisture during the process of exposure process. The ink used for drawing the design has to be opaque, not necessarily to be black. Each transparency will have fitting marks made on them and once all the colour separations are made they are kept on a lighted glass box one over another matching the fitting marks and checked for the accuracy of the design with overlapping if any. It is artists skill to make overlapping or without overlapping considering the order of the screen used in printing, coverage of the colour separation (A blotch may shrink the fabric after a print) so that ultimate printed design on fabric gives the same outlook of the original design.

Once colour separations on tracing paper has been checked and found fine in all respects these has to be engraved on to the sensitised screen.

The positive colour separations are laid flat on the sensitised screen matching the fitting mark on the screen and on the tracing paper and secured firmly using adhesive tape.A glass plate of suitable dimensions is then laid on top of the positive and securely clamped to the screen frame by means of suitable clamps. Some weight can be kept on the glass plate after keeping in position to expose from a light source, so that the glass plate is equally pressed at all places. Light exposure can be by sunlight, or artificial light. The time of exposure to get a satisfactory engraving, using different light sources is given below:

Light source	Distance from Light source	Time of Exposure
Strong sunlight		5 min
Bright daylight		10 min
Dull daylight		up to 2 h
Mercury vapour lamp	450 mm (18 in)	15 min
150 W lamp	450 mm (18 in)	3h
300 W lamp	600 mm (24 in)	up to 2 h

The temperature of the glass sheet should not exceed 300C, while exposing. During exposure the portion of sensitised screen where the light has fallen (Open portion in the tracing sheets) becomes hard and insoluble. The screen is removed and immersed in cold water for 2 min. Treat the screen again in water at 350400C, when all the PVA and Gelatin of the unexposed portion is washed off. In the case of Gelatin based screen coatings, the exposed portion gelatin even though it has become insoluble it is not hard enough to remain strong on repeated squeegee movements. Hence, it has to be hardened by treatment with

Quantity	Unit	Chemicals
25	Parts	Ammonium Dichromate
50	Parts	Chrome alum crystals
50	Parts	Formaldehyde 40% Solution
1000	Parts	

In case of polyvinyl based screen coating the hardening is done with the following treatment

Quantity	Unit	Chemicals
50	Parts	Acetaldehyde
50	Parts	Isobutyraldehyde or Butyraldehyde
20	Parts	Sulphuric acid 660Be (1680Tw)
1000	Parts	

Immerse in the above solution at room temperature (15-200C) for 45-600C.

In both cases, after the hardening treatment the screens are rinsed with cold water and dried lowly. The screen can be checked for any blockage, if found, may be cleared by brushing that area with 1:1 lactic acid 80%.

Screens can be further reinforced by brushing lacquer on one side and wiping from the other side with a dry cloth or cloth wetted with a solvent like butanol-xylene mixture. This may not be necessary for PVA based screens. Reinforcing can be done on one side or both as per requirements and dried. After drying the screens may be baked at a temperature of 120°-150°C for 3-5 min, the higher temperature requiring the shorter time. The screens are ready for use.

There are other methods of screen engraving which is explained under modern printing methods.

1.1.11.4   Printing tables

Textile printing was carried out with flat screens on long tables up to 100 m in length. The process is mainly limited to the production of high-quality prints today as it is more labour-intensive and less productive than screen printing by machine Tables are usually made of wood of lengths 50–75 meters or more according to availability of space and material to be printed. For example, saree printing tables is about 75–80 m and the length will be in such a way as to print multiple screen lengths after allowances. The same way the width of the table also depends on the width of the materials to be printed. Usually 55" so that 45" fabric can be printed after allowances. Manual printing widths higher than this is not easy. For ease of operation, the tables are made with a slope from the farther end towards operating side of the table. The usual gradient is 1" in 10". Tables are also made with other materials like concrete these days. It has advantages like steadiness and does not suffer from warping as in the case of wood. The top of the tables is covered with 2-3 layers of blanket or felt to give flexibility and finally with waterproof cover like Rexene with back cloth to absorb the spillages and penetrated paste.

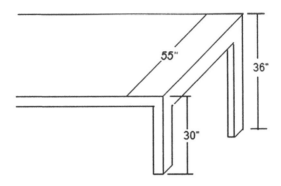

Diagram showing the usual sloping (gadient)

The fabric to be printed is gummed on the table with suitable adhesives like CMC, dextrin, cellulose derivatives, alginate, polyvinyl alcohol, etc. Sometimes a gluing carriage is used to glue fabrics to screen printing tables for printing. The liquid glue is applied by means of a roller applicator and the fabric is brought into contact with glue film by means of pressure rollers.

An electrical heater of any type is provided at the bottom of the tabletop so that the fabric dyes after printing in a short time.

Printing operation

Printing operation involves the following steps:

1. Pasting, or otherwise securing the fabric by pins to the tabletop.
2. The printing operation.
3. Stripping the fabric from the table.
4. Washing down the table to remove colour deposits (Washing is not usually necessary after each printing operation and takes up comparatively little time.)

The printing tables are 50–100 m in length as per the length of the print house and usually `1.2–1.6 m width. Longer length helps in higher production and time to dry the first print before starting the second print. The tables can be made of wooden frames or metal frames, the latter is being used more often now-a-days. Tables may have wooden tops with a woolen felt of about 6mm thick is laid above it to have a smooth soft surface. It is covered with a waterproof sheet cover. Fabric to be printed is stuck over this using an adhesive. (Some printers may pin the fabric over the table for printing, in such cases the waterproof sheet may be replaced by a backgrey).

The printing table is fitted with a rail on the operating side aligned well and firmly with the length of the table. The rail will have adjustable metal stops on them to exactly align each print position of the screen. The screen frame also will have screws or bolts fastened on them which make contact with stops fastened on the guide rail. The stops on the rail is first adjusted to match with the exact width of the design on the screen (as per the fitting marks).

The introduction of synthetic fibres, nylon and polyester as well as the replacement of wooden frames with metalenabled stable screen to be manufactured which maintained tension when wetted, and much higher print accuracy. Strong, stable screens enable the hand-screen printing process to be mechanised. The first development was the introduction of a moveable carriage, in which the screens are mounted one at a time. Various types of squeegees are available. Some are made entirely of wood while others have wooden handles with rubber or metal plates inserted. The length of the squeegee is approximately 40 mm (1-5 in) shorter than the internal dimensions of the screen.

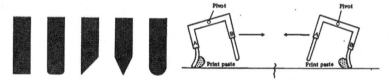

Different shapes of rubber squeegee tips   Action of a double squeegee

Rubber squeegee tips fitted inside wooden handle is more common. The tips can be of different shapes, the use of which is decided based on the material to be printed, type of table covering and the coverage of the design to be printed, etc. For large designs, and for fabrics that absorb a good deal of colour paste, a soft rounded blade is required. Patterns with fine detail that are to be printed on a flat, silk-like fabric require a sharp hard blade. Furthermore, a hard table covering requires a soft rubber squeegee, while a table with a soft underlay requires a harder blade. If the squeegee is held near the upright position, less paste will be applied than when a shallower angle is used. A more rapid movement of the squeegee or lighter pressure also results in less paste being applied. The edge of the blade must be perfectly level and clean.

The fabric may be fixed on the table by pins or sometimes combined on to the backgrey with special machines or stuck with the help of adhesive. The latter one being used more for printing continuous length of fabric. An adhesive like dextrin or gum according to the availability or semi-permanent adhesive can be used to fix the fabric on the table.

1.1.11.5 Printing

The screen is wetted out by printing on to a piece of spare fabric or absorbent paper and fixed on the carriage for the screen. A suitable amount of print paste is poured into the well of the screen frame and transferred through on to the cloth underneath, by drawing it to and fro across the screen with the squeegee, two to four passages generally being given. This facilitates quick printing with less fatigue. To reduce fatigue in screen printing, squeegees, bracketed in pairs and fitted with handles and adjustable weights are provided (see fig above and below). The colour pool is confined between the two blades. By the natural alteration of angle of the forearm between the pull and push strokes, the blade nearer the printer carries the load in the away motion, the further blade beingeffective on the strokes towards the printer.

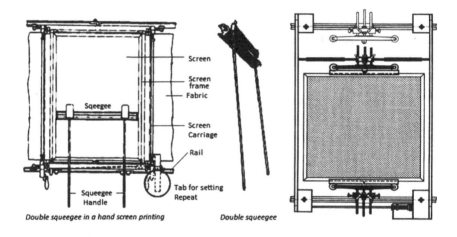

Double squeegee in a hand screen printing   Double squeegee

To avoid marking-off, the screen printer does not carry out his printing in a continuous order, but prints in alternative positions, that is, position 1, then misses a position to position 3, and on to position 5 and so on until the whole table length has been completed.

The drying of the printed material is assisted by blowing hot air over the table or by heated tables, so that provided the printing pastes are not too hygroscopic and the material is not too light in weight (and consequently limited in absorbency), the material should then be dry; otherwise, it is necessary to wait until the print is dry enough to avoid picking-up and marking-off. Printing is then continued in positions 2, 4, 6, etc., until the whole table is printed with the first colour. The same procedure is followed for the second and subsequent colours in the pattern.

To minimise manual effort in raising and lowering large screens they are counter-balanced in a suitable manner. Screens possess a greater surface area than any other method of printing. After printing, the thickening in the paste makes the screen sticky and it becomes little difficult to lift off from the printed cloth. For facilitating easy separation, a novel device is recommended known as Lift-off Process. The whole screen, in this case, does not rest on the cloth, but only the frame. When the squeegee is pressed by the printer during printing the screen under its edge is lowered by the pressure he exerts in scraping and forms a close temporary contact with the cloth. As soon as the squeegee proceeds forward the pressed part of the screen springs up due to the tension and elasticity of thecoated screen. The device to keep the screen apart is very simple. Four 0.8 mm thick strips of wood as wide as the thickness of the frame sides are screwed at the bottom of the frame so that when the screen is placed on the cloth for printing, they rest upon the cloth and the frame sides rest on them, while the gauze carrying the design is 0.8 mm above the surface of the cloth. After printing, the printer finds no difficulty in separating the screen, his work is hastened, and no smearing of the printed parts takes place.

With these types of squeegees developed Table screen printing was also developed, which gave much higher production than hand screen printing and better setting of designs. Settings were adjusted by the machines and only manual adjustment on the machines achieved perfect settings. High-quality, multi-coloured prints, e.g., in small runs on silk (e.g., scarves), are printed on tables with a moving film printing carriage [Fig. (a) below]. The carriage takes its direction from the repeat rails which run along the edge of the table [Fig. (b)]. The flat templates are printed through one after the other; initially one pattern is printed over another without intermediate drying. When all the individual colours have been applied, the fabric is stripped from the rubber conveyor and passed on to the vertical drier or a horizontal drier.

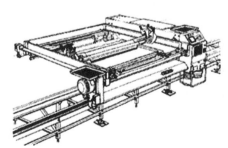

(a) Table screen printing   (b) Screen fitted on a table printing

After the printing, if necessary, it is dried again and taken for further fixing, washing off, etc.Screen printing is the most effective method for blotch printing effect. Uniformity of colour, accuracy of printing and solid effects is most successfully and easily obtained by screen printing. The printer can be an ordinary man.

In modern scenario the screen printing of the type explained here is maximum practiced in chest printing. Manual screen printing has given way to semi-automatic or fully automatic screen printing machines which are done much faster, and production is much higher manual printing. They will be explained in detail in the next section. Chest printing was practiced manual, but now machines are available which can produce prints at much higher rates.

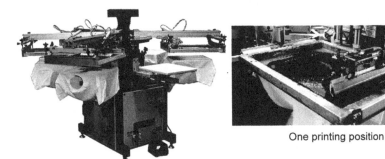

One printing position

A modern chest printing unit

The piece required to be printed is pinned on the tables arranged around a central turning device which is used for turning the screen around. Each screen carries one colour. It can be printed on the chest cut piece or on a stitched garment as per frequency required by rotating the screen around. This is a modern automatic printing machine but included here since the basic screen which has been used for so many years is used here.

2

Modern printing methods

Even though manual screen printing methods are still being used in special printings like Saree printing, Silk printing, etc. it has to be considered as an older technic as it is not suitable for longer print runs. To increase production and reducing the cost, the machine manufacturers mechanised printing operations automatic one by one. Thus, in 1930s screen printing was made partially automatic, which is called semi-automatic printing machines. Since these types of machines are still in use for bulk productions, we can consider from this point onwards Modern printing methods.

2.1   Semiautomatic screen printing

Semi-automatic flat-screen printing is still very popular even today where the scale of production is small. The screen carriage was one of the first to be made automatic whereby instead of positioning the screen by hand in each printing position the operation was mechanised. Raised slots were made so that while still on the rails the carriage could be lifted over and lowered into these slots. The second mechanisation was regarding the squeegee movement, which was manually pushed to and fro using long handle (explained above in manual screen printing section). This manual operation was slow and as the printed fabric width became higher it was tedious more difficult to perform forced it to be mechanised. In the automatic system the squeegee was driven in both directions across the table using a chain mechanism. The number of strokes could be pre-set and the pressure applied varied within appropriate limits until the optimum was achieved and then set for the remainder of the run. The mechanisation has also enabled us to use longer repeats which also

A semi-automatic screen printing machine   One screen position and squeegee

contributed to increase the production. Thus, it was able to make a production comparable to the production of roller printing machine which was the most commonly used bulk production machine. This has slowly reduced the importance of roller printing machines.

Semi-automatic flat screen printing is still done for the high fashion industry, especially high-quality silk prints with sometimes more than 20 colours. Part of this production will be done in future by the Ink-Jet printing method.

The Quality which can be produced by this kind of technique is outstanding long tables of up to 60 meters are used. The fabric is glued to the blanket either with water soluble glue (PVA) or by using thermoplastic glue.

The first colour is printed over the whole table length followed by the next colour, etc. The first printed colour is dried already before the next is printed, which produces very sharp prints especially in the case of fall-ons. ("wet on dry").

For example, most of the high-quality silk scarves for the well-known fashion houses are printed by this method. Unlimited No. of colours can be printed.

2.2   Fully automatic screen printing machines

In the 1950s fully automatic flat-screen printing machines entered the printing market-Buser-Stork-Johannes Zimmer. In these typesof machines, all operations were made automatic. It was found that in semi-automatic printing units the movement of the screen and printing at odd positions and printing again at even places makes the operation very slow. The operations become so slow when the no. of colours is more. The solution was found by incorporating a moving table and making the screen stationary. Movement of the table made it possible to bring the fabric to exact positions of the screen and print.

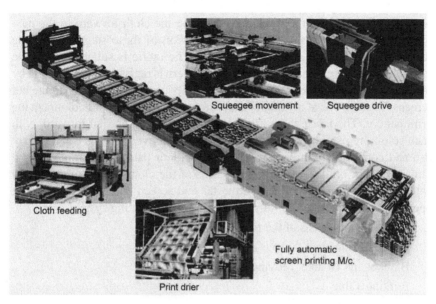

Squeegee movement

Squeegee drive

Cloth feeding

Fully automatic
screen printing M/c.

Print drier

The printing operation in a flatbed screen printing machine are made fully automatic by standardising the stages of preparation and producing the print including screen holding, addition of print paste, pressing the paste on to the cloth, lifting the screen and carrying the cloth forward to the next screen printing screen position. The automation of these stages makes the reproduction of printing results quite easy. The fabric is suitably fed to the machine in a crease free manner. The print pattern is registered on the fabric by pressing the printing paste through the specially engraved portions on the flat screens. There are as many numbers of screens as the number of colours in the print pattern. The fabric to be printed is conveyed, i.e., carried forward, with each colour, register by register, while the flat engraved screens continuously rise and come down, at each repeat of the colour pattern are printed at the same time, but on different printing places of the cloth. The entire coloured pattern will be printed only when the far end screen completes its printing operation. Given below is a fully automatic screen printing system.

The system consists of the following major parts:

Cloth feeding

Blanket and Blanket Moving system

Printing Table

Screens and Squeegees

Fabric lifting and Drier

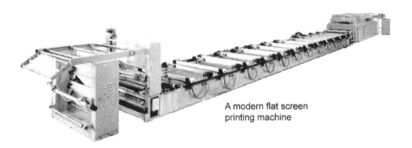

A modern flat screen
printing machine

2.2.1   Fabric feeding

As explained earlier the Blanket ('printing table') intermittently moves and the Screens are stationary in this type of printing system. For smooth running there has to be a system in place for avoiding unnecessary tensions on the fabrics or the blanket which will affect the printing. As far as the fabric feeding is concerned, this is accomplished by fabric accumulating system, or scray or such systems. Following drawing shows three such methods of feeding the fabric. Feeding from batches with high entry is one of the best

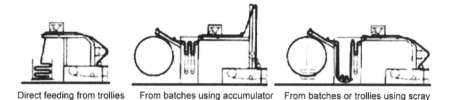

Direct feeding from trollies   From batches using accumulator   From batches or trollies using scray

systems where one can control the straight feeding with selvedges in line. The fabric includes a brushing and suction system to remove any loose fibre, or extraneous matters which may resist the prints. A very efficient cloth guider supports the straight laying of the fabric on to the blanket. A pressure roller presses the fabric firmly on to gummed blanket which takes the fabric forward along the blanket. In a different system called vertical feed system, where the fabric is stuck onto the blanket as it comes out of the washing system dry at the feeding end into a vertical floating movement of the blanket drum. It is remarkable for its simple and compact design. Owing to the vertical arrangement, the continuously running section of the belt is freely accessible. This gives the advantages of a very short path from batch to belt and continuous gumming of the cloth. It can be best explained by the following figure:

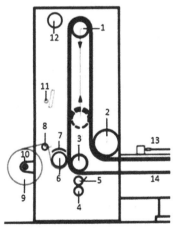

A feeding in unit (Schematic above and RHS actual) 1-vertically 'floating' belt drum, 2&3- Belt gude rolls, 4&5-Gumming unit, 6- The THERMOPLAST heating and pressing roller, 7- The heat shield,8-Guide roll, 9- Fabric roll,   10-The facis bearings with brake for the batch, 11- The selvedgeguides12- the cloth guide rollers, 13- First screen position, 14 Blanket 15- Fabric, 16- Blanket

The vertical feed-in unit contains all the equipment required for faultless feed-in and gumming. The intermittent movement of the blanket is compensated by the up and down movement of the blanket with the mechanism shown in the figure above. The fabric is stuck on the blanket before this mechanism. So, no tension problem due to the stop and start movement of the blanket.

2.2.1   The blanket and blanket moving system

Endless rubber coated fabric, blanket runs over the screen printing table, on which the piece good is fixed, wool felt is often used under the printing blanket as a flexible layer. The material to be printed is mounted on these blankets and repeatedly passed over the printing table, rhythmically moving and stopping. The screens can be lifted and lowered. Whilst the material is stationary, printing takes place using special squeegees. A correct printing blanket run is necessary for good repeats, and it is important for accuracy of fit of a design. The printing blanket run depends on the position of the rollers which are in contact with the printing blanket.

Another advantage of the blanket is the use of thermoplastic adhesive for sticking the fabric on the blanket. These adhesives which are effective from a certain temperature and which are used in film printing where they are heated to the necessary temperature by means of heated rollers or plates (explained below) Thermoplastic adhesives, vulcanised in rubberised continuous printing blankets, allow the preheated fabric to be printed without gluing and washing using the preheated printing sheet to position the pattern for printing. The temperature is adjusted and regulated by means of a thermostat. The fabric is transported over a convex plate allowing it to be bonded smoothly to the blanket (Fig.). Conventional types of glue can be applied to the thermoplast layer for other fabrics for which thermoplast adhesive are not efficient. After passing through the printing zone, the fabric can easily be pulled away from the printing cloth, even when cold. It is possible to print 200,000 to 1 million metres but it may be necessary to replace the thermoplast earlier depending on the printing process (See fig above showing the electric elements of the feeding units)

One screen for each colour is positioned in series accurately along the top of these blankets. Regarding the length of the printing table, there can be provision even up to 18 colours or so. The width of the gap between the areas printed by any two adjacent screens must be a whole number of lengthways design repeats. This need not necessarily be the same as the lengthways screen repeat as there may be several design repeats per screen repeat; for example, where there are three design repeats per screen repeat, the gap between adjacent screens need only be one third of a screen repeat. The fabric is gummed to the blanket at the entry end and moves along with the blanket in an intermittent fashion, one screen-repeat distance at a time. All the colours in the design are printed simultaneously while the fabric is stationary; then the screens are lifted and the fabric and blanket move on. When the fabric approaches the turning point of the blanket, it is pulled off and passes into a dryer. The soiled blanket is washed and dried during its return passage on the underside of the machine.

A driven roll controls the movement of the blanket. Once the fabric is taken off by the drier mechanism the blanket runs below the printing table where it passes through a blanket washing unit (of many different designs by different manufacturers) where they are cleaned off all the paste spilling and colour stains and dried comes up at the entry of the machine, if necessary gummed on the surface with natural adhesives ready for printing the next part of the fabric. It is still a practice to apply a water-based adhesive to the blanket at the entry end, by means of a brush running in a trough containing theadhesive solution, and to spread the layer more evenly with a rubber squeegee; the fabric is then pressed against the tacky blanket with a pressure roller. A hot-air dryer issometimes employed to partly dry the adhesive before the fabric is printed. The intermittent movement of the blanket poses some problem when the water-based gums are being used. It can cause a line of excess gum to be produced each time the blanket stops moving, which in extreme cases may affect the levelness of the print. The blanket, on its return passage, will also receive an irregular degree of washing. This problem is solved by moving the blanket continuously on the return passage under the machine and in the feed-in unit. Guide drums at both ends act as compensators to allow the simultaneous, intermittent and continuous running of the blanket. See the figure below the drum at the feeding end can move from A – B and at the delivery end C – D to allow the blanket to run continuously in the blanket washing and gumming area.

Even after conceiving the design concept of fully automatic screen printing by intermittent movement of the blanket, it was also difficult to maintain the accuracy, which was of utmost important in registering the repeats correctly. The blanket tended to slip and overrun, or stretching of the long rubber-coated, laminated neoprene blanket alsointroduced inaccuracies. In modern machines more often employ a series of electromagnetically or hydraulically operated clamps which grip the edges and move the blanket precisely one repeat distance without stretching it. The drive for the forward movement is usually hydraulic. (The exact limiting of the repeat advances is ensured by stops fitted to both sides of the machine. The stops act directly on the advance magnets. The repeat length is adjustable between 400 and 3000 mm. The setting is made by a system of spindles, motor gearing and brakes. In some designs coarse setting is done by pushbuttons, fine setting by hand-wheel. The clear-set counter has '/100mm graduations (or 1/1000 inch). Different designs are also used.)

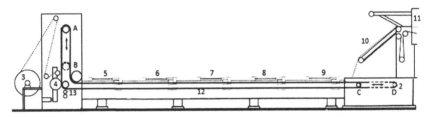

Schematic diagram of afully automatic screen printing Machine

1 and 2- These rollers move as shown, to maintain the lowerside of the blanket in constant motion, 3- Fabric roll, 4- Pressure roller, 5,6,7,8 and 9 - Screen positions, 10- fabric lifting and conveyor, 11- Drier, 12- Moving Blanket, 13- Temperory adhesive application unit

The more common practice nowadays is to apply a tacky semi-permanent or 'permanent' adhesive on the blanket. They are mainly based on acrylic copolymers, can withstand the washing necessary to remove excess print paste without becoming detached from the blanket. It is possible to print 200,000 to 1 million metres but it may be necessary to replace the thermoplast earlier depending on the printing process. These adhesives are coated on to the blanket and are only tacky when heated. Heat can be applied either directly to the adhesive layer or to the fabric, to achieve the required bond. The fabric is transported over a convex plate allowing it to be bonded smoothly to the blanket. The temperature is adjusted and regulated by means of a thermostat. Conventional types of glue can be applied to the thermoplast layer for other fabrics for which thermoplast adhesive is not efficient. After passing through the printing zone, the fabric can easily be pulled away from the printing cloth, even when cold (See fig above showing the electric elements of the feeding units)

2.2.3   Screens and squeegees

Screen development has been explained earlier. Each screen is fixed firmly onto a carriage positioned as explained in the blanket and blanket movement paragraph. (See fig below). Central adjusting devices hold the screens in the squeegee units. They ensure quick and accurate fine setting and quick changing of the screens. Narrow screens which do not occupy the full width of the printing table can be used in conjunction with telescopic tubes. Each carriage has its own adjusting units, mostly electronic and even remote controlled in modern machines.

In screen printing process whether manual or automatic, a print paste is applied to a textile fabric through a screen engraved with a design in the form of a "negative" by means of a squeegee. The screen frame serves the purpose of raising the screen whilst the textile fabric which is glued to an endless rubber blanket is transported the length of one repeat after which the screen is lowered into the printing position once again. Various squeegee systems are employed in flat screen printing:

1. Blade squeegees: single and double squeegees or tilting squeegees
2. Roller squeegees
3. Magnet-roll squeegees

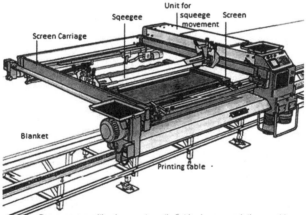

One screen position in an outomatic flat bed screen printing machine

Single screen position

In the case of blade squeegees the application of print paste and print paste penetration are controlled by the squeegee angle and the pressure. (See fig below (a) and (b)). In screen printing, wedge shaped area formed by the angle between blade squeegee and flat screen. The print paste wedge is a determining factor for print paste application. If the blade squeegee is set at a shallow angle then a long, narrow wedge forms (print paste application is correspondingly lower) and vice versa. The wedge is shorter in blade squeegees set at a steep angle. The wedge pressure created is decisive. Most popular squeegees today are the double rubber blade squeegee or magnetic squeegee. Movement of the squeegee can be along the length or width. The latter has been used more, possibly due to the advantage that no pressure is exerted on already printed areas at the end of the stroke. One, two or more passes of the squeegee can be made as required.

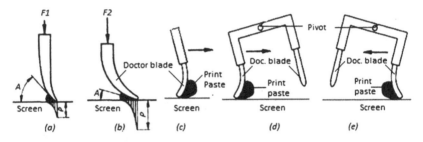

(a) & (b) Print paste penetration are controlled by the squeegee angle and the pressure - F1 & F2 - Blade pressure, A - Angle p - Paste penetration. (c) Operation of single blade (d) & (e) Operation of double blade in the direction of the arrow

In a double blade squeegee system, a pair of parallel rubber-blade squeegee is driven across the screen with the print paste in the gap between them. It acts as a single squeegee in each stroke (See fig above (c), (d) and (e)). Only the rear squeegee makes contact with the screen, the leading squeegee

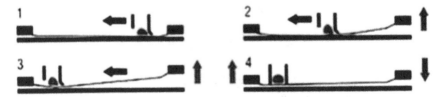

Movement of a double squeeze and screen lifting

being raised slightly above it. When the next stroke is made, the leading squeegee for the first pass becomes the rear one for the reverse direction. The double-blade arrangement is simpler to construct than one utilising a single squeegee that has to be lifted over the pool of print paste at the end of each stroke, although this type is found in some modern semi-automated machines. In the case of normal single squeegee units, they are raised in tilting fashion, rising first on the off-side and then on the operator's side to avoid jerky lifting of the screens and the risk of splashing.

The strokes of double squeegees are termed as flood stroke and print stroke:

Flood stroke:

The rear squeegee is slightly lifted and floods the paste in a thin layer over the screen length. The leading squeegee is raised above the flood squeegee. The paste can penetrate into the engraving. In flood stroke operation, a single stroke is made with the screens raised. This already fills the gauze openings with ink. After the screens are lowered, the usual squeegee stroke is made. Thus, in the short time of a single stroke, almost double the quantity of ink is applied to the fabric. With the flood stroke procedure, it is possible to greatly increase output in all cases where a single stroke is not sufficient. Further advantages of the flood stroke procedure are the sharper contours and the avoidance of all bubbling in the ink applied.

Print stroke:

The leading squeegee becomes the rear squeegee and presses the print paste through the engraving on to the fabric. The leading squeegee is lifted above the colour paste.

Working of a Rod squeegee (LHS Scematic, RHS Actual)- 1. Rod squeegee, 2- Print paste, 3- Screen, 4- Electromagnet, 5 - Rotational direction of the rod squeegee, 6- Direction of the movement of the Electromagnet, 7 - Fabric

Another type of squeegee used by machine manufacturers is a rolling-rod squeegee moved by an electromagnet, driven intermittently under the blanket. This squeegee is the same type used in Rotary printing machine (explained later in this book). The only difference in the system is that the magnetic field is stationary in the rotary printing machine in automatic flat-bed printing it is moving, so that the rod can roll in the lengthwise direction. Normally, one passage is all that is required for adequate cover and uniformity. The diameter of the rod is usually small enough to allow print paste to flow over and round it at the end of a pass. It is clear that screen distortion and wear are less where rolling rods, rather than rubber-blade squeegees, are used.

2.2.4   Fabric lifting and the drying

When the printed fabric approaches the turning point of the blanket, it is pulled off and passes into a dryer. The soiled blanket is washed and dried during its return passage on the underside of the machine. After the last print when the blanket moves forward the fabric also moves along (as it is glued to the blanket). Since the previous part of the fabric is inside the drier it blocks the light beam to the photocell which triggers the starting of the conveyor and slowly pulls out the fabric from the blanket. (See fig. below) By this

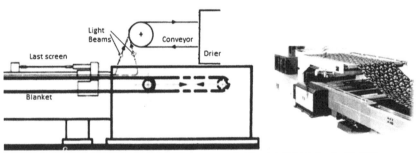

(a) Schematic diagram of the fabric lifting arrangement of the printed fabric before the drier (b) the actual printing machine part before the drier

time, the blanket stops for the next print and at that position the fabric again blocks the next light beam to the photocell which stops the conveyor system of the drier. Alternatively, a single diagonal beam can be used. The intermittent movement may cause problems of variable stretching and over-drying. Ideally the fabric should move through the dryer at a constant speed equal to the average linear velocity of the blanket, but this is not always possible. The limits between which this process continues and thus the drier runs in tandem with the printing blanket, the fabric can be lifted off the blanket are the end of the last screen and the end of the top linear surface of the blanket.

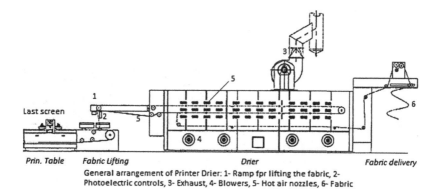

General arrangement of Printer Drier: 1- Ramp fpr lifting the fabric, 2-
Photoelectric controls, 3- Exhaust, 4- Blowers, 5- Hot air nozzles, 6- Fabric

The cloth runs continuously and completely tension-free on the conveyor belt through the drying chamber. The fabric is transported by the conveyor through the drier and taken out at the delivery end while the endless conveyor returns to the feeding end of the drier. The drier can be of different design, it can be a single, double or triple pass as per the speed of the printer and length of the drier. If polymerisation of pigment is to be done in the drier their driers with this provision like extra runs for polymerisation. There are chances of wet prints staining the conveyor. Once the conveyor stains in turn stains the fabric the conveyor may be washed and used. In some designs of the drier the initial part of the fabric and conveyor does not touch each other to avoid wet staining and after the initial run for a length where the fabric is partially dried it is carried by the conveyor.

The entry ramp (see above fig.) can be raised or lowered according to repeat length and type of fabric. With knit goods, this fact can decide on success or failure, as the small distance from printing table to dryer belt prevents curling and distortion of the fabric.

The continuous advance prevents local over-drying. The exhaust fans draw damp air from the dryer. The resultant slight negative pressure in the dryer chamber prevents gases from escaping and ensures the inflow of fresh air.

2.2.5   Blanket cleaning system

Once the fabric is taken off from the blanket by the drier, the endless blanket runs through under the printing though various units to make the blanket clean and fresh for the new printing at the feeding end of the blanket explained earlier. The blankets bets stained with print paste in many ways like printing over the width of the fabric, spillages, print paste penetrated through the fabric especially thin open weave fabrics (e.g.,Saree prints), etc. This stain has to be washed off in order to avoid staining the fresh fabric for printing. This is done by passing the blanket through a blanket washing unit. There are different variants of this unit designed by different machine manufacturers. They are

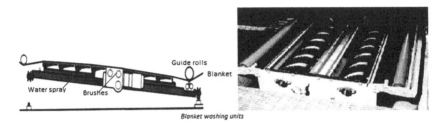

Blanket washing units

washed with water jets and rotary brushes and then dried, passes under the printing table and is ready to be used for another printing cycle. The soiled surface of the belt is sprayed with jets of fresh water or recycled water. The belt then passes through a range of water channels and steel blades, which remove the fluff and any residue of ink and conventional gum. Nylon brushes driven by an electric motor assist the washing process. Before the belt leaves the washing unit, it is rinsed with fresh water, which also serves for continual renewal of the washing water. A squeezing device clears excess water from the belt and dried. Two hydraulic cylinders lift the washing unit into working position. When the machine is stopped, the washing unit returns under its own weight to the lower position. The whole washing unit is stainless. It is mounted on wheels and can easily be run out sideways for maintenance purposes.

The ultimate production of an automatic flat-bed screen printing machine depends on various factors. One may generally think it depends on the speed of the blanket butthe speed at which the blanket is advanced is much less important, as the blanket is in movement for considerably less time. Speed of advance has to be restricted, as at high speed the system would have so much forward inertia that there would be a tendency for the fabric to overrun the blanket. One of the factors affecting the production is the no. of squeegee strokes required. All the screens may not need more than one stroke, but if one screen needs more stroke the other has to wait as all the screen print at the same time. Usually blotch screen needs more strokes due to the quantity of past to be transferred to the fabric is higher, and requirement of uniformity of blotch, penetration, etc. The no of strokes also depends on the thickness of the fabric and irregularity of the surface of fabric like corduroy, etc. In some machines the squeegees can be used to make a stroke while the screens are in the raised position (known as a 'flood stroke'). This fills the mesh in the printing areas of the screen with paste before the screen touches the fabric, and more colour is applied with the first stroke when the screen is lowered. The bigger the repeat and squeegee movement is widthwise, blanket movement forward will be longer and hence the production can be higher. If the squeegee movement is lengthwise, as thesqueegee will take longer to move along the screen where the repeat distances are large. The drier capacity (quantity of fabric that can be dried in a specific time) can limit production as the speed of production has to be adjusted to this which becomes especially important when the print coverage is higher.

The automatic flat screen printing is still used in saree printing in India. It is best suited to print the pallu, the design of which will be different than the body and the repeat comes after almost 5 m or so after the body print. In an automatic flatbed screen printing machine, the pallu screens can be adjusted in such a way that it will be printed only after certain repeat of the body or print the pallu first and the continuously print the body while the pallu screens are in lifted position and when the body design is being printed the pallu screens can be kept in lifted position.

A saree printing in progress in an automatic flatbed printing machine

2.3   Rotary screen printing machines

Automatic flat-screen machines cannot be described as continuous because of their intermittent printing action. The possibility of continuous printing could be possible only if continuous movement of the fabric has been achieved by moving the screens along with the fabric while printing. Many efforts by the manufacturers produced different machines. Fully continuous printing is best achieved using cylindrical forms. Many attempts were made in the early 50s to form flat metal screens into cylinders, despite the necessity of a soldered seam. The idea was first proposed in 1947 in Portugal; however, it was only put into practice in the 1950s by Almerindo Barros of Portugal(Aljaba). They were constructed on the principle of conventional roller printing machineswith rotary screens made of rounded bronze wire with internal squeegees, in place of normal copper rollers. These were mainly used for carpet printing.

But the initial commercial machine was first introduced by Stork (Holland) at the ITMA show in Germany in 1963. In concept, the idea is to take a flat screen and simply shape it into a roll by sealing the ends of the flat screen together. The simple modification converts a semi-continuous process to a continuous one. However, initially there were many technical hurdles to overcome before rotary screen machines became practical.

2.3.1   Development of rotary screen printing machine

In the historical development of Rotary printing machine, was a Schwed patent where first movable screen idea was brought, even though it was used for bigger repeat, again mostly used for carpet printing it was the first idea of continuous movement of screen and fabric was brought.

Schwed patent   Rotary printing for upto 4 m repeat

The successful use of closelywoven nylon fabrics in flat screen lead to use these materials for cylinder screens as well. The "two-ply" screen consisted of metal mesh inside and a woven nylon sleeve outside.

Aljaba rotary screen printing machine was more a representative of today's rotary printing machine, but its design was similar to the roller printing machine. However, unlikeroller printing, it is the printing cylinder that is driven, while the rotary screens revolve only with slight pressure. They are stabilised and adjusted at the ends by a free-running cog. The product, together with the back cloth, is fed in at a constant tension, but not stuck on.

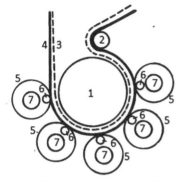 

Rotations printing machine Aljaba   *Aljaba machine*
1. Pressure bowl, 2. Guide roller, 3. Printing blanket, 4. Fabric to be printed, 5. Screen, 6. Roll squeegee, 7. Middle axes

The invention of seamless screens of electro-deposited nickel was the significant step which led to the rapid expansion of rotary-screen printing Peter Zimmer introduced the galvano screen in 1961. In 1963 Stork introduced the lacquer screen. The Stork machine based on lacquer screens introduced at ITMA 1963 was an immediate success. Machines using magnet rod squeegees as by Peter Zimmer have been very successful in printing wide width fabrics.

One of the first rotary printing machine

Today's rotary printing machine consists of a printing table on which an endless conveyor made of resin-coated rubber on which the fabric to be printed adheres, after passing through a spreader and a warp straightening unit controlled by photocells. The printing stations are arranged along the whole length of the conveyor belt (min. 8, max 12-16 or even more in spl. cases) printing stations depending the maximum no. of colours to be printed on which are assembled the engraved rollers. A squeegee blade of rod is introduced inside the photoengraved roller together with the feed unit of the printing paste, which is fed by the distributor and the squeegee forces the past on to the fabric as in the case of flat screen printing, while the fabric moved along with the conveyor and the engraved roller rotates on its own axis. At the end of the printing table the conveyor belt returns, and the fabric is taken off from the belt and dried in a drier more less the same manner as in the case of the automatic flat-bed printing machine. The cylindrical screens can be much closer together than it is possible with flat screens and so the blanket is shorter for a given number of colours. The fabric dryer, however, must be longer to enable the dryer to dry the printed fabric at a much higher running speed. Speeds up to 30-50 m/min are used depending on the design and the fabric quality:

Infeed

Printing Table and Blanket

Screens and Squeegees

Drier

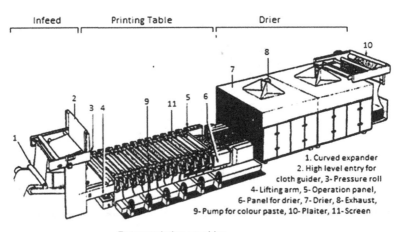

Rotary printing machine

2.3.2   Infeed

The infeed is almost same as we have explained under automatic flat-bed screen printing machine. Infeed unit will have a system from batch or trolley Since the rotary printing machines run much faster than the flat bed machines it should have a much more efficient dust extraction unit and selvedge guider to deliver the fabric to the blanket. The dust extraction unit is installed just before the printing unit to remove lint, loose fibres and other loosely adhering matter from the surface of textile fabrics to be printed (which could impair printing quality, e.g., local resists and screen blocking problems, etc.) a dust extraction unit is installed, e.g., immediately before the printing zone. Various high speed cloth guiders are available which can suit the requirement of the printing machine. Normally there will be a lifting arm with curved expander roll and a pressure roll to assist the fabric feeding from a 'A- Frame' without tension. Fabric then goes to the dust extraction unit where the extraneous matters fluff, etc., adhering to the fabric is brushed and removed by suction. There are many different designs of the dust extractor. After dust extraction fabric is passed through a compensator and high entry to cloth guider and then to a curved plate fabric heater. The fabric is pressed on the gummed blanket with the help of a pressure roll.

2.3.3   Printing table and blanket

Printing table is where the actual printing takes place. Printing table has the following parts on it (1) Drive, guide rollers, tensioning facility, deflection roll, supporting rollers for the blanket, (2) Blanket washing unit, (3) Printing Heads and screen suspension, (4) Gumming unit, (5) Fabric gluing device, (6) Electromagnet (in case of magnetic squeegee provision).

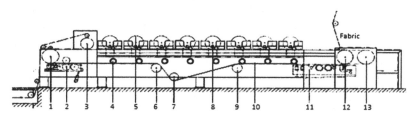

Printing table : 1-Deflection roll for blanket, 2-Gumming unit, 4-Patern repeating wheel, 5-Screen, 6&9-Supporting roller for blanket, 7-Correction, roller, 8-Squeegee, 10-Blanket, 11-Blanket washing unit, 12&13-Drive and tensioning rollers.

   Printing table is the most important part of the printing machine supporting the Blanket on which the printing takes place. The tabletop is an endless blanket, which like a conveyor made of firm resin coated rubber with minimum extension under tension. The blanket runs flat on the top of the printing table through a heavy drive roll which also can be adjusted for tension. After the drive roll the blanket goes under the table passes through the washing and drying units, gumming unit (same as explained under Flatbed printing machine) The gummed blanket (without gumming if thermoplastic layer for gumming is used) comes to the top via a deflection roll (guide roll cum deflection roll for straight running of the blanket). Both blanket washer and gumming units can be taken out from its position for cleaning purposes.

Printing table

The length of the printing table depends on the number of printing heads, which is decided on the basis of maximum colour printing is planned.The printing stations are arranged along the whole length of the conveyor belt or Blanket (min. 8, max12-16 or even more) printing stations depending upon on the length of the printing table and the customers need), on which are assembled the engraved screens. The screens are mounted above the blanket supported by the screen holders with bearing which rests on the side framework of the printing table. The whole unit is called a screen suspension (screen holder) which serves to position, tension, adjust and drive the screen. Each screen has a supporting roll below the blanket. There are open bearing or closed bearing system (as per manufacturers, some manufacturers supply both systems as per customers requirement) for the screen supports which turns the screen while printer runs.

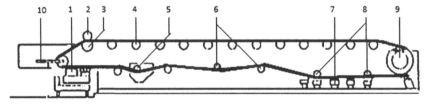

Path of Blanket on the Printing table   1- Gumming unit, 2- Cloth pressing Roll, 3- Blanket supporting Roll, 4- Screen supporting rolls, Balnket control, 6- Guide rolls of Blanket, 7- Blanket washing units, 8 - Pressing roll for the blanket washer, 9 - Main-drive roller with drive 10- Blanket tensioning device

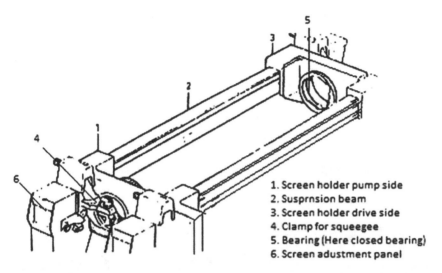

1. Screen holder pump side
2. Susprnsion beam
3. Screen holder drive side
4. Clamp for squeegee
5. Bearing (Here closed bearing)
6. Screen adustment panel

Single screen holder

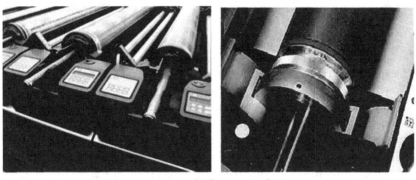

Open bearing system of different manufacturers

Closed bearing system of different manufacturers

The screens will have end rings fixed at each end to help to fix it with the screen supports. The squeegee, which is fitted with a flexible coating blade, is inside the screen. The continuous adjustment of the squeegee pressure makes precision adjustment of the print paste coating possible for each individual screen without making it necessary to stop the printing machine. For long-pile fabrics requiring complete print penetration, an "air blade" which increases the penetrating pressure three-fold has been developed.

The print paste is pumped, from a container at the side of the machine into the screen trough a flexible pipe. Inside the screen the pipe is rigid as it also acts as a support for the squeegee. Holes in the pipe allow the paste to run down to the bottom of the screen. A level control actuates the pump when the past level falls below a pre-set height.

Endring fixed on and engraved screen   Pumping systm for print paste to each screen

There are three units under the printing table which are in contact with the outside face of the blanket. First is a blanket washing unit, with scrapers or doctor blades, and water spray, brushes, to wash efficiently to remove all the paste remnants, fluffs gum, etc. from the blanket. The unit is fixed on a trolley with wheels which is locked in position. It can be removed by pulling out on the wheels after unlocking for cleaning. The gumming unit is positioned at the feeding end but below the print table. The blanket after washing and drying comes to the gumming unit. There are many designs of gluing units. The glue can be applied by a roll half dipped in the glue and the top part of the rill touching the blanket (like a nip roll) or by a doctor blade or can be pumped onto the printing blanket and dosed with the aid of an airflow squeegee. Whichever method it should ensure uniform thickness of the layer of glue and is reproducible. It should be possible to set the width and the position of the glue on the printing blanket to avoid unnecessary wastage of glue. Normally, the gluing device could be pulled out (see fig below) sideways from under the machine. Filling, emptying and flushing are done with the integrated pump or manual. Drying of glue during prolonged stops (weekends) can be prevented by emptying the reservoir and filling it with water. The glue application squeegee will be completely immersed. Instead of gumming the blanket can be coated with a thermoplast which is not washed off in (non-washable) blanket washer. Each time the blanket comes to the feeding end it is heated when it melts and acts as glue. The fabric for printing also is heated by passing over a heated plate before coming into contact with

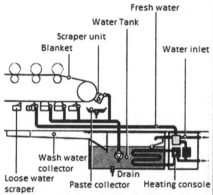

Schematic drawing of the blanket washer

Blanket washing unit (actual)

Blanket gumming unit

the thermoplast to help in sticking on the blanket. Thermoplast, which is a semi-permanent glue is mainly used for synthetic fibres. It can print up to 200,000 to 1 million metres but it may be necessary to replace the thermoplast earlier depending on the printing process.

As the name indicates, thermoplast is activated by heat. This heat is supplied by:

    * infra-red radiators on the blankets
    * the heated plate via the cloth

The advantages of thermoplast are:
• good adhesion on synthetics
• no penetration of the glue in the case of thin qualities
• the cloth immediately sticks after the cloth pressing roller
• frequently cheaper than water-soluble glue for qualities that are difficult to glue

Disadvantages:
• pigment dyestuffs and lint may cover the thermoplastic layer, so that it no longer has any adhesive power.

Note: If so desired, it is possible to apply a water soluble glue on top the thermoplastic layer, except polyvinyl alcohol adhesives which contain methylated spirits (C_2H_5OH) which will dissolve Thermoplast.

2.3.4   Screens and squeegees

Rotary screens for screen printing are hollow metal cylinders containing fine pores whose walls have a thickness of 0.08–0.15 mm and which can be more than 3 metres in length as per the requirement of the fabric width to be printed. The printing table width also will be as per consumer's requirement of fabric width to be printed, screen holders adjustable to a range from a maximum to minimum. Although the seamless nickel screens are only about 0.1 mm thick, they are strong enough provided they are put under lengthways tension. Light aluminium alloy end-rings are fitted to the ends of the screen, care being taken that the plane of the ring is at right angles to the axis of rotation of the screen. Some rotary screens are driven from both sides to avoid the danger of twisting and buckling.

The screen engraving details are discussed under the chapter 'Design Aspects'

2.3.4.1   Rotary screens

In rotary printing thin nickel perforated rollers are used to print the design on to the fabric. These rollers on which the designs are engraved are called screens.

There are two types of screens used in Rotary printing:

1. Normal Lacquer Screens
2. Electroplated Screens

Normal lacquer screens

Rotary screens or Lacquer standard screens, introduced by Stork in 1963 but now available from several manufactures, have uniformly spaced hexagonal holes arranged in lines parallel with the axis rotation of the cylinder and offset in alternate lines, as in a honeycomb, for maximum strength. Now there are many types of screens are available according to the mesh size – stork screens are named as Standard, Penta, Nova, etc. but in the case of other manufacturers there may be different nomenclature. The details of the screens discussed here are based on stork nomenclature. Other manufacturers will have different terminology for these types of screens or a different technology screen itself. Careful observation of the various screen hole shapes shows some differences – The side shapes of the perforations and the width of the non-perforated solid part between the holes. There are many manufacturers of the screens. They have different standards of screens produced. Different manufacturers will have different standards and terminologies for different type of screens. We will explain this reference to the Rotary Screen Brand Stork.

Standard screens

The walls of the perforations have a near-conical, profile. The outside dams between the openings are relatively wide. The paste supply with Standard Screens is shown on the left of the diagram. The hole shape is almost conical, whilst the dam between the holes is relatively wide. This means a relatively

high pressure is required to make the paste flow over the nickel dams. Higher pressure passes the paste through the substrate the backside of the fabric which is not actually required. The result is that more paste is wasted, and blotches are uneven.

The hole size of the screens is designated as mesh. In standard screens there are different meshes like 60, 80, 100, etc.

  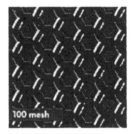

Standard screen enlarged to show the hole size

Penta screens

Different types are introduced to enable the printer to print more intricate designs where fine stripes, and outlines, etc. are involved. The main difference in the PentaScreens compared to the standard screens is that the holes are of Biconical shape compared to conical shape in Standard Screens and the dams are narrower. This shape requires much less pressure to pass the paste through the holes because additional pressure is built up in the middle of the hole where the shape is narrowest. Once the paste passes through the narrow portion a venturi effect is created section of the hole shape is required for printing. Thinner blade squeegees, reduced squeegee pressure, lower paste wastage, higher production speed, etc. are the advantage.

Dams are considerably smaller than in the Standard Screens. These are more easily bridged following reduction of squeegee pressure and increase of paste viscosity.

As well as providing better edge sharpness and a more uniform surface blotch print; this brings paste savings due to the reduced penetration.

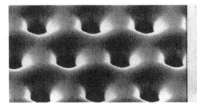 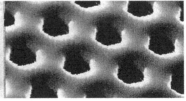

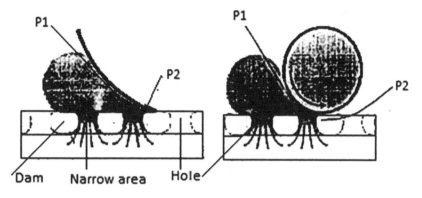

Penta screen

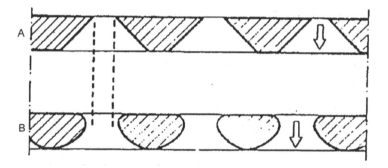

Comparison of Hole shape of Normal Screen (A) and PentaScreen(B)

NovaScreens

The holes are even more favourable, with an optimised bi-conical shape. The outside dams between the openings are even narrower. The wall thickness has increased considerably. The optimal parameters of NovaScreens lead to unique advantages:

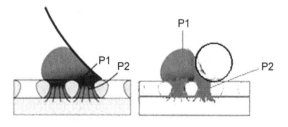

Nova screen section

Higher mesh counts combined with wider hole diameters not only result in also offer considerably higher open areas and a far lower risk of screen blockages.

There are a different group of NovaScreens (ED) which has a higher wall thickness for better stability. These screens have all the advantages of the other type of screens along with stability. Please note that the paste quantity inside the screen is lower in the new technology screens compared to standard screens which helps with better design clarity and stability of screens. The bi-conical hole shape positively influences the print paste flow through the perforation. Here again, the reduced widths of the outer dams result in a remarkable surface print quality, if required, and excellent edge sharpness. This means that a single NovaScreen is able to print tones, fine details and even blotches. In addition, the optimised hole structure (small dams) less penetration is needed to bridge the dams to achieve an even color distribution. This enables us to save print paste better. Thicker qualities can be printed by adjusting print recipe little and applying little higher pressure on the squeegee to get deeper penetration. NovaScreens have following advantages:

* less blocking
* less wear and tear of the squeegee
* save print paste
* higher print speed
* possible to print with higher viscosity for the sharpest results
* less rejections
* fine outlines
* fine geometric designs (weave imitations)
* quality half-tonal designs
*constant evenness of blotches

Special screens with denotation ED are for special purposes additional to wall thickness. 135ED are for blotches, White pigments, metal prints, thicker fabrics, etc.

165ED are for combination of design types (halftones, blotches, geometrical designs, fine lines)

195ED: highest mesh count, relatively large holes, very fine lines, geometricaldesigns and tonal effects.

Hole shape of one manufacturer (enlarged)

Specifications of a screen

A screen has many specifications with which is identified. A printer selects a screen for his particular purpose taking into consideration all these specifications. It should be noted that it is not necessary to select same type screen of each colour of a design. One can use one type screen for blotch, one type for fine details, one type for broad sections of the design, etc. Following are the specifications of a screen, which is explained with reference to Stork screens.

Mesh count

Mesh count is the most important feature of a screen. Basically, the screen is identified with this number and to some extent all other characteristics can be derived from this. The "mesh count" indicates the number of screen perforations per linear inch. This determines the fineness of the screen itself arid indicates the achievable print result. The holes in the Stork screens have a honeycomb cell shape. This may vary with different manufacturers. The hole shape is normally hexagonal.

Thickness

This refers to the thickness of the screen wall. It is needless to say that it is the maximum contributor towards the stability of a screen. The thickness of the screens is indicated in microns. The NovaScreens version is the thickest screen that provides greater stability in the printing process.

Hole diameter

In the previous enlarged diagram one can see that the holes are not directly drilled on to the screen. There are different structures for different typesof screens. Design of the hole structure contribute in various factors in the actual printing operation which is described earlier. The NovaScreens has relatively large holes because of the optimised bucolical structure.

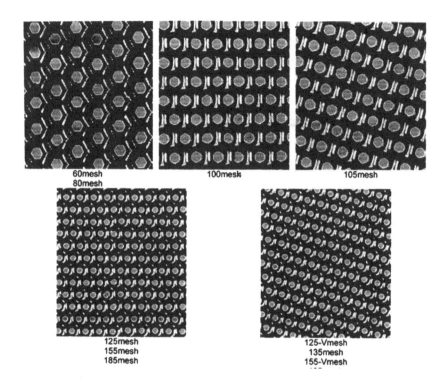

E.g.: compare a PentaScreen 155 with a NovaScreen 165 ED. The NovaScreen has more holes and these are also larger. This means a finer print and less blocking in the printing process.

Number of holes

It is evident that the number of perforations per cm2 determines the fineness of a screen and, to a large extent, the fineness of the printed result. A higher mesh count also results in more uniform paste dispersion in solid motifs. Consequently, it can create considerably smoother blotch prints (especially in combination with a higher open area).

Open area

The open area of a screen is determined by the number of perforations and the hole diameter. The indication of the open area is expressed as a percentage; The NovaScreen 165 ED with an open area of 19% comprises 81 % solid nickel. The percentages between the various screen types are very different because of the different screen structure. This is one of the most important factors in the determining the amountof paste supplied onto the fabric.

Repeat (mm)	Length (mm)	Mesh	Open area %	Thickness (mm)
640	1280-3500	60	19-21	0.115-0.120
		80	16-18	0.110-0.115
		100	14-16	0.105-0.110
		105	16-18	0.105-0.110
		125	14-15	0.105-0.110
		125V	14-15	0.105-0.110
		135	13-14	0.100-0.105
		155	13-14	0.100-0.105
		155V	13-14	0.100-0.105
		165	12-13	0.100-0.105
		185	11-12	0.100-0.105
688	1280-3500	60	19-21	0.115-0.120
		80	16-18	0.110-0.115
		100	14-16	0.105-0.110
		105	14-16	0.105-0.110
		125	14-15	0.105-0.110
		125V	14-15	0.105-0.110
		135	13-14	0.100-0.105
		155	13-14	0.100-0.105
820	1280-3500	60	19-21	0.115-0.120
		80	16-18	0.110-0.115
		100	14-16	0.105-0.110
		105	14-16	0.105-0.110
		125	14-15	0.105-0.110
		125V	14-15	0.105-0.110
		135	13-14	0.100-0.105
		155	13-14	0.100-0.105
914	1280-3500	60	19-21	0.125-0.130
		80	16-18	0.120-0.125
		100	14-16	0.115-0.120
		105	14-16	0.115-0.120
		125	14-15	0.115-0.120
		125V	14-15	0.115-0.120

		135	13-14	0.110-0.115
		155	13-14	0.110-0.115
1017	1280-3500	60	19-21	0.125-0.130
		80	16-18	0.120-0.125
		100	14-16	0.115-0.120
		105	14-16	0.115-0.120
		125	14-15	0.115-0.120
		125V	14-15	0.115-0.120

Example of one manufacturer's different type of screens – Repeat, Length, Mesh, Size, Thickness of screen and Open area percentage

Theoretical finest line

The finest lines we can "theoretically" print (because, as well as the screens, many other printing and paste conditions have a great influence) are shown in the table. The finest line that can be printed in an ideal situation is mathematically determined by adding the effective hole diameter and the thickness of the next following dam of nickel, which is also called the "pitch". Thus, for example, the theoretical finest line for a NovaScreen 165 ED would be 0.15 mm (= 25.4 mm:165 ED). Taking this value into account, one can determine the finest continuous and non-interrupted printable line. This requires TWO perforations to be covered by the line on a conventional film or in a digitally created image used for the engraving process.

Screen Geometry

The geometry of the screen perforations reflects their hexagonal shape and distribution in the screen matrix. Perforations are always aligned in two planes, positioned at an angle of 60° relative to one another. However, to accommodate different types of application, two alternative arrangements relative to the horizontal plane of the screen are employed.

The 60° perforation line is clearly related to a horizontal perforation line. As indicated, all the Standard Screens, as well as the 105 mesh and 125 mesh PentaScreens, belong to this category. The basic 60° relationship remains in the lower drawing, but the 'horizontal' perforation line is tilted at 15° so the ascending line is at 45° to the true horizontal. This 45°/15°-structure is employed for all the other PentaScreens and all NovaScreens But what do these different screen structures really mean to the printer?Generally, screen types with a 60° ascension can cause the following, very typical faults: (a) Where horizontally orientated lines have to be printed, they will show interruptions of variable lengths, which increases as the line becomes finer. The reason for this effect is that such fine lines never correspond exactly to the horizontal perforation line of the screen. Consequently, in areas where the engraved line intermittently falls between two of the perforation lines, no paste can be transferred to the substrate.(b) The same effect can occur with geometrical designs including straight lines ascending and/or descending at an angle of approximately 60°. This leads mainly to the notorious saw-tooth effects, where the engraved line also jumps from one perforation line to the next, causing a stepped, discontinuously printed line. (c) Even with thicker outlines, this phenomenon can cause a similarly disturbing effect when an engraved line covering two or three screen perforation lines momentarily extends to three or four lines. In such cases, the outline will be printed with regularly repeating thicker-than normal sections. (d) For the same reason, geometrical motifs with straight edges, squares, rectangles, rhombuses, etc. - can exhibit irregular, uneven outlines.All screens with 45°/15° structures have a totally different behaviour in this respect. The fault types mentioned under (a) to (d) generally cannot occur, because the angles of film or file lines and those of the perforation lines are far removed from each other. The two structures tend not to follow one another; instead, the engraving line cuts across the screen's perforation lines.On the other hand, this does mean that 45°-structured screen types will show the same defects for designs containing lines at a 45° angle to the horizontal.Another criterion for the selection of 60°or 45°structured screens can be to avoid fabric moiré effects. The 60° screens more often show fabric-related moiré. This problem can usually be solved by employing a different mesh-count and/or a 45° screen (such as NovaScreen).

	Standard screen 80 ED	Penta screen®		Nova screen®		
		125 mesh	155 mesh	135 ED	165 ED	195 ED
Stability index	106	100	104	104	106	112
Ascending/descending performation lines	60° / 0°	60° / 0°	45° / 15°	45° / 15°	45° / 15°	45° / 15°

Stability index
In the table the mostly subjective evaluation of screen stability is replaced by mathematically calculated values. The higher the stability index, the more stable the screen - indicating that less creasing or bending will occur in the production process.

It is the hole-dam geometry and the quality of the nickel that determines the stability index. The values are based on PentaScreen 125 mesh. One can clearly see that the Standard Screen 80 ED type is approximately 6% more stable than the PentaScreen 125 mesh, while the NovaScreen 165 ED has an approximately 6% higher stability than the PentaScreen 125 mesh. (All Storks specifications)

Given below specifications of screens of a particular manufacturer (Stork):

Standard Screens					PentaScreen					NovaScreen				
Repeat	Mesh	Thickness (µm)	Open area %	Hole Dia (µm)	Repeat	Mesh	Thickness (um)	Open area %	Hole Dia (µm)	Repeat	Mesh	Thickness (µm)	Open area %	Hole Dia (µm)
53.7	60	100	14	161	51.8	125	100	15	79	53.7	165ED	115	19	67
53.7	80	87	12	111	51.8	155	100	12	58	64	135	105	24	92
64	40	118	23	305	53.7	125	100	15	79	64	135ED	120	22	88
64	60	100	14	161	57.4	105	105	15	101	64	165	100	21	71
64	70	100	13	131	57.4	125	100	15	79	64	165ED	115	19	67
64	80	87	12	111	60.9	125	100	15	79	64	195	100	18	55
64	80ED	95	10	101	64	105	105	15	101	64	195ED	115	16	52
64	80T	87	12	111	64	125	100	15	79	68.8	135ED	120	22	88
64	80H	87	12	111	64	125v	100	15	79	68.8	165ED	115	19	67
64	100	95	8	74	64	155	100	12	58	68.8	195ED	115	16	52
68.8	80ED	95	10	101	64	155DLH	105	7	44	71.6	165ED	115	19	67
68.8	80ED	95	10	101	64	185	90	11	47	81.9	135ED	120	22	88
68.8	100	95	8	74	64	215	90	7	31	81.9	165ED	115	19	67
72.5	60	100	14	161	66.8	125	100	15	79	81.9	195ED	115	16	52
72.5	80ED	95	10	101	68.8	125	100	15	79	91.4	135ED	120	22	88
80.1	80ED	95	10	101	68.8	155DLH	105	7	44	91.4	165ED	115	19	67
81.9	40	118	23	305	72.5	125	100	15	79	91.4	195ED	115	16	52
81.9	60	100	14	161	72.5	155	100	12	58	92.3	135ED	120	22	88
81.9	70	100	13	131	80.1	125	100	15	79	92.3	165ED	115	19	67
81.9	80ED	95	10	101	80.1	155	100	12	58	101.8	165ED	115	19	67
81.9	100	95	8	74	81.9	105 ED	120	15	101					
91.4	40	118	23	305	81.9	125	100	15	79					
91.4	60	100	14	161	81.9	155	100	12	58					
91.4	70	100	13	131	81.9	155DLH	105	7	44					

91.4	80ED	95	10	101	91.4	105ED	120	15	101
91.4	100	95	8	74	91.4	125	100	15	79
101.8	40	118	23	305	91.4	155	100	12	58
101.8	60	100	14	161	91.4	155DLH	105	7	44
101.8	80ED	95	10	101	92.3	125	100	15	79
101.8	100	95	8	74	102	125	100	15	79
168	60	100	14	161	107	105ED	120	15	101
168	70	100	13	131	107	125	100	15	79
182	70	100	13	131	114	105ED	120	15	101
					119	105ED	120	15	101
					119	125	100	15	79
					120	125	100	15	79

Note: ED – Extra thick, 125v – same as 125 but specially made for Geometrical Designs and horizontal lines, T – Transfer Version, DLH – Small open area, suitable for transfer prints, H Suitable for Half tones

Repeat size is the dia of the screen. The Repeat size is denoted by the cm value, but the accurate repeat size is measured in exact mm which is very important for the design maker and the engraver. Film size is decided based on the repeat and the thickness of the film, etc., as the dia of the film is higher than the screen repeat size.

Another manufacturer's different screen comparisons (Stork)

The most common repeat size is 640mm; other sizes are 819, 914, 1018. 1450, 1680. Bigger repeats are used with special attachments. There are now screens of 2m dia which is able to print one bed sheet as one repeat. These are mounted on the printing machine with open bearing and sleeves to hold the screen in position. These screens are useful for printing bed sheets, scarves,

Sleeves for loading the big repeat screens

Kanga (an apron cloth used by African ladies) which has big repeat designs covering the size of the garment, or bordered designs (like in Kangas).

Big repeat screens loaded on the machine

Theoretical paste volume:

The theoretical paste volume of a screen indicates the sum of all hole volumes in ml/m^2 which equates to the "paste carrying capacity" of the screen. The principal factors influencing the paste volume of a screen are the number of perforations per cm^2 or per m^2, together with the hole size and wall thickness.

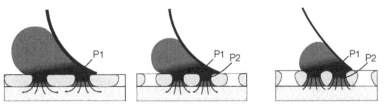

Standard Screen   NovaScreen   PentaScreen
Comparison of Printing (Blade squeegee)

	Standard	Penta		Nova		
	80ED	125 mesh	155 mesh	136ED	165ED	195ED
Thickness in µm	95	100	100	120	115	115
Hole Diameter in µm	101	79	60	88	67	52
Open area in %	10	16	12	22	19	16
Theoritical finest line in mm	0.31	0.21	0.16	0.19	0.15	0.13
Number of holes per cm^2	1128	2776	4141	3367	4880	6633
Theoritical paste volume in ml/m^2	9.5	16	13	26.4	21.9	18.4

When printing delicate designs with critical printing pastes, storks Screens tend to get blocked. Some type screens block more than others. Usually, the blocking is observed only when one observes in the printed designs only. This happens only after a no. of holes get blocked. At this time, the printer has to stop the machine and clean the screens which means increase in stoppage time and water wastage. But NovaScreen of stork (or equivalent screens of other manufacturers) is designed in such a way that these types of stoppages are largely reduced, 6 – 7 times. Thus, longer lengths can be printed without any interruption.

Paste filtration

Another reason for screen blockage is the particle size of the print paste. It is needless to say, that mesh size for filtering the print paste has to be selected judiciously to avoid screen blockage. As with printing screens, the fineness of the stainless steel filter gauze is indicated in "mesh counts". This can be very confusing, as the woven structure of the filter gauze is very different from that of an electroformed rotary screen. It must not, therefore, be assumed that gauzes and screens with the same mesh-counts are compatible with one another. The important factor in filtering is that, to avoid screen blocking problems during printing, the holes in the filter gauze must be smaller than those of the rotary screen. A 125 mesh rotary screen, for example, has a 79 µm hole diameter - whereas a 120-mesh gauze has a 120 µm hole. In this case, we therefore would advise the use of a 200 mesh filter gauze, with a corresponding hole size of 73 µm. What the printer really needs in his colour kitchen is several different filters, so that he can choose the gauze with an opening width at least a little smaller than the hole diameter of the screen being used. The most common screen types and the corresponding filter gauzetypes are listed in the table, together with the technical data for both.

Given below a guideline specification of different screens (Stork):

Mesh	Hole Dia. (μm)	Filter No./Mesh	Opening Dia. (μm)
Standard			
40	320	80	175
60	161	120	120
70	131	120	120
80	101	200	73
80ED	101	200	73
100	74	200	73
Penta			
105	101	200	73
125	79	200	73
155	60	325	42
155DLH	44	325	42
185	47	325	42
Nova			
135ED	88	200	73
165ED	67	325	42
195ED	52	325	42

Impact of paste volume: The volume figures in the table show that the PentaScreen 125 mesh has approximately 500/0higher paste volume than the comparable 80 ED mesh Standard Screen - and the Nova Screen 165 ED again has almost 500/0higher paste volume than the Penta Screen 125 mesh. Some may think that a higher paste volume would be associated with higher print paste consumption but, in fact, the reverse is true. With correct adjustment of the paste and printing parameters at 1igher volume screen can produce considerable paste savings. The table also shows many in practice well known print-technical correlations - especially the widely proven ones between the StandardScreens and the PentaScreens.

Theoretical Paste Volume indicates the mathematically calculated carrying capacity or filling quantity in ml/m2 of the screen type concerned. (TPV = open area % x thickness)

Real Paste Supply defines the paste quantity effectively transferred to the substrate, e.g., the 120 g/m2 quoted although the actual value always depends on all the relevant paste and printing parameters, and on the fabric itself:

Description	Standard	Penta		Nova		
	80ED	125 mesh	155 mesh	136ED	165ED	195ED
Theoritical paste volume of screens (ml/m2)	9.5	15	13	26.4	21.9	18.4
Suppose your need 120 (g/m2) paste onto substrate	120	120	120	120	120	120
Print condition factor (e.g., dye class/ qualiy, paste recipe/ viscosity, squeegee type/pressure, fabric sort/preparation, print speed/paste level etc.)	12.5	8	9	4.5	5.5	6.5

Print Conditions Factor is the effort needed to achieve the required paste supply. The PCF-value is determined by all the important print and paste-related variables, such as dyestuff class and quality, the printing paste recipe and viscosity, the squeegee type and pressure, the kind of fabric and its pre-treatment, the printing speed, the paste levelinside the screen, etc.

1. Penta screen

From a TPV of 15 ml/m^2 and an RPS of 120 g/m^2 we derive a PCF value of 8. This embraces all the above-mentioned variables. For the sake of simplicity, however, it can be considered as a function of e.g., squeegee pressure only.

2. Standard screen

With the same RPS of 120 g/m^2 and a TPV of 9.5 ml/m^2, the PCF becomes approximately 12.5; i.e., about 50% higher than in case 1. One or more paste and/or printing conditions must therefore be adjusted in order to achieve the same paste supply of 120 g/m^2. This may require a lower viscosity, higher pressure, reduced printing speed, a more thorough fabric pre-treatment, etc. which not only means a decrease in quality, but also material losses and lower productivity, possibly resulting in substantially increased costs.

3. Novascreen

NovaScreens' lower PCF value of about 5.5 for a 165 ED (low effort) clearly confirms that the electro-forming developments enabled Stork Screens to introduce performance improvements that offer many advantages to the printer. This means you are able to print with a lower squeegee pressure, smaller diameter magnetic roller, higher viscosity-, higher speed, or a combination of these variables. Accordingly, as proven in practice over many years, the NovaScreen, should not just be seen as the top precision and quality rotary screen, but also as the easy-to-use and cost-saving screen. With correct adjustment of the printing parameters, they provide highly advantageous price/performance ratios in the print shop.

Printing of geometrical design

When selecting the screens for geometrical designs, lines, stripes, etc. the screens have to be selected diligently so that you get perfect lines and outlines without breaks in printing.

Screen geometry

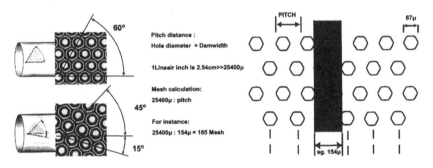

Up to 125 mesh, the angle is 600 and above 125 the angle is 450.

Guideline for Mesh Uses

Mesh 40, 60, 70 – Suitable for Blotches (On heavy grades of cloth), Mainly used for penetration.

80, 80ED – For use with Outlines, suitable for Blotches (On normal grades of cloth), Adequate penetration.

100 – Surface Printing. For use with outlines.

105, 110 – Suitable for Blotches, Excellent results in Halftone, Discharge and Resist printing Combination Printing.

125, 125v – For all applications, suitable for Blotches, surfaces print and penetration, sharp outlines, fine details.

155 – Fine Outlines, Discharge and Resist printing, mainly suited for sharp Geometric designs with horizontal and vertical lines.

For a 60° line in the printing design, one should use screens with 45° angle and 45° angle lines, one should notuse screens with 60°.

Stork penta screens

Penta 125 mesh: Universally applicable, blotch printing is even more consistent, surface and penetration printing, contours and fine details up to 0.2 mm.

Penta 155 mesh: Suitable for blotch printing on light material qualities, contours up to 0.16 mm, half tone printing, discharge and resist printing, excellently suited for designs with horizontal and vertical lines, 155 DLH: Specifically, for transfer printing and half tone printing.

Penta 185 mesh: For extremely fine half-tone printing, contours up to 0.13 mm, designs with horizontal and vertical lines.

Penta 215 mesh: Half-tone printing, contours up to 0.11 mm.

Electroplated screens

During manufacture (nickel plating) the design is simultaneously applied to an electro-plated screen by leaving holes open only in those places where the design requires this to be done.

Making an electro-plated screen demands rather complicated equipment and skilled personnel.

The choice of a mesh no of screen depends on the cloth to be printed, the design to be printed in combination with the desired printing effect and the composition of the printing paste.

Given below the guideline for selecting mesh no. as per fabric:

Mesh Nos	Fabric Qualities
40	For terry-cloth, corduroy qualities, needle-felt and other qualities
125, 60	For coarser qualities, such as textured fabrics and knits and generally for blotch printing.
125, 80	The most 'widely used types of screens for normal contours and plain weave qualities
185, 100	Foe light Qualities and design with fine contours

Selection of screens as per design- Guideline

Mesh Nos	Fabric Qualities
185	Used for fine contours and well defined figures
125	Combination of small surfaces and lines and for blotch printing in which the cut off figures must also be well defined
Penta v screens	V-screens have a special raster which makes them suitable horizontal and vertical lines without causing a *moire effect*.

| 255 | By combining special raster films with 255M beautiful raster effects can be obtained. |
| T screens (80) | For transfer printing T-screens (SOH) with a very smooth surface are recommended,we do not recommend the application of a thick coat of lacquer here. |

Notes:

1. The holes of the screens are small. The higher the mesh numbers the smaller the opening.

2. Proper filtration of the printing paste is required to avoid clogging.

3. In the case of normal coats of lacquer, the contour may at some places be interrupted by the space between the holes. When a thick coat of lacquer is Applied, there will actually occur a channel in the coat of lacquer. The dyestuff is forced into this channel the result of which is an uninterrupted line.

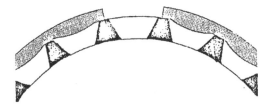

4. For big blotches we recommend a coarser mesh type to facilitate penetration of the dyestuff. Blotches consume the greatest quantity of printing paste, of course, which is essentially filtered by the screen. By selecting a larger mesh size for a larger quantity of printing paste, the risk of clogging and stripe formation is reduced.

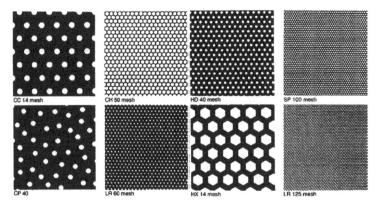

Different rasters on Stork rotary screens

Selection of screen as per composition of printing paste

Though in principle the composition of the printing paste has to be adapted to the cloth to be printed and to the screen suitable for this purpose, it often occurs in practice that as a result of the use of various types of printing machine and standardisation in the colour kitchen the ideal composition of the printing paste for a certain article is not obtained. Low viscosity pastes will more easily penetrate through the screen than very thick pastes. If such deviations exist, they can be partially adapted by the choice of the screen.

By varying the mesh number, the composition of the printing paste and the type of squeegee blade it is possible to obtain a very great variety in quality. These factors together determine the final appearance of the print, and they should be so adapted to each other that extreme viscosity and/or squeegee pressure are avoided.

First step of printing is to fix the screens. Screens can be fixed in two ways- Dark to light or light to dark screens. In ideal case, both methods can be adopted in any designs there are some prerequisites for a particular sequence.

When Dark to Light sequence is selected, we should have constant width of the fabric, proper sealing of the screens, an appropriate overlap (say 0.2 mm), fabric is properly fixed on the blanket, etc. By this method sharpest prints, and most beautiful overlaps can be obtained.

In certain special cases this sequence may not work and thus the light to dark sequence has to be adopted strictly in such cases. This method demands a good absorptive power of the fabric, and print pastes to be generally thin. This method is most suitable where poor adhesive power when the blotch is printed in the first printing position; a highly varying width of the cloth, so that the dark colours smudge the light colours via the blanket; designs with large overlaps, as a result of which the lighter colours also mix with the dark contours.

2.3.4.2   The squeegees

The squeegee performs several functions in the printing process, but must make compromises in completing each in order to accomplish them all. The most important of these functions include the following:

   * forcing the screen and substrate into contact, thereby allowing paste transfer to occur

   * providing hydrodynamic pressure to force the ink through the screen

During the print stroke, forces exerted on the squeegee include downward pressure to bring the screen and substrate into contact at the squeegee edge, as well as friction forces (drag) that cause the squeegee edge to resist movement across the screen's surface. These forces combine to deflect the squeegee from its original angle to its actual printing angle, commonly referred to as the contact angle. The size of the contact angle has a critical influence on ink shearing and the amount of hydrodynamic pressure applied to the ink--the smaller the contact angle between blade and mesh, the greater the hydrodynamic forces and the poorer the shearing action of the squeegee. Controlling this contact angle is difficult because so many variables exist. The most important of these variables include the following:

* Amount of squeegee pressure set by the operator

* Paste system used and how it affects the frictional resistance between the squeegee and screen

* Speed of the squeegee stroke

* Type and durometer of blade

* Free blade height between blade edge and holder

Blade squeegees

This is the conventional squeegee system where the curvature of the blade, and therefore the angle of the contact between the blade and the screen, changes according to the applied pressure, which can be readily altered by adjusting the bearings at both ends of the squeegee assembly. With traditional squeegees, the effect of screen tension across the print width also is an issue. The tensioned mesh creates additional resistance and deflection to the squeegee toward the ends of the blade closest to the frame edges.

Once the screen is fixed in position the next the squeegee has to be fixed inside the screen in position. The function of a squeegee is to press controllable quantity of the printing paste through the perforation of the screen on to the cloth. The paste is supplied into the screen by pumps through a distribution pipe of the squeegee which distributes the paste uniformly across the width of the fabric before the squeegee blade.

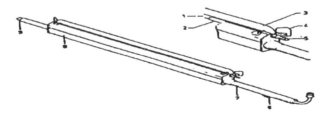

A squeegee assembly. 1- Clamping Hose, 2 – Clamping Pipe, 3 – Squeegee Blade, 4 - Scraper, 5 – Air supply valve, 6 – Suspension cam, 7 – Paste supply pipe, 8 – Housing 9 – Squeegee cap

The squeegee consists of a rigid stainless steel housing which contains the printing paste supply pipe and the squeegee blade clamping strip.

The squeegee blade comes in strip rolls of different width and gauges (see Table below). As per the requirement of the print the printer cuts the strip and blade is clamped by means of an inflatable clamping hose and a clamping pipe. This permits rapid and easy assembly or replacement.

Thickness of the Squeegee Blade (mm)	0.1	0.1	0.15	0.15	0.2
Height (width) of the Squeegee Blade (mm)	40	40	50	55	

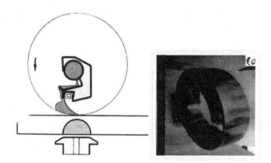

Squeegee Blade strips

When the squeegee is lowered in the printing position the squeegee blade assumes a curved shape, resulting in a "wedge" between squeegee blade and screen. Adjustment of the squeegee pressure and the type of blade used are decisive for the shape as well as the size of the "wedge". Every type of steel blade has its own specific range of applications.

In addition to normal blade squeegees, special blade squeegees with synthetic tips are available to reduce the frictional force between the blade and screens which can be used in the absence of the airflow squeegees and special requirements.

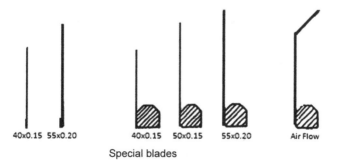

40x0.15 55x0.20   40x0.15 50x0.15   55x0.20   Air Flow

Special blades

Blade squeegees, rod squeegees and combined blade and rod squeegees are used in rotary screen printing.Traditionally blade squeegees are used in special cases blade cum rod squeegee is also used by printers. But nowadays the magnetic rod method is also becoming increasingly important in its use in textile screen printing.

The squeegee assembly is held by squeegee suspension on both sides of the screen holder. The assembly is passed through the screen and on the drive side the squeegee moves into eye. Whereas on the pump side it moves into one cut out and clamped. The paste supply pipe will be projecting pout at the pump side and each colour paste supply hose is clamped to each proposed blade. There is a provision to adjust the applied height of the squeegee blade and it can be locked in position.

How to fix/replace Squeegee Blades
Place the squeegee in two brackets. Cut squeegee blade to correct length and apply rounding to the ends with the aid of the jig which has been supplied. Use fine abrasive cloth to finish the sharp edges that result from cutting.

Length of the squeegee blade Maximum printing width of the screen in general:

1340 mm	1280 mm
1680 mm	1620 mm
1910 mm	1850 mm

Place the squeegee blade at the bottom of the groove of the squeegee. Mount clamping hose.

First tighten the screw on one side. Pull squeegee blade taut on the other side and secure it in position with a screw. Now inflate clamping hose (5 - 6 bars). Check squeegee blade for any burrs and for straightness. The squeegee is now ready for use (The method may vary for different manufacturers).

Notes:
1. The plastic baffles on both sides of the squeegee have to be present to prevent the dyestuff from flowing behind the squeegee blade.
2. If M.L.R. squeegee blades or Air-flow squeegees are used the squeegee blades are supplied in the required size per printing width. So, in that case no preparation of the blade is necessary.

Machine 'stopped' position the squeegee is raised and not in contact with screen. When the machine is started the squeegee assembly is lowered to printing position and the pressure of squeegee can be adjusted as per printer's requirement.

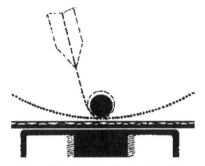

Magnetic rod squeezee in comparison blade squeezee

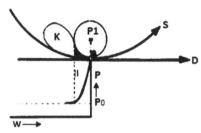

Action of magnetic rod squeegee

The printing which is applied using the blade squeegee is more superficial, rod squeegees.

Generally, press the coating into the substrate more. Blade squeegees require a relatively coarse screen and printing pastes with relatively low viscosity for deep penetration. The rod squeegee is characterised by the type of printing (the rod behaves in a similar way to a roller printing and the contact pressure is relatively high) which resembles roller printing. Fine pattern details can be printed more clearly using the magnetic rod squeegee because its rod squeegee characteristics make it possible to use finer screens and printing pastes of higher viscosity. Large- or full-area patterns or ground prints can be printed more evenly by using a rod squeegee than by using a blade squeegee. It is preferable to install two rod-squeegee stations in existing blade squeegee machines – one for the fine details (contours) and one for the large areas (blotches).

Selvedge to selvedge variation chances are more prevalent in blade squeegees especially in wider fabrics. In such cases and blotch prints it is better to use Magnetic rod squeegees. Magnetic field attracts the rod uniformly across the width of the substrate gives a better chance for uniformity of the print. In blade squeegee contact pressure produced mechanically (pneumatic, hydraulic or spring pressure and lever effects); risk of distortion – the force required for producing the contact pressure is introduced, or thesqueegee is supported, via the ends of the squeegee suspension unit. The capacity for adjustment and the reproducibility of the contact pressure depends on several factors.

Adjustment can be corrected, or correction must be applied, on both sides of the machine. As the pressure increases the friction between the screen and blade increases. In the case of rod squeegee contact pressure produced electromagnetically; distortion-free printing; contact pressure is produced by a magnetic field parallel to the rod withoutapplying any load to the squeegee-unit mountings or the ink-paste feed pipe. The contact pressure is very easy to adjust and reproduce – adjustment is carried out by means of a printing pressure step switch at each printing station or continuously by using a rotary potentiometer.

When printed light-weight fabric a metal blade is often preferred to a rod. In modern printing mills very often both systems can be found.

Airflow squeegee
The airflow squeegee is similar to the conventional type, but has an adjustable tip angle and pressure, offering a wide control range for application rate and penetration while the machine is actually running. The air-flow squeegee can be used within the entire pressure range and consequently it is a universal squeegee. The air-flow squeegee is a further development of the scraping squeegee. There are two characteristic features:

1. The wedge angle which is decisive for the quantity of print paste to be applied and the mechanical pressure (bellows pressure) on the tip of the squeegee are independently adjustable.

2. The tip of the air-flow squeegee is made of synthetic material so that there is only minimal friction between the squeegee and the screen.

In practice this means that the air-flow squeegee gives the following advantages:

- Universally applicable.
- Each type of squeegee blade has its own specific range of applications. However, the air-flow squeegee can be used for the whole printing range - from fine contours to blotches. At high bellow pressure you will have the benefit of a rigid blade (40 × 0.15 and 55 × 0.2) and at low bellows pressure the benefit of a slack blade (50 × 0.15).

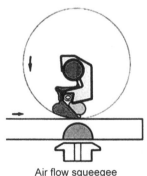

Air flow squeegee

The bellows pressure may be increased and the necessity for this depends on the question whether the cloth to be printed contains many knots or slubs or whether there are creases in the screen. Higher squeegee pressures give higher counter-pressure of the blade (resilience; as a result the total mechanical strength on the squeegee blade is also increased (resilience of the squeegee blade + bellows pressure). If an air-flow squeegee is used in combination with steel blades, the arc formation can be influenced by varying theBellows pressure. At higher squeegee pressure the bellows pressure can be decreased. Low bellows pressure results in more surface print which sometimes gives more brilliance whereas high bellows pressure gives more penetration. Using the airflow squeegee the quantity of dyestuff you want to apply is regulated by the squeegee angle (Wedge angle). Due to a different formation of the wedge angle between the squeegee and the screen the application curve of the air-flow squeegee shows a marked difference. In comparison with 55×0.2 and 50×0.15. The angle of the squeegee can be adjusted as per printer's requirements. From experience, for normal printing a lower angle is preferable and for heavier qualities and blotch printing higher angles are used. Along with the angle of the squeegee horizontal position is to be adjusted and generally the air-flow squeegee is set to position 5 when the slant is lowest (as with blade 55) and to position 6 when the highest. All these adjustments are guidelines only the printer has to do his own combination adjustments to get the best results.

Advantages of using airflow squeegees

1. Centre selvedge variation: If the squeegee is adjusted backwards (in the direction of the drier) and up to a maximum of 3.5 the sides will be lighter or dark sides can be eliminated. Adjustment in forward direction (in the direction of the infeed) will lead to darker sides.

2. Better run properties: The air-flow squeegee's tip of synthetic material guarantees only minimal friction between the squeegee and the screen. In this way vibration of the squeegee is avoided and the risk of creases in the screen is greatly reduced.

3. Adjustable mechanical pressure. The mechanical pressure (bellows pressure) on the tip of the squeegee can be adjusted while it is independent of the wedge angle, and it is uniform across the full width. As a result: contours are printed better; the holes in the screen will less easily get clogged and knots in the cloth are printed with more coverage. This also helps in smoothening the creases in the screen and they are therefore less visible in the printing result.

4. Contact surface between the blade and the screen: Due to the preset angle of the air-flow squeegee there will be a 'line of contact' between the squeegee tip and the screen. When a steel squeegee is used at higher squeegee pressures there will be a contact surface between the squeegee and the screen. Low friction and the 'line of contact' give more uniformity of print when the air-flow squeegee is used.

Rod (Roller) squeegee

Principle

The rod pushes the paste through the open areas of the screen onto the substrate. The substrate is glued onto the printing belt. A table supports the printing belt where it touches the roller squeegee. Magnet coils have been fitted into this table; the pull of these magnet coils is steplessly adjustable by one potentiometer.

The printing result depends on different factors:

• cloth
• rod diameter
• magnetic force
• viscosity
• mesh number
• printing speed

Together these factors define the result. If these factors are correctly combined, an optimal effect will be achieved.

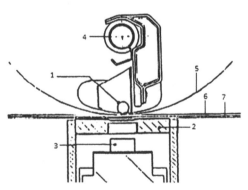

Schematic drawing of a Magnetic rod system 1- Rod, 2- Table
3. Magnet, 4- Paste distribution pipe, 5- Screen, 6- Fabric, 7- Blanket

Magnetic Squeegee with closed bearing system

The variables introduced by squeegee pressure, friction between the squeegee and mesh, and screen tension are the basis for many of the repeatability and consistency problems that plague the screen-printing process. To achieve the higher print speeds demanded, the effects of friction and applied pressure must be made much less significant in the process.

The best current squeegee options for increasing production speeds are to use squeegees that feature special support layers or use rigid back plates to prevent deflection. These specialised blades and accessories tend to limit your latitude in set-up but lead to more consistent and repeatable setups once you get used to them. The downside to these solutions is that the higher printing speeds they promote significantly increase friction at the contact point, which can reduce the life of the screen and/or squeegee.

When looking at the technologies used in other print processes, the obvious solution that emerges is to replace the squeegee blade with a roller squeegee. This device allows greater speed to be achieved while maintaining an excellent and consistent ink-film thickness. The roller squeegee is a concept that particularly suits web and cylinder press formats, and it may even prove viable for flatbed presses.

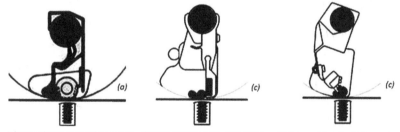

(a) Normal Magnetic Rod Squeegee   (b) Rod sqeegee for low Penetration   (c) Ros squeegee for high penetration

It has been proved that the rotary-squeegee concept worked, achieving 50% greater ink deposits than a traditional squeegee and powering the squeegee's rotation (rather than allowing it to rotate freely in response to screen motion) was the best way to minimise the effects of frictional drag on ink transfer.

Roller-squeegee systems actually achieve twice the hydrodynamic force of the conventional squeegee, and provide a controlled and highly supported contact region and very little deflection. Such squeegees provide consistent contact across the width of the image during printing. A limitation of the roller squeegee is that it requires a heavier construction than the conventional squeegee blade, especially because it must be independently driven to rotate. This also makes it more difficult to set up, which is why its application in flatbed presses may be limited.

The rod pushes the paste through the open areas of the screen onto the substrate. The substrate is glued onto the printing belt. A table supports the printing belt where it touches the roller squeegee. Incorporated in this table are permanent magnet cores whose force of attraction can be set incrementally. The printing result depends on various factors like cloth, rod diameter, magnetic flux, viscosity, mesh count, printing speed Together these factors define the result. If these factors are correctly combined, an optimal effect can be achieved.

The magnet roller squeegee is exceptional ease to operate with relatively few control functions. The diameter of the roller and the viscosity of the paste mainly determine the amount of print paste applied.

Rod diameter and type

With the blade squeegee, one can vary the amount of printing paste to be applied during printing. With a roller squeegee, the choice has to be made in advance since the rod diameter determines the amount to be applied. With roller squeegees one should also take the desirable speed of production and viscosity of the printing paste into account.

Smooth and knurled rods

When printing on certain types of web and/orwith printing pastes, spots may be formed caused by a kind of aquaplaning. By using a knurled rod this problem may be solved.

The magnetic flux (normally 4-6) serves to keep the roller squeegee in its place and squeegees the printing paste. Too high a magnetic flux may cause creasing of the screen and crushing of the colours. The amount of paste applied cannot be varied by means of the magnetic flux.

Printing width

Combined printing with roller and blade squeegee can take place up to printing higher widths

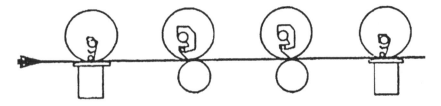

Lower width → → Higher width

Use of Magnetic and blade and rod squeegees for different width substrates

The contact time of the paste while printing operation is determined by the wedge area formed during the operation. In the print diagram, this area corresponds to the zone stretching from the initial rise in impact pressure, through the penetration resistance zone, to the squeegee contact point. The

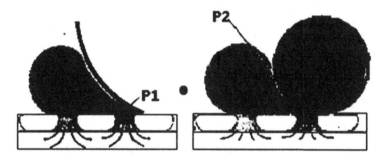

pressure behaviour in this area, which is known as the contact zone, has a significant effect on the amount of printing paste applied. A prerequisite for a uniform coating is a constant and defined level of ink paste over the entire printing width to keep the length and ratio of the two wedge areas constant. The discharge speed is dependent on the size of the impact pressure and the flow resistance of the screen. The size of the impact pressure is determined by the geometry of the squeegee, the squeegee force, the print speed and the flow properties of the printing paste. The flow resistance is determined by the percentage of the open screen area.

For each printing position there is a paste supply system consisting of a pump, paste level control and a pump cleaning device. As explained earlier the pump supplies paste through the supply pipe along with the squeegee assembly. If the pump is continuously pumping the paste may overflow. Hence the level of paste inside the screen has to be controlled. This is done by the paste level controller, which of the pump when the level is reached and as the printing is going on and the paste is consumed the level comes down and the pump is started, and allthese actions are automatically controlled by the level control. The level controller is a sensor rod which is fitted on a holder tip of which is the level of the paste inside the screen. This rod can be lifted or lowered at the printer's requirement.The pump cleaning device serves to clean hoses and pumps.

2.3.5   Drier

The driers incorporated in Rotary printing machines are of the same type as in an automatic flat-bed printing machine with some provisions of accommodating higher speed of the rotary machine. The drier should have a much higher drying capacity i.e., for a running speed up to 80-100 m/m. Generally, steam, thermic oil heating or direct gas fired driers are available. To avoid undue stretching of the wet printed fabric, especially knitted goods, the fabric is dried on conveyor and last stages may be without conveyor for woven fabric. For knitted fabrics there are dyers with two conveyors so as full drier path is on conveyor. Drying is done by hot air medium, by air passing though the steam or oil heated heat exchangers or radiators using a blower which delivered to the face of the fabric through nozzles. If gas fires the hot

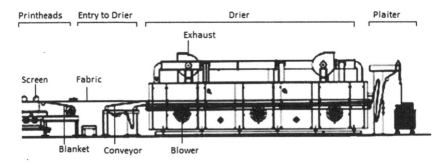

hot air can be directly blown to fabric through nozzles. Fabric is moved without tension on the conveyor belt which moved with it drive and once the conveyor ends the fabric is pulled out of the drier by the motor of the exit device. The conveyor belt is driven by a motor which runs synchronically to the drive of the printing blanket through a photoelectric cell. Conveyor is allowed to run in place by corrections made automatically by means of a probe, guiding rollers and motors. Part of the evaporated moist air is taken out by the exhaust.

2.3.6   Printing machine operation

In Rotary printing machine the printing operation is much simpler than other printing machines. The print pastes are placed in each screen position and the hose connecting to the pump is inserted into the paste. The screen fixed in position by locking to the screen drive on both sides and tensioned length way. Screen has to be positioned making precise adjustments in longitudinal, lateral and diagonal registers. The automatic repeating of round screens in rotation printing requires a fixed optic for each screen and a fixed point on the screen (register cross). The register cross on the screen can then be detected (e.g., by means of a laser beam). The position of the carrier peg for the screen can also be located in the screen hole of the printing group, which also occurs optically by means of suitable sensors. This relies on there being an end ring gluing device. By means of a simple optical sensor the position of the register cross in relation to the end ring and thus also in relation to the point of application of the carrier peg on the screen opening of the printing machine can be set precisely and reproducibly (see Fig.). Precise cutting of the screen stocking and the control of thermal problems when gluing are further requirements, so that less corrections are necessary in the repeating of the screen in the printing machine. In manual adjustments finer adjustment has to be done visually as

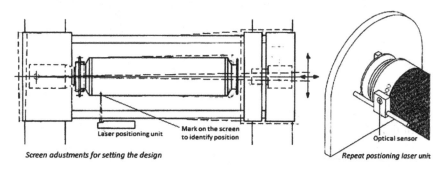

Screen adustments for setting the design   *Repeat postioning laser unit*

the printing starts. The position of the screen with respect the print head is decided by the experience of the printer and the coverage and overlap of the engraving. As a rule, the blotch screens are placed as the last screen The squeegee unit is fixed inside the screen and the level control is positioned taking into consideration of the coverage of the screen print and print

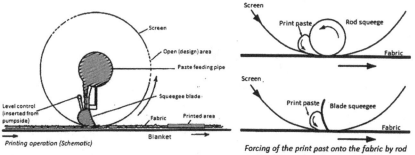

Printing operation (Schematic)   *Forcing of the print past onto the fabric by rod and bladesqueegees during machine running*

requirements and the pump started. Level controllers are a series of measuring systems that use the sensing device to compare should/is values (microprocessor) and subsequently control an on/off device to adjust the actual level of the paste.Machine is started slowly and the fabric is fed on to the blanket, which carries it along with it as the fabric gets attached to the blanket due to the gum. The design setting is visually checked and adjusted, and once design fitting is perfect the machine is run in full speed, confirming the fitting of the

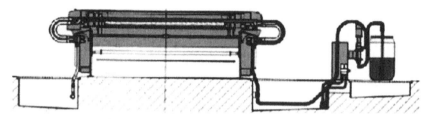

Colour paste pumping to the screen in an on machine screen washing system

design has not changed. During the running the squeegee forces the print paste through the mesh where it is opened (i.e., engraved design area). At the end of the print table the blanket goes down below the print table whereas the fabric is taken away be the drier, tension being controlled by a photocell unit synchronising the speed of the drier and the print table. Dried fabric plaited or batched at the delivery end of the drier is taken for further processing. It is better to plait the fabric which is going for steaming to allow it to cool and absorb maximum moisture possible whereas one can batch the fabric which is going for curing or dry heat treatment.

2.3.7   Variations from the basic machines

There are variations of the machines available from the manufacturers for special requirements, to increase the production, etc.

2.3.7.1   Larger repeats

The normal machine screen head can accommodate repeats size of 640-1020mm. If there are requirements for bigger repeats special provisions has to be made on the machine. These combine the advantage of a high print speed offered by the rotary screen with that of a large format offered by the flat screen. For reasons of stability, the perimeter of the screen can only be increased to a maximum of 2 m. These increase the design possibilities but at an extremely high cost. Screens are significantly more expensive and sometimes the components of the printing machinery have to be changed bring the circumference of the screen to 2 m. A further development is electronic controlled intermittent printing by two partially engraved complementary screens which lift and print in a reciprocating manner.

Large repeat screen fitted on the printing heads   Colourway changing in stork peagasus (Heat no. 2)

2.3.7.2   Quick colourway change

As print runs become shorter and shorter, change-over times are playing an increasing important role in determining the efficiency of print machines. Often these machines have more downtime than actual printing time. Print + Change system (Pegasus) can put an end to this screens and squeegees of previous designs can be dismantled during the printing process without any problem (see fig. above). Printing can continue at full speed; there is no risk of cloth contamination and the screens for the new design can be installed at the same time. Stork Pegasus has introduced a system whereby one set of screens can be dummied or cutoff from printing in a special way when the other sets of screens can be fitted in the remaining print head (Provided there are the minimum twice the number of heads available on the machine than the number of screens for that design) and the next colourway can start running without much delay. Once the print run is complete, the shutters beneath the used screens are closed and the ones beneath the new ones opened. The pervious screens are taken out and washed and fixed ready for the next set of colourway while the present screens are in production, The system consist of several mobile shutters (one for each print position) that can be inserted between the screen and the printing blanket. Once a printing position has been raised, the shutter can be positioned at the touch of a button. It makes a complete seal, so that screens and squeegees can be removed without any risk. Leaking paste will drop onto the shutter, not the fabric. The shutters can be opened again once a new design has been installed. They slide into a cassette at the bottom of the printing machine; any paste leaks are scraped off during this process. The shutters can be operated on a position by position basis. This is a very undesirable and wasteful situation. Printing can recommence immediately when the repeat is completed, the used screen can be removed at your leisure.

2.3.7.3   Screen washing on the machine

The on machine washing system is supplied by many manufacturers (e.g., Stork, Ichinose). Removing the screen from the screen holders, saving the paste left in the screen, washing of the screen and again fitting it back also is time consuming and labour intensive and often on repeated handling of the screen can cause damages to the screen. Hence the on-machine screen cleaning was proposed. It is an extension of quick colour change on stork Pegasus where the printing machine continues to print specific positions while other positions are cleaned. The squeegee holder will have an additional water spray pipe in addition to paste delivery pipe

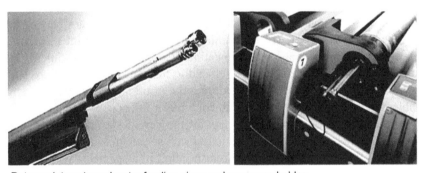

Rotary print paste and water feeding pipes and squeegee holder

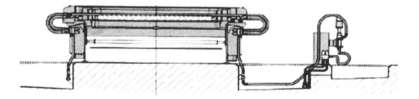

Ichinose Samurai

Printing operation is Samurai model printing machine is a little different than other machines. The printing table is as usual in any other machine where printing blankets are horizontal during the preparation phase. All preparatory and maintenance work is carried out in this position. But once ready to print, and at the push of a button, the printing units are raised around the longitudinal machine axis into an inclined position Automatic registration and automatic colour feeding starts up.

The colour paste is pumped from above into the screen, the volume is controlled via pump speed. The colour past in the screen is supplied by a way of a gravity-driven, pressure-less flow. As claimed by Ichinose this supply system provides outstanding uniformity; the colour flows continuously in a closed circulation. The rheological properties remain extremely constant.

The Samurai provides on-machine washing of screens and squeegees in a time-saving, simultaneous and automatic procedure. Once printing has been completed, the colour residue is extracted at the same time from screens and pipes. This process is highly effective, as the colour remaining in the screen will follow the law of gravity and flows of its own accord to the point of extraction. After the colour residue has been extracted, service water is fed into the system. This water flows through the screen and is conveyed – together with the water flowing out of the screen –into the service water tank. Final rinsing is effected using a small amount of fresh water. Compressed air is fed into the screen together with the washing water. This air first enhances the washing process, then finally expels the remaining water droplets.

Colour changes are effected without removal and re-installation of the screen. They remain on the machine, safe from damage. The result is less misprints, no re-registration is required, fitting is perfect, right from the first meter.

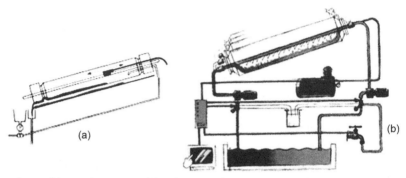

On machine pasterrecovery (a) and screen cleaning process (b)

A special use of Rotary printing machine is for Transfer Paper printing. These papers are used in transfer printing which is explained in this book.

2.3.8   Strike offs (Sampling)

In the commercial production process of printing, the printer has to show how the design appears on the fabric with colour matching of each screen for the approval from the customer (equivalent to labdip approval in dyeing). Initially it is done on a small piece of fabric using strike off machines. In many units, it is done by flat screens printing one screen after another manually or on a small screen printing unit. (See fig below)

Strike off machines   Strike off unit iwth rotary screens

Single head rotary machines are also used for obtaining samples closer to actual bulk production. Here the screens have to be made even before the approval of the design by the customer, which is a costly affair, poses problems. Hence this method is not popular. Some manufacturers make sample machines which can use small rotary screens for making strike offs. (See fig. above)

3
Design Aspects

In textile processing industry the printing designs are developed either by own artists or a design supplied by the customer is adopted for the print for the method used (if the design supplied is not possible to be used for the prints). In the latter case, for example a design repeat has to be adopted for a rotary screen either by repetition or by shrinking a the design (or as such) to accommodate the circumference of the rotary screen (There are exceptions like a single print, often as large as the garment or made up, printed nowadays made possible with the advent of digital prints which was often difficult with the limitations of screen sizes used.

3.1   Developing own designs

The basic unit in a design is called a repeat. Thus, as explained earlier whether a handprint done in small scales or a bulk print in a factory, fabric design involves the seamless repetition of this basic unit called repeat over the length and width of the cloth.

3.1.1   Direction and orientation of the repeat

A basic unit can be repeated in a print in different ways such as directional and non-directional.

Directional prints are usually in the lengthwise direction of the fabric. The motifs will be parallel to the selvedge of the fabric. Directional prints can be one-way or two-way.

In one-way prints the motif will be in one direction, but the position of the motif may or may not be at the same position in different line. The fabric can only be turned one way; Otherwise, the print looks upside down or sideways. In two way prints, the motif can be in two different directions (say, vertical and horizontal).

One-way prints can only be turned one way; otherwise, the print looks upside down or sideways. Even if most motifs are two-way or non-directional, if just one motif in a design is oriented one way, the entire design is considered one-way. *Two-way prints* look the same whether they're turned right side up or upside down. Non-directional prints can be either tossed or four-way. *Tossed prints* can be rotated in any direction, including on the bias, and look the same. *Four-way prints* look the same whether they are oriented at 0°, 90°, 180°, or 270°, and the eye will typically travel along these straight lines.

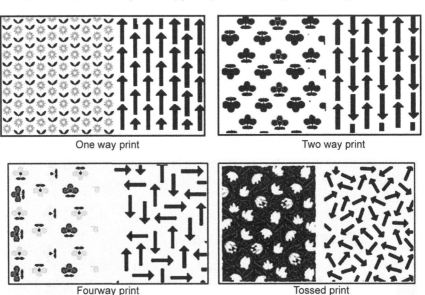

One way print   Two way print

Fourway print   Tossed print

Other than the above types there are many types of prints, some of which are explained below. *Railroaded* prints mainly used in upholstery fabrics, are ones with motifs running parallel to the selvedge of the fabric. *Border prints* are railroaded by nature and are designed with specific end uses for example Indian sarees, African kangas and can be used for pillow covers, tablecloths, etc. The decorative border travels along one or both selvedges (the two borders can be same or different) and normally the inside body of the fabric is filled with a quieter print or can be a solid shade. Usually, designs are developed with specific uses in mind.

There are designs created with even more specific end uses in mind such as craft panels, all-in-one prints, and cheater prints. *Craft panels* include all the ready-to-sew pieces for a project, such as a stuffed toy, printed directly onto the fabric. They might be printed directly onto a finished product, such as a T-shirt or tote bag, scarves, towels or aprons. Another commonly used prints are called *engineered prints* which is a different version whereby the design is directly printed on the garments. An *all-in-one print* whereby different motifs are artistically mixed in a design and is popular in some countries. A *cheater print* is developed after quilt pattern and is created as a mock patchwork, which might be simple squares or complicated. Like all-in-one prints, they are a great way to incorporate several prints into one length of cloth.

3.1.2   Prints named after motifs

Even though, motif refers to an element in a design but for a designer it is a repeated element.

Based on motif the design can be classified into many groups. It can be shrunk into three broad categories - geometric, floral, and novelty or into two: abstract (geometric) and representational (including both floral and novelty).

3.1.2.1   Name of some prints based on motifs

Geometric prints can include polka dots, stripes, plaids, checks, herringbone diamonds, etc. Floral prints have generally flowers, leaves, fruits, etc. Novelty, or "conversational," prints can be anything which is not included in the above two categories.

Paisley

Plaid

Stripes

Floral

Animal

Camo

Houns tooth

Polka dots

Plaid

Checks

Herringbone

Tribal

Geometric

Ikat

Engineered design

3.1.3   Repeat

A repeat is the basic unit of a design. The repeat is spaced in a design in different ways to fill the space in such a way that the repeat is easily identified or sometimes it is camouflaged so well in the print that it is difficult to identify. Hence the repeat types can be classified into   groups.

In *Square Repeat* (also called *Block Repeat, Straight repeat*), the motif or motifs are built within or overlapping a foundational rectangle (or square or parallelogram), and that rectangle is repeated as a simple grid. The grid may be invisible or incorporated overtly as part of the design. All printing methods except digital require that the basic design repeat eventually be built up into a rectangle, so really, all repeating patterns are variations of a square repeat.

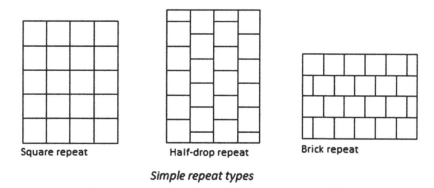

Square repeat   Half-drop repeat   Brick repeat

Simple repeat types

Whereas *Half-drop Repeat* is made by taking alternating columns of the square repeat grid and pushing them down a fraction of the block height—a quarter, a half, three quarters—and you have a drop. Half-drops, where the design is pushed down half of its height, are the most common type of drop and the most common type of repeat overall. On viewing the print one can see the motifs diagonally in both directions. It's also easier to camouflage the repeat this way, because it breaks up motifs, so that the straight line repeat is which would cause unintentional striping or tracking is avoided.

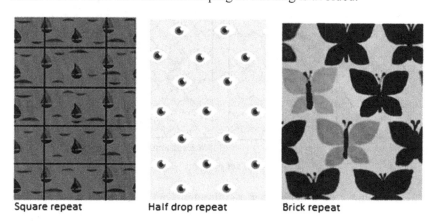

Square repeat   Half drop repeat   Brick repeat

Examples of designs with different repeats

In *Brick Repeats* is actually follows the same concepts as in drops, but the alternating pattern rows, instead of the columns, are staggered.

3.1.4   Random versus set layouts

Motifs can be laid out I the prints in such a way that they look randomly scattered over the cloth or in an arranged pattern. As explained earlier in random layouts the repeat is not obvious or in other words, it will be difficult to locate. This is achieved by strategically overlapping motifs over the edges of the repeat boundaries, using square and drop repeats or even brick repeat as well.

In a set lay out the motifs can be easily identified and due to the set arrangement, the layout also can be recognised. Here also Square, drop, and brick repeats can form the underlying grid of a set layout. Other interlocking shapes, sometimes referred to as tilings or tessellations, are often used as the basis for set layouts as well. For example, take a square grid, turn it 45°, and you have a basic diamond pattern. By examining the designs below one can understand the set layouts:

Diamond/Lozenge   Hexagon   Scale or Clamshell   Ogee

Examples of set patterns

According to the customer requirement or specific end use of the printed fabric (e.g., different garments, made ups, etc.) the motifs can be packed together, showing little to no background,or spaced, showing lots of background. Allover print or allover pattern is often used to refer to packed designs.

3.1.5   Design development

For a textile printer, either he has to develop the design (probably with the help of designers or artists) or develop a design supplied by the customer for the printing machines used for printing. When a motif is accommodated exactly inside a square or rectangle it is easy to find out a motif from a print. But usually, it is not done like this. Even though all patterns are built on a foundational rectangle or square, it is usual (for random and some kinds of set layouts) to lay some motifs over the top and one side of the rectangle and then make sure that the parts that fall outside the rectangle enter back into it on the opposite side in perfect alignment.

To create square repeats and half-drop repeats manually by the following method can be followed.

3.1.5.1   Square repeat

1. Cut a rectangle or square of parchment paper of about the repeat size. Draw the motifs in the middle away from all the four sides. Divide the square (rectangle) into four quadrants and number the 1 to 4. Cut the paper into two equally and horizontally.

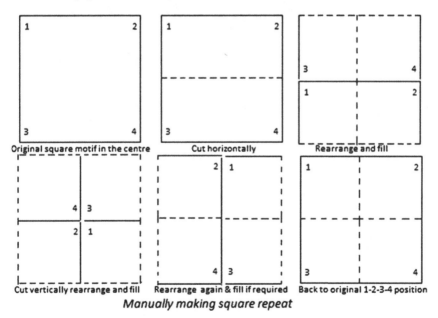

Original square motif in the centre   Cut horizontally   Rearrange and fill

Cut vertically rearrange and fill   Rearrange again & fill if required   Back to original 1-2-3-4 position

Manually making square repeat

2. Interchange top and bottom pieces (horizontally) and tape them. Continue the design into the middle portion which is blank covering

the horizontal joint. Don't do anything at the left and right which may also look blank.

3. The paper is then cut halfway vertically and interchange the pieces (sideways or vertically) or in other words move the left-hand section to the right side, and tapes the sections together. As done earlier fill design in the blank portion seamlessly. Again, fill in the middle.

4. Some small blank portions along the top and bottom middle edges. To complete this detach horizontally through the design on the previously cut lines, reassemble and tape the quadrants together. Fill the blank portion drawing the design in continuity.

5. The repeat pattern is ready. One can still rearrange the 4 pieces into the original 1-2-3-4 position, or even without doing this also the repeat pattern remains the same. But the starting place of the pattern repeat is different.

3.1.5.2   Half-drop repeat

1. Follow steps 1 and 2 above.

2. Cut vertically at the middle either cutting through the design or around the motifs but starting and ending points should be at the middle. Either way the repeat will still work.

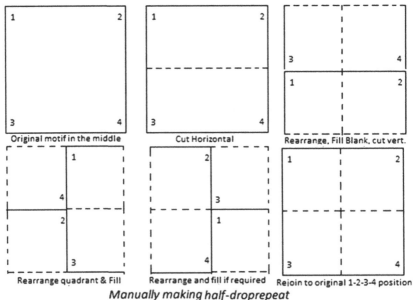

Manually making half-droprepeat

3. Separate the quadrants and reassemble and tape the quadrants together, lining up the square centre corners. This configuration makes it so that motifs falling off the top right (side) edge re-enter on the lower left edge of the repeat, and motifs falling off the bottom right (side) edge re-enter on the upper left edge.

4. Fill in the middle of the drawing over the centre vertical seam. Cut the quadrants apart again and reassemble and retape them together. Fill in the remaining middle space if required. Reassemble and tape the quadrants back into the original 1-2-3-4 configuration and the repeat is ready.

3.1.6   Design development by using a computer programme

Manual development of design is a time consuming process and the accuracy of the repeat, etc. dependant on the artist. Now a days the designing is completely done using computer programs. There are textile specialised computer programmes are available with special features. But common Adobe Photoshop and Adobe Illustrator can also be used for making textile designs. Since it is a specialised area, we won't go into details of design making by these programmes.

Photoshop is a raster image editing program—it builds images from pixels, whereas Illustrator is a vector image editing program - it creates images from points (vectors) and lines. It can be used as a fabric designing tool for developing from scratch or manipulating a scan. The latter case is more suitable for developing from a basic design or an idea for a design given by the customer. There are some negative points for photoshop software also - unless special settings are selected, artwork loses clarity when scaled up and down (which is normally required to accommodate within the screen repeat / size). But a seasoned artist can maintain the integrity and expression of hand-drawn lines using Photoshop. Contrary to this Illustrator images can be scaled up and down infinitely without any loss of quality and its colouring tools are far more robust. Illustrator is best suited to very graphic, sharp lines. Square and half drop repeats can easily be done on computer using Photoshop or Photoshop Illustrator. Other than photoshop there are some other open source software like GIMP is available for designers. Other alternatives are specialises designing software like Design Suite (NedGraphics), SymmetryShop (artlandia) which are plug-ins for Photoshop and Illustrator. These programs build dozens of patterns and repeat types from editable "seed" motifs at the touch of a button directly onto the screen; there is no need to create individual pattern swatches and proof the patterns by drawing and filling new shapes.

Other than design manipulation computer software can be used spot designing grids, colouring using Colour Palettes and Schemes and many other aspects which may not be as easy for an artist who does work manually. For example, colouring of the designs. You can colour from a colour palette, pantone colour system, colour wheel or digital colour.

3.1.7   Digital colour

An artist who is using computer software does not have to touch paint, he can do colouring computer. When using computer for colouring normally one of the colour models or colour spaces are used. they are theoretical models that quantitatively define colours based on different components. The commonly used colour models are RGB, CMYK, HSV, HSL, and HSB and Lab (or L*,a*,b* and CIELAB).

RGB: This is a model based on emitted light in the primary colours of red (R), green (G), and blue (B). Coloured light acts differently than paint or ink. Red, green, and blue light combined in equal proportions makes white; while an equal mixture of red, green, and blue pigments creates black. Computer screens, TVs, and other electronic devices use the RGB model. Hex colour codes are RGB colours in a code that is readable by web browsers. There are no absolute standards for these colours; the model merely tells the device how to make the colours, but they end up looking a bit different on every screen.

CMYK: Also known as process colour, this model is based on the application of mixtures of inks—cyan (C), magenta (M), yellow (Y), and black (K)—to a white surface. It also refers to the printing process that uses this model. Tiny dots of each colour are laid on the paper in various proportions, and the eye reads them as solid colours. You can work with CMYK colours on a computer, but what you see are approximations of CMYK colours rendered via RGB.

HSV, HSL, and HSB: These models define colours by their relative hue (H), saturation (S), and value (V) or lightness (L) or brightness (B). They are derived from the RGB model but based on more familiar and intuitive colour-mixing concepts.

Lab, also known as L*,a*,b* and CIELAB: These models are designed to approximate all colours perceptible to the human eye. They are based on lightness (L) and on two colour-opponent dimensions (red-green and yellow-blue), labelled a and b respectively.

In Adobe Photoshop, the Colour Picker allows you to supply codes for HSB, RGB, CMYK, Lab, and hex colours. In Adobe Illustrator, the Swatch

Options dialog does the same thing. (Colours untranslatable from one model to another will be indicated as "out of gamut.") More budget-conscious software (such as Photoshop Elements) and free, opensource image editing programs (such as GIMP) typically support only HSB, RGB, and hex colours.

The Pantone Matching System (PMS) is one of the colour systems widely used by designers in the textile industry. Pantone provides colour swatches on different materials like paper, cotton fabrics and even blends. It is better to provide the cotton or blend swatches which is easier to achieve in printing. The idea behind the Pantone Matching System, which is basedon formulas for mixing inks and dyes, is to allow everyone in the chain of design and production to match specific colours among different devices, equipment, and substrates. (If they don't use the system, you can purchase individual chips or swatches from Pantone to send, rather than going through the trouble of painting or printing your own or assembling your own fabric swatches, in which case you might not be able to find a specific colour.) Since most of the processors use the pantone colour guides, the communication can be done conveniently by quoting the number of the colour which has to be specified, instead of passing on the swatches. No matter which Pantone system products you use, you can cross-reference the colour numbers for free on their website.

There are other digital colours and software resources which can be used for selecting colours, for example Kuler (kuler.adobe.com), ColourLovers (colourlovers.com), ColourSchemer Studio (colorschemer.com), etc. A designer can supply the design by painting the colour manually as a hard copy or as soft copy coloured by digital colours. Another way is to supply colour swatches for each colour in a print. Transferring of a design to screen can be understood in the following part.

3.1.7   Manipulation from an artist's design (Design supplied by the customer)

When planning to produce and market a print, the design and fabric quality must first be selected. The artist would not have probably not drawn the original design in repeat (as required for a printer); it may moreover require amendment of scale and of the number of colours as per requirement of the market.

At this point it is necessary to decide which printing method will be used, as the lengthwise repeat is subject to the limitations of the number of screens chosen. No repeat limitation for flat screens; for rotary a usual (most common) repeat of 64 cm (or as per special screens used) should be used.

3.1.8   Repeat sketch

One of the most important technical parameters in commercial textile printing is repeat, or the repetition of a pattern along the length and width of the fabric in mathematical precision. Many designers, particularly those working in design studios for fashion, may not consider repeat in a technical way at all, although designs for interior textiles are nearly always created in repeat. However, even if your designs are not going to be completed in repeat it is important that you are aware of the visual impact of repeat on your design and consider the continuity of the image across the edges of the design, as this gives an indication of the repeat. If the design is not in repeat, the printer and his designing studio will ensure that the repeat is completed prior to production.

Every print pattern is based on a drawing of the design. The dimension of the pattern repeat, or of the screen repeat, is an important factor which must be considered when the drawing is drafted. It depends on the kind of pattern, respectively on the print style, on the width of the goods, and on the available screen printing equipment; in some cases, technical aspects and local working conditions and requirements will also exercise an influence. A design unit which is repeated at regular intervals is known as "repeat" or "fitting".

The individual design units (repeats) must fit together perfectly to avoid the appearance of discontinuities that may become visible on long length of a fabric. To this end, the original sketch is redrawn to give a modified version known as a repeat sketch.

When thinking about repeating, you need to consider whether you wish to develop a motif-based design or whether you would like to create a more complex image with a range of visual elements. Very few textile designs use a single motif; while one motif can create a very strong design statement, using several individual elements will result in a more complex and sophisticated design resolution. One might also consider what kind of rhythm and flow you would like to create in your design through the structure of the repeat.

Most commercial textile designs are produced to be printed in repeat onto rolls of fabric that are relatively large. A rotary screen is 64 cm (25 in) in circumference and therefore 64 × 64 cm (25 × 25 in) or 32 × 32 cm (12½ × 12 ½ in) repeats are standard. With the increased use of digital design, it is also common for designs for fashion to be printed on A4 - sized paper, which is standard letter paper 210 × 297 mm (8 × 11½ in) or A3 paper. which is double A4, 420 × 297 mm (16 × 11½ in). Digital, or inkjet, printing can also be used to produce non-repeating patterns, engineered prints and patterns that are printed directly on fabric shapes or garments. The repeat of the design is delineated by an exactly rectangular square ("repeat rectangle"), the sides of which, the repeat lines, are determined by the length of the repeat (in warp direction of the goods) and the width of the repeat (in weft direction). The repeat of the design is identical with the screen repeat if the design fills the screen completely in most cases, however, the design repeat is part of the screen repeat and is repeated regularly for several times within the screen, the dimensions of the screen repeat being a multiple of those of the design repeal. The repeats can be placed side by side (adjacent repeats) or they can be staggered like any of the manner shown below:

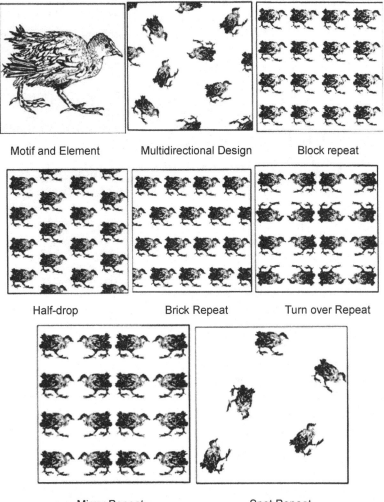

Motif and Element   Multidirectional Design   Block repeat

Half-drop   Brick Repeat   Turn over Repeat

Mirror Repeat   Spot Repeat

In any case, the design drawing must be true to repeat, or it must be modified in such a manner that it becomes true to repeat. This is best done by dividing the repeat rectangle into four parts which are then interchanged diagonally, so that it will be easy to detect any inaccuracies within the repeat, which can then be removed by suitable retouching or modification. In the case of samples with adjacent repeats, the division is carried out by two lines intersecting at right angles. These lines need not go through the centre of the repeat. In the case of staggered repeats, the division is in the form of an upright "Z" if the repeat is staggered in warp direction, or of a reclining "Z" if it is staggered in weft direction. The upper and lower transverse lines forming the letter "Z" must always be half the repeat length (See Fig.). Dividing and diagonal exchanging can be done either by cutting up the repeat drawing and refitting the cut parts, or by transferring the design to transparent paper or transparent sheeting, and displacing the design in stages.

Putting a design in a repeat that has not been thoroughly thought through will result in the amplification of any mistakes, as they will be repeated across a whole width of fabric. Some mistakes are common: strong horizontals, verticals or diagonals can emerge from a composition when motifs have been inadvertently placed in a line; 'tramlines' can appear as a linear space around me design if you do not think carefully about working across the repeat: or ungainly spaces can emerge if you do not consider the composition sufficiently during the design process.

The boundary of the repeat in the final version is usually disguised by following the edges of motifs as much as possible. It is very important that the boundary should not run through the middle of a blotch, as it would show up as a dark line in the final print.

3.1.9   Colour separations

Once the adjusted repeat sketch has been accepted, the next step is to separate the colours in the design. This is done by reproducing the design areas of each colour separately on a clear (translucent) dimensionally stable film in such a manner that the coloured areas are impervious to light. This can be done by different methods e.g.,

- by pointing with opaque ink
- by cutting using cut films
- by direct copying

On photographic transparent paper or film by using reproduction (process) cameras.

Illustration of a colour separation of a four colour design is given below. A is the design and B, C, D, and E are the colour separations and F is the background. Some colours can be made by overlapping also. The designer separates or mixes the colour by overlapping to get the final design appealing or as per consumer's requirements. Sometimes a designer can make colour separations in such a way as the number of printing colours can be less than the actual number of colours in the design. Extra colours are made by overlapping two or more colours. The colour separations (or diaspositive) are usually hand-painted positives produced using opaque paint. This laborious process has now been automated by using programmed photoelectric scanners.

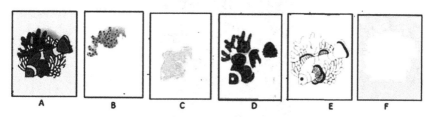

This separation may be applicable when the printing is a direct printing style. But when these colour separations are made the designer should also keep in mind the method of printing or printing technique. Thus, the printer should decide on the printing method (Direct, discharge or resist printing) beforehand. For example, if the design has to be done by discharge the screen F may not be necessary. Also, the designer should know about the precision obtainable by the printing method decided. Also, it is necessary to know whether overlapping designs or colours, half tones, and so on. Contours, as well as the use of dot screens, etc., will also exercise an influence. For example, overlapping by ¼ to ½ mm will be sufficient to prevent 'blank' spaces when printing adjacent colours on accurate printing equipment, but on equipment of poor precision the overlapping is often required to be as much as 1 mm or more. Small differences are to some extent covered by adjusting pressure on the screen or adjusting the thickness of paste, etc. by the printer also. This additional width is usually only provided for the paler of the adjacent colours.

In case of fine outlines, or contours and small areas, the colour separations are normally a little, more 'tight-fitting' because they will mostly turn out a little wider when they are printed. In overlapping designs care should be taken to obtain good mixed shades by suitably covering up the coloured areas when the individual colour separations are being prepared. The same applies to half-tones and semi-tone effects. In this case, all the colours which are to be reduced must be combined in a common colour separation for printing the half tone resist.

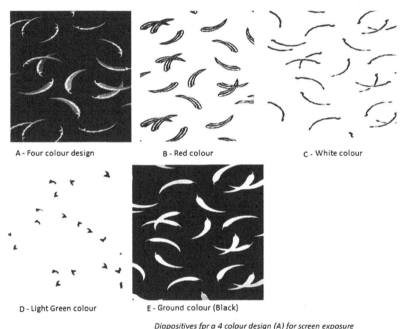

A - Four colour design   B - Red colour   C - White colour

D - Light Green colour   E - Ground colour (Black)

Diapositives fpr a 4 colour design (A) for screen exposure

3.1.10   Cut film in screen making

When producing hand-painted colour separation designs, smooth surfaced patterns may be produced through screens. Transparent film is used which has a similarly transparent laminate coating and the pattern is traced with a knife on the laminated side. The laminated layer is then removed, without damaging the base film, wherever the pattern should be absent when printing.

3.1.11   Half tone Printing

Half tone printing is a technique used for giving shaded effects. This is achieved by printing closely spaced small dots which cannot be detected by naked eye but give impression of a shaded effect. (See the fig below). The effect can be impressive only if the dots do not collapse while printing or fixation and after treatments.

How half tone is printed   How half tone is seen with naked eye

Due to the structure of the fabric to achieve this effect we require a considerably smaller number of dots, i.e., a coarser cross-line compare to paper which has a smooth surface. Screen with a maximum density of 30 dots/cm (or 900/cm2) or, for screen printing, only 12–20 dots /cm = 144–400 dots /cm2. By interposing a cross line screen of this type and colour filters, the negative of the original design is obtained with the reproduction camera and, after copying on to film, can be joined together to produce large diapositives from which the design may be transferred to the screen with a printing frame (for flat screen printing). By using different halftones, variable print effects are obtained. For example, a seven-colour print effect may be achieved with just three printing screens if two of these screens contain, in addition to the full tone, a further two half-tone gradations in each case. Half-tone printing therefore allows the production of multicolour prints with various tonal gradations. Half tone screens are prone to mechanical damage and blockage easily. To get moiré free screens the mesh density (no. of individual thread/ cm) should be 1.5-2.5 times or 3-4 times greater than the half tone density (see fig. below).

(a) Normal Gauze	(b) Screen gauze with reduced open area

Threads/cm	68	68/68
Thread dia	43μ	43
Mesh size	100 μ	60
Free printing area	49%	17%
Fabric thickness	70 μ	82
Volume of paste cm³/cm²	34.2	14

The angular position between the half-tones and the thread system of the screen gauze should be adjusted to 4–5° but never to 0°, 45° or 90° (which nearly always results in moiré formation). In 4-colour printing, the angle of the half-tones in the colour separations is adjusted to, e.g., 5°, 35°, 58° and 81° in relation to the printing screen mesh. Satin and taffeta weaves have proved particularly suitable for half-tone printing.

Net half-tones (completely uniform mesh) are used for more precise shading, circular grain and worm grain half-tones (which produce an image in irregular dots with a "hammer blow" effect) are also used for somewhat coarser prints. When the design repeat extends all over the screen embossed film with the design can be used directly as a dispositive for copying on to the screen.

Notes:

1. High-tension screens are best for printing halftones because they enable better registration, controlled off contact and excellent "snap-off" behind the squeegee.

2. While making screens stretch all the frames with the same procedure and tension

3. Large areas of the same colour or dark areas will be the most likely to cause moiré patterns, use stable, metal frames of the same dimensions and use the same mesh on all the screens of a design.

4. In printing use very sharp, clean edged squeegees and try to use high viscosity pastes with halftones.

5. When making film, angling controls the moiré pattern created between the halftone lines of each individual colour film separation. Halftone angling may be given in two ways:

    (a) Within 90° for rulings with two axes of symmetry such as checkerboard & dot rulings. This would be for round dots or squared dot patterns.

    (b) Within 180° for rulings with one axis of symmetry such as bead rulings. This would be for elliptical dot patterns.

6. Colours like cyan, magenta and black should be at an angle of 30° from one another. Yellow, being a weaker colour, can be set at a 15° angle from a darker colour. When different screens are combined, several distracting visual effects can occur, including the edges being overly emphasised, as well as a moiré pattern. This problem can be reduced by rotating the screens in relation to each other. This screen angle is another common measurement used in printing, measured in

degrees clockwise from a line running to the left (9 o'clock is zero degrees).

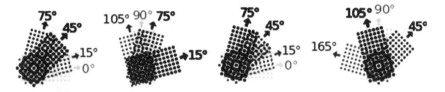

Images with high black content and deep tones:

Colours	Within 90°	Within 180°
Yellow	0	0
Magenta	15°	15°
Cyan	75°	75°
Black	45°	135°

For images where yellow and magenta are dominant like in skin tones or orange colours:

Colours	Within 90°	Within 180°
Yellow	0	0
Magenta	15°	135°
Cyan	75°	75°
Black	15°	15°

For images where yellow and cyan are dominant like in greens and water or aqua colours:

Colours	Within 90°	Within 180°
Yellow	0	0
Magenta	15°	75°
Cyan	45°	135°
Black	75°	15°

7. The most dominant colours should be at 45°, (within 90°), and at 135°, (within 180°). For a one colour halftone the angles should be at 45° for both dispositions of 90° and 180°.

8. While printing do not over flood the halftone screen on the flood stroke. Printing four colour process, duo tones, and one colour halftones on a manual textile screen printing press can be challenging. Halftones are most often done on automatic machines due to their consistency and accuracy. Manually printing halftones takes the skilled hands of an experienced screen print.

9. When printing halftones, 230-count mesh is the commonly used, though some printers prefer an even higher mesh count. In colour prints the shade of the biggest dots will be predominantly seen.

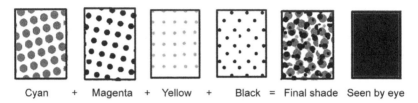

Cyan   +   Magenta   +   Yellow   +   Black   =   Final shade   Seen by eye

10. Higher mesh counts generally cost more and are harder to stretch without tearing or popping.
11. The method of making half tone on rotary screen is different, which will be explained under rotary screen engraving.

3.2   Flat screen engraving

3.2.1   Screen making

Metal Frames must be cleaned first, degreases with fat dissolving agents, freed from any rust. It is also better to roughen the surface on which the gauze is to be glued using a steel brush or an emery wheel. The gauze is fastened to the frame in stretched condition using soluble adhesives. There are several methods and manual, semi-automatic and automatic auxiliary equipment are available for screen manufacture. The basic requirement is to fasten the gauze tightly (as tight as a drumskin) without any distortion and with same tension all over. Outfitting may occur if the screens are not stretched and fastened tightly or if they stretched too tight while fastening. Normal stretching for a proper screen making may be 2% for silk, 5% for stabilised polyamide, 3% for polyester.

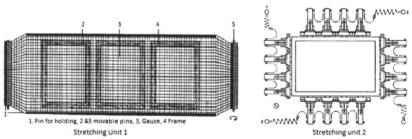

1. Pin for holding, 2 &5 movable pins, 3, Gauze, 4 Frame
Stretching Unit 1   Stretching unit 2

Some of the stretching units. Unit 1 is a semi-automatic unit, Unit 2 is fully automatic unit with electric stretching parts

There are stretching and gumming units available (see fig above). These units allow even stretching and gumming in the stretched state. One can stretch one, two or more frames as per size of the frame.

When manually stretching mainly two methods are used. But it is quicker, easier and more perfect is possible with stretching devices. Manual stretching of the screen has been explained in the history of textile printing, which were the methods adopted in the beginning of screen printing efforts. But here we explain manual methods which are even practiced today, small scale manufacturers and for making screens for strike off.

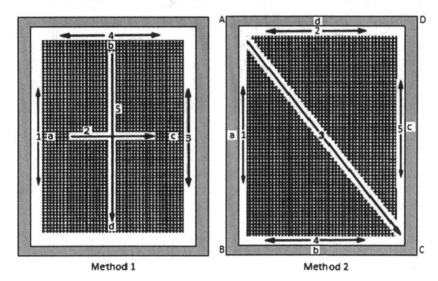

Method 1   Method 2

In manual stretching the bolting cloth is cut in to slightly bigger than the frame size giving allowance for handling. Wooden frames are covered working from centre to corners to prevent formation of folds (method 1) or from one corner across the corresponding longitudinal or traverse side to the opposite corner and other side to the opposite corner and the other sides of the frame (method 2).

Method 1: The gauze is fastened in the centre of the longitudinal side of the frame (a) or outside of the frame (1) and stretched by hand or with a pair of stretchers to the opposite side c (2) and fastened there as above (3). Next the same way the other part (4) is stretched from b to d (5), working being continued alternately in direction of the corners progressing by a few inches along each of the opposite pair of sides. (See fig. above)

Method B: The gauze is adjusted as far as possible without any bow or skew) fastened at the corner A/Side a, followed by a similar procedure at

A/d. The fabric (1) is drawn towards B, where it is pinned in the centre of a, whereupon the entire longitudinal side a is finally stretched. The same operation is now followed on side d (2). After this, the gauze is drawn diagonally from A to C (3), and then fastened on side b, beginning at C/b and then at B/b, pulling all the time (4). Finally, the gauze is stretched over c (5) and fastened from D

Gauze stretched and fastened to the frames   An exposed screen ready for printing

The stretched gauze is fastened on the frame whether wooden or metal, using adhesive lacquers which are soluble in solvents, or adhesives which can be cross linked or thermoplastic adhesive compounds or lacquers. The framed wood or metal is cleaned, degreased and freed from any traces of rust in case of metal frames before using for stretching the gauze. It is also better to roughen the side of the frame where the bolting cloth has to be fastened, with a metal brush or emery. Adhesives can be applied through the gauze in stretched position or the soluble adhesive can be applied on the frames beforehand dried and the gauze can be stretched tightly over them and the adhesive is moistened through the gauze and again allowed to dry. When using thermoplastic or cross linkable adhesive the setting has to be done by heat sealing or cured suitably.

3.2.1   Cleaning of the blank screens

After stretching and fastening, it is necessary to degrease and clean the screens or gauzes thoroughly with a proper solvent. Insufficient degreasing is often the cause of poor adhesion of the lacquer or light sensitive coatings to the gauze. The cleaning method differs with the material of the bolting cloth. Nowadays commonly used is Polyamide.

A stronger alkaline soda wash at 50°C can be used. Soft brush or sponge can be used for rubbing both sides of the screen. Screens are thoroughly washed and if necessary, with 3-5% acetic acid and rinsed again. Highly stained screens may be washed with detergents which contains solvents. There are special cleaning chemicals and other screen engraving chemicals supplied by special manufacturers.

3.2.2   Coating

Flat screens are usually coated by means of hollow or trough shaped doctor blade, or with an angular doctor made from wood, metal or plastic. The screens are kept in an upright but slightly inclined position for coating. The filled doctor is the drawn upwards, exercising a certain amount of pressure, the speed of movement being selected as required and the doctor is slightly tilted during this operation. At the end of the movement the blade is restored from the tilted position to its original position so that the excess solution is again taken up. To make the coating thinner the doctor blade is pressed against the screen and quicker the movement of the doctor. Screens are also coated by mechanical means using special units meant for this.

3.2.3   Transfer of designs to screens

A design can be transferred to a screen by painting or by photochemical process. Painting is seldom followed now as it needs real skill and time consuming. Now it is done by photochemical process.

3.2.3.1   *Procedure for painting method*

1. Painting with Lacquer Method

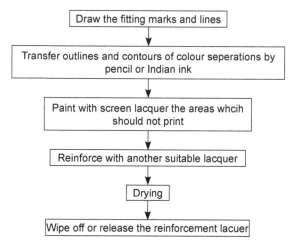

2. Painting with Preliminary Lacquering

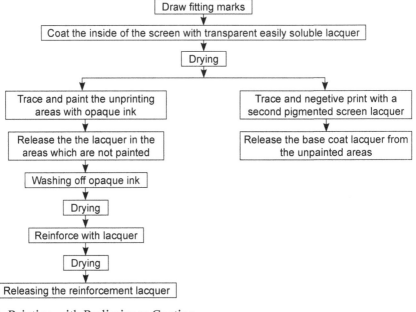

3. Painting with Preliminary Coating

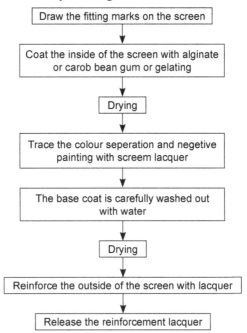

4. Resist method

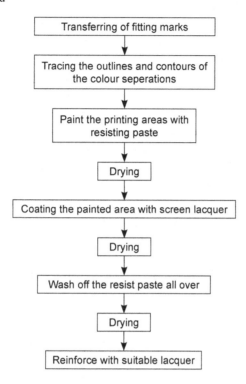

3.2.4   Photochemical process

Photochemical process is the most accurate method of transferring a design on to screen. This does not need a skilled artist is the most commonly used today. There are many methods in this process but it involves commonly, is the application of a light sensitive coat usually to the outside of the screen and after drying covering with a diapositive of a colour separation and then exposing to light. The exposed areas become largely insoluble in water, while in the unexposed areas the coating can be easily removed by washing out with water.

3.2.5   Light sensitive coating.

Nowadays, light sensitive coating material is available from manufacturers who are specialised in this area but the basic coating is made as per following recipe (example)

3.2.5.1 Recipe

A	B	C	D	Unit	Additions
170	90	80	100	g	Gelatin
450	510	580	600	ml	Dissolved in water
5			5	g	Zirconium Oxide
45			45	g	Dispersed with water
	10	8		g	Ammonia 25%
8		8		g	Glycerine
	5	4		g	Defoamer
	90			g	Kaolin
			80	g	Titanium Dioxide
45	15	18	20	g	Ammonium Bichromate
277	120	122	220	ml	Water
1000	**1000**	**1000**	**1000**	**g**	

Gelatin Solution is used as a common base. Any protein substance Albumen, Pearl Glue, fish glue, etc. were used. Now polyvinyl Alcohol or its derivatives are mostly replaced this.

Ammonium bichromate, Potassium dichromate or Sodium dichromate are used as the sensitisers. The sensitivity to light of the light-sensitive coat is influenced by these sensitisers in the stated sequence, i.e., Ammonium bichromate gives the best results. If dichromate is more the coating may become brittle. An addition of 5g/kg zirconium oxide as catalyst will enhance the reactivity of the chrome gelatin coating. Pigments titanium dioxide helps to improve the adhesion to gauze. The sensitiser has to be added last.

Modern light sensitive coat is largely replaced by polyvinyl alcohol mixtures. Some guideline recipes are:

Recipe 1

Qty.	Unit	Additions
910	g	Polyvinyl Alcohol 10%
90	g	Ammonium Bichromate solution 22%
1000	**g**	**Total**

Recipe 2

Qty.	Unit	Additions
700	g	Polyvinyl Derivative Emulsion 20%
40	g	Ammonium Bichromate solution 22%
260	g	Water
1000	**g**	**Total**

3.2.6   Application of light-sensitive coatings

The operations including mixing of the components and coating, etc. are all done in a dark room or in subdued yellow or orange coloured light. The air and equipments used must be free from dust. Chrome gelatin is usually applied lukewarm (30 – 350C), while chrome /polyvinyl preparations are applied cold or ambient temperature. Coating methods have been explained earlier. It is important to ensure a homogeneous, thin coating and to avoid streaks, drops, bubbles or stoppage in the coat. Thinner the light sensitive coating, finer or sharper the reproduction of the image. As the coating thickness increases, the required exposure time during copying process also increase. Also, there is risk of the coat detached from the gauze or from the lacquer coat during developing.

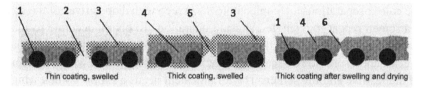

1 – Warp or Weft of the Gauze, 2 – Clear portion in the Design, 3 – swollen portion of the coating during washing, 4 – Basic Coating, 5 – The clear portion after swelling, 6 – The clear portion getting covered after washing.

Automatic screen coating unit

After the light sensitive coating has been applied, the screens are dried in the horizontal position in a light proof drying chamber with circulating air of about 28 – 300C. Some of the hardening reinforcement lacquers are very sensitive to moisture before hardening. For this reason, it will be useful to observe a sufficiently long drying period, or even to dry for a brief period at a high temperature or with infra-red driers as a final operation.

3.2.7   Engraving designs on to the screen

The designs can be transferred on to screen by large size diapositive or by step up method.

3.2.7.1   Using large diapositive:

The colour separations are copied onto the screen with diapositives of the size of a screen repeat in a single exposure operation. To obtain accurate results, it will be necessary to mark the glass plates of the copying devices, as well as screens, and to provide an accurate coincidence of these marks when adjusting the diapositive and the screen. The arrangement needs a support block which is smaller in length and width than the inside of the screen but slightly higher than the sides of the screen frame a woollen felt, or a soft foam or foam rubber of 10 – 15 mm thickness is placed on this block.

3.2.7.2   Exposure to light

The screen is placed on the block in such a manner that the coated surface face upwards. The large size diapositive is then fastened with a strip of transparent Viscose sheet, coating facing coating, and a glass plate is placed on top of it and weighted, if necessary, by suitable weights. After this, the exposure to light is done with a light source which is suspended above the arrangement fig. A. The exposing arrangement also can be made as shown in fig. B and C. To prevent undesirable reflections, it is better to keep the distance between a spotlight lamp and diapositive at least equal to the diagonal length of the diapositive.

3.2.7.3   Light source

The choice of light source depends on many factors, including the technicalities connected with reproduction, as well as the type of light sensitive coating employed. In contrast to chrome-gelatin recipes, polyvinyl alcohol and related products require a light source which is rich in short (ultraviolet) waves. Arc lamps of 30 – 200 amps, especially spot light three phase lamps, are suitable as light source, but mercury vapour lamps and incandescent lamps, as well as fluorescent lamps and neon lamps are also used. Electronic Flash lamps which are introduced can give the exposure in short time. For gelatin coats incandescent lamps of 500 watts each will be sufficient, but since they are having very low UV content long exposure times will be necessary. Powerful photoflood lamps will be better. Incandescent lamps of this type, as well as mercury vapour lamps, which require only short exposure times, develop intense heat, which can cause thermal hardening of the light sensitive coating even those areas which are protected against light, due to the excessive temperature increase. This can be avoided by providing for satisfactory ventilation. Fluorescent lamps have the advantage relatively low temperature at their surface so that the distance between the lamps and copying table can be reduced considerably. Presently UV lamps with very low heat are universally used with low exposure time.

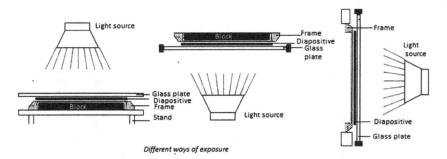

Different ways of exposure

3.2.7.4   Time of exposure

It can vary from 3-0 seconds to 10 minutes or even lower, depending on various factors like the light source, distance of light source and the screen, recipe used, amount of heat developed during exposure, the complexities of the design especially of the fine details, etc.

3.2.8   Step up method

The design can contain several repeats or in some cases a single repeat. The design is reduced or enlarged with the approval of the customer, in such a way as several repeat or one repeat. Then each colour in the design is separated and drawn as single repeat and repeated several times if the design having several repeats or as a single repeat. (See below)

The designer (In the case of big designs, or a single repeat chest print, single repeat African designs, etc.) fills the screen size (in the case of the flat screens a suitable number of repeats or single repeat will cover length, of the screen (or the width of the fabric) and width of the screen (each print of one screen will print a set of repeats or a single repeat and subsequent prints it will repeat the design exactly in such way as the design looks continuous.) In case of rotary screens, the screens are of standard size, the number of repeatsise made in such way (shrunk or enlarged as mentioned above) that it exactly fits in the circumference of the screen. (see 'screen sizes' in the section giving the details of screens) lengthwise repeats are made as per the width of the cloth (usually slightly higher than the actual width of the fabric to give allowance for the width way movement of the fabric, if any, while printing. Extra length of the after the design exposure is blocked.

In each design a designer has to find out the actual single repeat and check the way in which it is repeated in the design. In the following design two different repeat is taken out and shown on the RHS (any number of the single repeat can be derived). Once the repeat is decided the designer has to find out how the repeat is repeated in the design. In the shown design the repeat is arranged in the brick type alignment to achieve the full design. Another way is

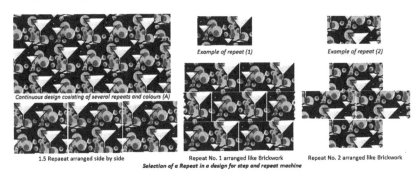

Continuous design cosisting of several repeats and colours (A)

Example of repeat (1)

Example of repeat (2)

1.5 Repaeat arranged side by side   Repeat No. 1 arranged like Brickwork   Repeat No. 2 arranged like Brickwork

Selection of a Repeat in a design for step and repeat machine

to take 1 ½ times of the single repeat can be arranged side by side to get a continuous full design. Once the repeat is decided the colourwise separation is done and repeated on a step and repeat machine to get a diapositive on a sheet as big as the screen to be exposed. In flat screen exposure, the sheet is laid flat on the screen, kept in exact position by aligning the fitting marks on the screen and the sheet and exposed. In case of rotary screen exposure, the diapositive made exactly as per the circumference of the screen and wound around the screen placed exactly by aligning the fitting marks one the screen and the sheet and exposed.

The length of the flat screen or the rotary screen will cover the width of the fabric. The width of the flat screen or the circumference will repeat the design lengthwise of the fabric (with continuity). There can be very complicated designs where the single repeat is arranged in complicated manner to camouflage the repeat and thereby covering full area perfectly without blank spaces.

A colour separation can be transferred to the screen by exposing a small repeat several times carefully. By using a step and repeat machine the diapositive of a small pattern repeat is shifted each time and the already exposed portion is and not exposed areas are carefully covered with material which is impervious to light. This can be done either manually or with the help of a machine. Manual method really needs skill, and it is time consuming.

Step up repeat method is applied on copying tables and similar devices. In this method a network of lines is drawn on the screen, in accordance with the repeat crosses on the individual design repeats and filling dimensions. The dispositive of the colour separation is then pasted on the glass plate of a copying table (depending on the type of copying equipment used) or if a support block is employed it is placed on to the screen or pasted on the glass plate of a step up copying plate.

A step-up copying plate consists of a plate – Glass of 8 – 10 mm thickness, which is framed by suitably dimensioned plywood plates the surface of which lies about 0.5 mm lower than that of the glass plate. Each colour diapositive is made with fitting marks on each sheet so that the screens are exposed at the exact position so that the prints repeat with continuity showing the design is continuous. If the ground colour is not white a screen is made for the ground colour also. Earlier days these diapositives were made manually, but now the design can be colour separated using software and the repeats can be printed using step and prepeat machine (see below) or using special software the repeats are made on the computer and directly printed on transparent paper using opaque ink on a printer.

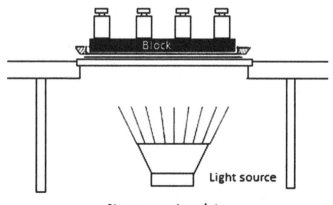

Step up copying plate

In this manner, a sufficiently large surface is obtained on which the screen can be moved, and at the same time a close contact has also been obtained between the light-sensitive coating of the screen which is to be placed on the surface, and the diapositive of a repeat or a repeat strip, the diapositive having been pasted on the glass plate are covered up to make them opaque. After the screen has been placed in position, a soft layer of felt or foam, as well as a plywood plate, are placed on the screen and weighted with suitable weights before exposure. The surface of the table is supported by two jacks. The light source is placed under the plate of glass. The position of the diapositive remains unchanged during the step-up copying operation. To expose next repeat the screen is shifted until the corresponding fitting marks coincide once again, provided the copying device is large enough. The diapositive and the light sensitive coating on the screen are then pressed together again and the exposure to the light is again effected. This procedure is repeated till the screen repeat is filled.

3.2.9   Step up method with prepared screens

In this method, the copying arrangement consists of two completely plane plate-glasses and two support jacks, one of which is fitted with a repeat guide rail. The fitting marks are only marked on one of the screens which are required for the print pattern. Subsequently, all screens are coated either directly with the light sensitive coating, or first of all with a transparent screen lacquer which is then coated with the light sensitive coat on the inside of the screen. The screen on which the fitting marks are drawn, is then placed on a plate-glass, the bottom of the screen facing the glass. This plate-glass is supported on jacks and is illuminated from below with a weak yellow light. The diapositive of the main colour (the contours) is then placed on to the light-sensitive coating in such a way that the fitting marks of the diapositive coincide with the glass plate is placed on top, exposure then being done from above. This procedure is repeated after the diapositive has been shifted by the length of one repeat. After all exposure is over full screen is developed and is ready for use.

3.2.10   Using a Step and Repeat Machine

Once a single repeat of the design has been redrawn in a suitable form it must normally be replicated in the correct arrangements so as to fill the screen, and cover the full width of the fabric to be printed. When flat screens are being photo-patterned multiple exposures of the positive, a single repeat is multiplied carried out direct on to the coated and dried screen, using a "screen step and repeat" machine.

Manual repeating process, however careful is done, can give some mistakes and hence not very dependable. These days, it is done with the help of a Step and Repeat Machine which can give accurate step up copies. The main feature of these machines are an illuminating device, a copying window, and a horizontal or vertical mounting support for the screen. There are machines available which are semi-automatic or fully automatic. Only one repeat diapositive for each colour separation and the machine has the provision to move this one repeat transversally or longitudinally to any extent accurately. The mounting support for the screen, respectively the screen itself, is fixed, while the illumination device is movable. The illumination device can be moved on one direction, while the mounting support for the screen can be moved at right angles to the direction of movement of the illumination device. The repeat cross of the diapositive is then made to coincide accurately with reference mark on the glass plate. The free part of the copying window is then covered up so that they are impervious to light. When the screen is fitted into the copying apparatus ensure an accurate agreement between the axis cross of the diapositive with fitting marks on the screen. On the inner surface of the screen a pressing plate is placed over the copying window. This pressing plate has a foam rubber covering, and it is pressed down either by means of weights by spring pressure, or with compressed air. In vacuum copying equipment, the air is suctioned off, this producing a close contact between the diapositive and the light sensitive coating of the screen.

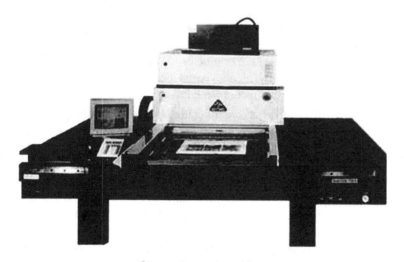

Step and repeat machine

Copying is done in series if the repeat is continuous. When a new series begins, this is often started from the last copy, the direction of movement then being reverse. If the repeats are staggered, the amount of repeat is adjusted according to the amount of stagger. If the staggering does not amount to a full repeat, but only to a half repeat, every second repeat field of a series is , and the gaps are completed when the staggered adjacent repeat series is exposed. Three phase spotlight arc lamps, mercury vapor lamps, or xenon high-pressure lamps are used as light source.

A step and repeat machine can be used both to produce large size diapositives (useful in the case of Rotary Screen engraving) and of part diapositives, as well as for the transfer of the design on to the screen as described above.

This method cannot be used for a cylindrical screen, and so for rotary screens the step-and-repeat process is carried out at the film stage. The single-repeat colour separation positive must first be converted into a negative, after which multiple exposures are made on to a large piece of photographic film.

3.2.11   Developing of the screens

After the exposure the screen is developed by treating with water, or in other words the areas which are not exposed are uncovered (which remained water soluble). The screens are dipped in the developing tray and moved to and fro, at first not too violently and where they are simultaneously or immediately thereafter given a hand spray shower which must not be too severe especially in the case of chrome gelatin or glue coatings. Developing usually dine in subdued day light. Temperature of the water used will depend on the coating used. For example, if it is gelatin or glue based the temperature can be around 40 - 500C and if it is polyvinyl alcohol based it can be 60 – 700C. It is advisable to follow the manufacturer's recommendations.

Small faults in the still wet screens caused by partial blocking of the blank areas in the design can be corrected as follows: Treat the developed screens with a suitably diluted sodium hypochlorite solution (if the bolting cloth is made of silk treat with a saturated solution of oxalic acid. After an adequate reaction time, another treatment is carried out with a 2% solution of sodium hydrosulphite, followed by rinsing with cold water. After the development has been completed, the screen is briefly rinsed cold, and then pre-dried as quickly as possible with warm air of maximum 350C (using a hair drier) and drying may be completed at ambient temperature on in a drying chamber at 25 – 30 °C.

3.2.12   Other methods of exposing

3.2.12.1 Older methods

Polyvinyl coatings can be hardened by reaction with aldehydes (acetalysation)
Treatment is carried out for instance, in a weakly acidic range with

Quantity	Unit	Additions
50-60	g	Acetaldehyde
50 - 40	g	n-Butyl Aldehyde
20	g	Sulphuric Acid 660Be
880	g	Water

Treat for 30 – 60 min at ambient temperature

3.2.12.1 For gelatin Glue Coatings

Quantity	Unit	Additions
50	g	Formaldehyde
10	g	Bichromate of ammonium, potassium or sodium
940	g	Water

The whole process can be summarised as below:

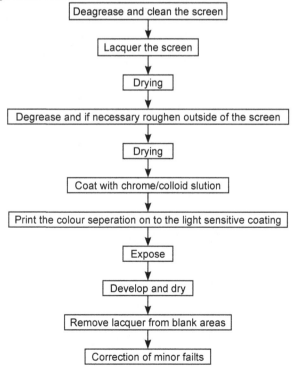

There are other methods to transfer designs on to screen but not practiced widely. Hence, it is not explained in detail.

Cut film method

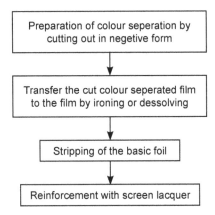

Resist Method (1)

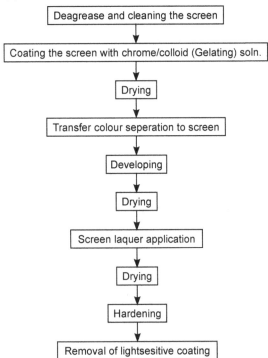

Resist Method (2)

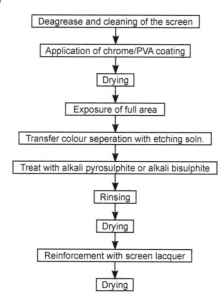

Resist Method (3)

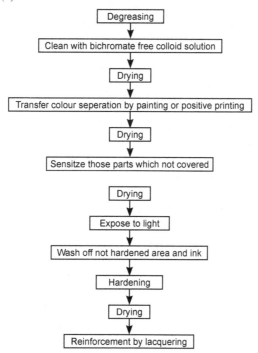

3.2.12.2   Paper Cut Method

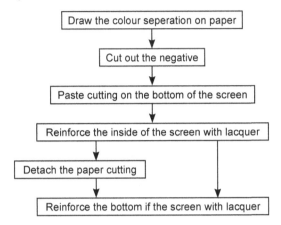

3.2.12.3   Wax paper method

A thin sufficiently transparent wax paper is coated one side by means of a broad, soft brush with approximately 10% solution of shellac in methylated spirit. After drying this process is repeated followed by another drying operation. This paper is now placed on the design drawing, which is true to repeat, the uncoated side facing the drawing. After the fitting marks have been applied, the contours of the colour separation are engraved into the shellac film of the individual wax paper sheets by means of a sharp needle. If the pattern repeat and the screen repeat are identical the colour separations are cut out in negative form, either after the engraving operation with the needle, or directly. If not, the required number of repeat copies is made. For this purpose, the necessary number of wax paper sheets placed on top of each other on a plane plate which serves as a support during cutting, the lacquer coated side of each sheet facing upwards. The contours of the colour separations have been engraved on to the topmost sheet before these sheets are piled up. A warm iron is then placed on the corners of the pile, so that the wax and the shellac can melt, and the sheets stick each other and also to the base plate. According to the engraved contours, all paper sheets are cut simultaneously with a sharp knife. Furthermore, the fitting marks are transferred from the topmost sheet to the lower sheets by piercing these sheets with a needle. The individual repeats are pasted with wax true to repeat on the mounting table the lacquer side facing upwards. The frame with bolting cloth is placed on the top. The inside of the screen is then ironed with a warm iron so that the paper combines with the bolting cloth. After this the screen can be strengthened further as in other cases.

3.2.12.4 Pigment paper method

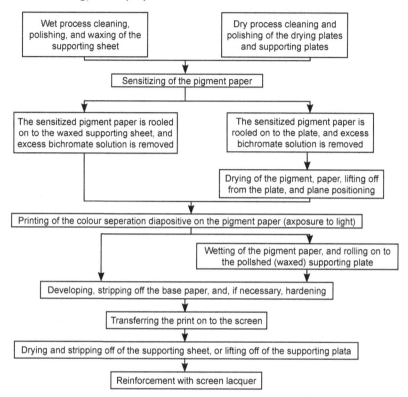

3.2.13   Stripping of the screens

Removal of light sensitive coatings is necessary for either corrections of designs (partial stripping) or for reuse (complete stripping). The screens made of polyamide or polyester can be stripped with a treatment with caustic soda (200Be, or 32.2 0Tw or 15% approx.) and rinsing and neutralising. A second method is by treatment with hypochlorite (about $20 - 30$ g/ available chlorine) solution and rinsing and dechlorination using sodium bisulphate (20g/l). Silk screens with gelatin coatings can be stripped with 50% Lactic Acid. Stronger coatings may also remove (polyvinyl coatings) by this method (dip the screen in 50% Lactic acid solution for 15 min and the rub with sponge and then rinse.

Poly amide and Polyester screens with very strong coating (not removed by previous methods) may be treated with a saturated solution of potassium permanganate. The solution may be brushed on to the screen and after a while it may be rinsed and treated with a strong solution of sodium or potassium bisulphate or pyrosulphite to remove the brown colour. Permanganate treatment should not be given for too long, lest it will damage the screen itself. If complete removal is not effected a repeated treatment can be done (common for all methods)

3.2.14   Removal of lacquer coats

Wash the screens well with water and if necessary, also a desizing agent to remove any possible thickening residues. Rinse and dry. Place the screen on an absorbent paper and their inside is covered with pieces of cloth wetted with solvent (for lacquer) in such a manner that the entire lacquer coating is covered. If necessary, plastic foils can be placed above the screen frame to prevent the solvent from evaporating. After some time, amajor part of the lacquer has been taken up by the paper support and the cloth. The screen is wiped clean of the solvent. One can use a suction device also to clean. Lacquer coat is possible to be removed only if it is soluble in a solvent. If light sensitive coating remains it can be removed by one of the methods described above.

3.2.15   Faults and defects on engraved screens and remedial actions

1. Lengthwise stripes
Various reasons are Emulsion not mixed properly, lacquer viscosity too low, Poor squeegee quality, tape not applied correctly, solid particle on the squeegee, cone, or in emulsion (Filter the emulsion and clean the cone), screen is not cut straight, screen is not rounded properly or screens are not clean. Coater is not perpendicular.

2. Cross stripes
Vibration in the coater, screen is not degreased properly, screen is not cut properly, air current around the coater, emulsion too thin, wrong ratio of screen type/lacquer viscosity/coating speed. As remedial action check the quality of the reservoir, readjust the coater, Degrease screen thoroughly, cut screen straight, protect coater against draught, adjust dilution of the emulsion, check coating speed / viscosity against the mesh count.

3. Fish eyes
Emulsion (sensitised) not good, emulsion (not sensitised) too old, Emulsion too fresh not properly deaired, atmospheric interferences, oils, solvent,

chemicals, screen is still greasy. Remedial actions: Make new emulsion, use new batch, ensure deairing for at least 12 hrs ensure cool clean air in the coating section, degrease screen properly

4. Dripping

Wrong dilution, emulsion too thick, polluted screen, wrong tape used, creases in the screen. Remedies: adjust emulsion dilution as per specification, adjust coating speed if necessary, Degrease before use, check screen.

5. Sticky after drying

Air humidity too high, drying time too short, unsuitable emulsion
Remedies: Use good climatizer, or climatize complete room. Try longer drying time.

6. Emulsion

Emulsion cannot be developed at all or only very poorly 'bleeding' of emulsion Wrong sensitiser and/or wrong ratio, emulsion is too old, temperature drier too high. Air humidity is too high, film density is too low, exposure time is too high. No good contact between film and coating, coated screen stored for too much time and/or at higher temperature. Water pressure in developer too high. Remedies: Use original sensitiser and prescribed ratio, use freshly mixed emulsion, set drier at 280C, provide climate control system, Film must have minimum density of 3, Choose correct exposure time by executing exposure test, Follow work sequence- mounting pressure, main pressure, place coated side of the film on the screen. In manual developing, use a hand shower to rinse the screen. In auto developing check the pressure of water.

7. Cloudy picture

Lacquer dilution is too high therefore lacquer pigment deposits. Remedy: Do not exceed the maximum dilution stated.

8. Poor adhesion of Lacquer coats and of light sensitive coats to the screen

Poor adhesion of the light sensitive coat may be due to unsatisfactory cleaning of the gauze (or omission of cleaning) or soiling of the screen after cleaning or before coating. It can also happen due to drying at higher temperature after coating whereby the lacquer becomes brittle and peeling off. Take precautions in the following: the moisture content of the bolting cloth should not be too high when the lacquer is applied. Lacquer solution should be sufficiently elastic in relation to the gauze material (especially in case of Polyamide material). Screen lacquer should be sufficiently resistant to chemicals. The light sensitive coating can peel off if the coating is not fully dried before it is exposed to light or if the humidity is too high. Excess quantity of chrome and or the water for dissolving the coating chemicals is too hot can cause the coating to chip off.

9. Porous light-sensitive coating

If the coating solution is too thin, or if there is air bubbles in the solutions, it can produce porous coatings. Chromate solution can partially crystallise if the concentration of chromate solution is high and can cause minute holes in the coating. Porous coating can result due to the presence of dust in the solution or in the room where coating is done or during drying. If the drying is done above 400C the coating may become porous.

10. Engraving is not sharp

The usage of too much solvent in the dissolving of the lacquer it may not give sharp images. The unsharp images can be due to the defect in the diapositive, insufficient contact between the diapositive and the screen, to thick light sensitive coatings, over exposure or undesired reflection during exposure, insufficient rinsing during developing.

11. Haloing

Can occur if the light sensitive coating was exposed to light during drying before engraving, or if the period between coating and engraving was too long or if rinsing during developing was insufficient. Insufficient opaqueness of the covered or blackened parts of the diapositive or hardening caused by the action of heat also can cause haloing.

12. Hardening of the photosensitive coating

It can be either due to an excessively high temperature during drying of the applied photosensitive coating before the exposure or due to the heat radiated by the light source (may be due to the distance between the light source and table is too low or poor heat circulation and ventilation)

13. Clogging of the mesh

Poor development of the photosensitive coating after exposure to light, contaminated developing bath, too slow drying after developing, inadequate wiping off of the screen lacquer. If the atmospheric humidity is very high or of the gauze is not completely dry, some of the hardening screen lacquers may tend to flocculation and hence cover up the open areas of the design.

14. Fitting problems

Fitting problems during printing can occur due to insufficient tension of the bolting cloth, or due to wobbling or bending of the frames caused by the manufacturing of the frames. Other reasons may be careless mounting of the diapositive before exposure, use of the diapositive material, which is affected by the humidity, non-maintenance of the screens during washing, drying and storage.

15. Haze after development

Regarding Photo Emulsion: Photo emulsion too old, poor storage conditions. Wrong mixing ration emulsion/sensitiser, wrong dilution. Remedy: Use fresh emulsion, store emulsion in a cool dark place. Do not expose the emulsion to temperature below freezing point. Dilute and use sensitiser as per according to manufacturer's specs.

Regarding screens: Screen is not dry before coating. Rough or creased screen Remedy- Store screen in climatizer after drying (afterdegreasing).

Regarding Coating: Uneven emulsion layer in the circumference. Remedy- Degrease and round the screen according to specs. Use clean squeegee and emulsion reservoir.

Regarding Drying: Air humidity too high in the engraving department, drying time of emulsion too short. Poor drying conditions, exposure of the screen which was stored too cold. Remedy- Climatize the engraving department, Dry at 300C in a good climatizer with air circulation and relative humidity of 30 – 50%.

Regarding Exposure: Exposure time is too short. Polluted film. Too high temperature during development. Remedy- Expose as per manufacturers recommendations with light source of the correct spectrum. Use only clean UV light transmitting films. Ensure proper cooling of the exposure lamp.

Regarding Developing: Insufficient manual Development. Insufficient machine development. Poor water quality Remedy: Develop Penta and Nova screen preferably in machine, check developing programme, post develop with clean water for at least 2 min. Clean or replace spray head, replace felt on the developer, soften the water prior to usage.

16. Pin holes:

Greasy polluted screen, Dust in the coating room, polystyrene from packaging sticking on screen, wrong emulsion, Emulsion prepared wrongly, Polluted lacquer reservoir, air pocket due to incorrect filling of the reservoir, too much talc powder used, talc powder grains coarse, polluted film, wrong exposure time, water too warm during development.

Remedies: Round and degrease as per specs, work in dust free environment, Use good quality emulsion, dilute emulsion as per specs., Filter and dear for min 12 hrs., Clean lacquer reservoir properly, pour deaired lacquer over a spatula to the reservoir, dry screen for longer time till the screen is not sticky, use fine micro talcum powder, use clean, dust free film with min density of 3, make film antistatic, determine optional exposure time, Use water with maximum temperature of 30°C.

17. Lacquer dissolves on a larger surface

Screen not properly grounded. Wrong lacquer type used. Lacquer exposed incorrectly. Printing paste or detergent is too aggressive. Poor adhesion of the lacquer layer to the screen. Remedies: Use prescribed temperature for rounding off, Check the temperature of the polymeriser – above and below temp. indicators, Use lacquer of better resistance, Follow prescribed work procedure, Use milder products in the printing paste.

Note: Donot round off and polymerise in one polymeriser or in other words, use separate units for rounding off and polymerising.

3.3   Engraving of rotary screens

The main differences between rotary screen and flat screen in the engraving point of view are:

The rotary screens are cylindrical, they are of a fixed size and the screens are rigid compared to flat screens which are flat, can be of any size and the mesh is flexible around 0.08 mm. The perforation is in the form of a dot screen. The walls of the holes through the thickness of the screen are sloping, so that the holes of standard screens are larger on the outside of the wall than on the inside. The patterns, in the traditional methods are transferred by photo-chemical means, preferably with a sensitisable and hardening screen lacquer or according to the method of preliminary lacquering. In the latter case, preliminary lacquering is done with a suitable screen lacquer and the light-sensitive coating is applied after drying. The patterns with a continuous blotch design or with longitudinal stripes can be reproduced by this method without any visible seams. In the case of patterns with a gradual and continuous lessening of the intensity of shade from full depth to pale tints, the separation of the image of colour separation into screen dots can lead to moire effects if the dot screen has not been adjusted to the perforation. The preliminary lacquering process is easy to carry out, and requires much less experience than etching or screen manufacture by galvanoplasty, besides being more economical.

The method of introducing the design on to the screen is similar to that used for flat screens, except that the shape of the screens necessitates modifications in the details. Lacquer standard screens, introduced by Stork in 1963 but now available from several manufactures, have uniform spaced hexagonal holes arranged in lines parallel with the axis rotation of the cylinder and offset in alternate lines, as in a honeycomb, for maximum strength.

Engraving of Rotary Screens involve following steps mainly:
Engraving steps

- Unpacking screens
- Stabilising screen
- Degreasing screen
- Coating screen
- Drying Screen
- Exposure coated, dried screen together with full size film to light.
- Developing screen
- Checking screen
- Polymerising screen
- Gluing ending to screen

3.3.1   Unpacking the screens

Removing the screen from the Manufacturers Packages is done using an unpacking trough. First measure the height of the screen package. Adjust the bracket on the left side of the measured height. Adjust the bracket on

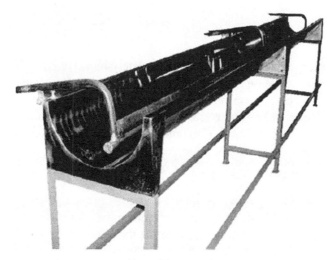

Unpacking trough

the right side 1 cm lower. Lift the screens by the edges, by two people and place the screen on the left hand (raised side) side of the unpacking trough carefully lower the left hand pressure bar on to the screen and slide one screen from inside 20 – 20 cm outside. Lower the second right – hand side pressure bar on to the extended part of the screen which has been slid out. Slide the screen fully out. Take the RHS pressure bar up. Insert the split auxiliary ring provided on both ends of the screen. After inspection, if necessary, the screens are rounded off by heating in a chamber at 160 – 180 0C for one hour in a polymeriser. This way the screen becomes fully round and repeat size from the kidney shaped screen from the unpacking.

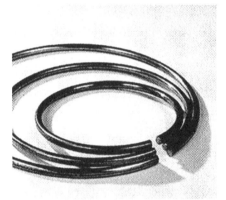
Split rings for different size screens

Inserting the rings to the ends of the screen

3.3.2   Degreasing

Next the screens with the rings on are completely immersed in a degreasing tank to remove the grease and oily matter on the screen due to manufacturing and packing process or suspend so that it is fully immersed for about 10 min. Take out the rings and rinse the screen with water keeping the screen upright by spraying water from the top. (Do not dry the screen in the polymeriser)

3.3.3   Manual coating

Depending on the engraving technique and coating method the engraving lacquer must be sensitised or diluted as per recommendations from the supplier. It is preferable to prepare the lacquer a before the coating to be done and keep it safely covered to allow full deaeration. The photographic emulsion must be applied in a thin and uniform layer. Coating by hand is done by placing the screen in an upright position with the help of two conical auxiliary rings, and by applying the emulsion with a circular rubber doctor blade or sleeve with rubber ring, in upward direction. Every screen is given 2 – 4 coating (circular coating technique. Make sure that the squeegee is clean, and the edges are round and smooth. During manual coating there are chances of foaming of the engraving lacquer. Therefore, the lacquer level has to be kept as low as possible and add fresh lacquer frequently as required.

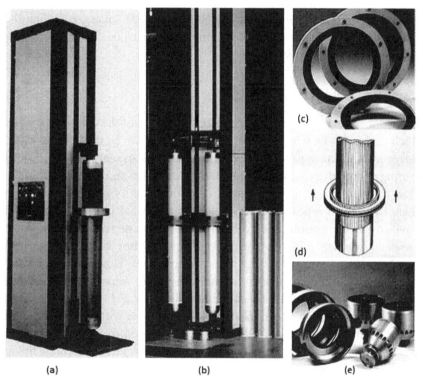

(a) Machine Coating (b) Double coating (c) Manual coating squeegees (d) Manual coating (e) Squeegees for Machine coating etc.

After each coating, the screens are dried in a drying chamber which is protected from light, at a low temperature of 18 – 200C, low relative humidity of about 60% and sufficiently intense air circulation. 60 – 100 mesh screens may be dried for 30 min. while for intermediate drying after second to fourth coating 15 min. may be sufficient. 40 mesh screens will need longer time of drying. It is always better to have the drying chambers fitted with an air conditioning plant, fan, heating units, a thermostat and a hygrometer.

3.3.4   Machine coating

First the screens are taped at the ends in accordance with the coating machines used. The machine is vertical with the bottom end fixed and the top end movable so that the screens of various widths (length) can be coated. Coating machine with single squeegee (dip coating) top to bottom should be taped top and bottom. In case of coating machine with double squeegee taping depends on the coating speed. (For e.g., > 1 min. do not tape, < 1 min Tape). No taping is required for coating machine with blade squeegee. Filling the photographic lacquer solution and running of the machine one should follow the manufacturer's instructions.

3.3.5   Engraving

Engraving is usually done with large size diapositive of the colour separation which can cover the whole screen. The diapositive can be made using a step and repeat machine or printing directly using CAD or special software. Instead of using a large diapositive it is possible to use a step and repeat machine to effect the manual adding up of parts of cut films and it especially advantage where the pattern repeats are very small. This method is more economical provided a suitable printing device is available. When step and repeat machine is used the seamless joining of the prints is assured. The coated circular screen must be fastened to a supporting roller for exposure. This supporting roller is positioned by means of a fixed and a movable, respectively adjustable conical mandrel. These two cones support the roller and hold it in position. For adjusting the diapositive, these must be equipped with longitudinal and transverse fitting lines or with fitting marks arranged like cross hairs. The colour separations of multicolour designs must be checked for accurate fitting before they are used. The step and repeat method is determined by the repeat size and if necessary also by the staggering repeat.

3.3.6   Exposing

After the axes of the repeat have been transferred to the coated screens, the screens are loaded on to the exposing machine and the corresponding colour separation is carefully wrapped around the screen and fastened for instance with transparent adhesive tapes. The exposure to light can be effected on any suitable exposing unit preferably light with shorter wavelength. The exposure time will depend on several factors like, the kind, intensity and distance of light source, the transparency of the foil on which the design is printed, photo emulsion used, etc. In case of doubt, a longer exposure is preferred to a short

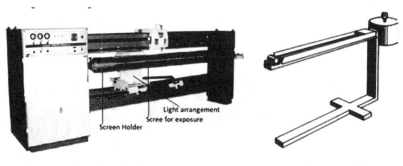

Exposing Machine   Suctioning device

exposure. In any case, the screen surface should not be exposed to excessive heat (radiation heat) to eliminate the chances of hardening due to heat which can cause problems during developing. The maximum temperature on the screen surface should not exceed 300C, hence the exposing room should be designed with good ventilation and provision for heat convection.

3.3.7   Examples of exposure time

Light source Philips HPR-125 W (High Pressure Mercury Vapour Lamp)

Mesh no.	Time (min)
100	6
80	7
60	10
40	20

3.3.8   Developing

After the exposure to light the screen is placed in running water of 10 – 250C for about 15 – 30 minpreferably in a rectangular containerin which the screen rests on two semi-circular blocks and is completely covered with water which flows from below to the top. This will prevent on the one hand the formation of an oilylayer on the water surface and on the other hand washing out of the bichromate from the light sensitive coating is accelerated without getting

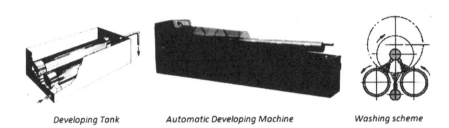

Developing Tank   Automatic Developing Machine   Washing scheme

accumulated in the water. During this period the unexposed area begins to swell. Developing is completed when the coating dissolves under haze formation in the unexposed areas. The cylindrical screen is now taken out from the developing tank and thoroughly sprayed with cold water to remove all the dissolved material. It is then placed in a drying chamberat 650C and left there till the screen is completely dry (about 30 min.). There are automatic developing machines are also available. The applied hardener should be removed from the open areas of the screen. This is done by a suctioning off device. The cylindrical screen is guided carefully over the suction slot of the evacuating device. The screen is then dried.

At this stage the screen has to be checked for any faults and retouched if the holes are in the blocked areas or removed if there are any blocks in the open area. This is done on a retouching (correction jack). The retouching may be done after hardening also.

Next the developed screen has to be hardened. The screens are reinforced both ends by inserting the rings. The suitably diluted (as per instruction of the manufacturer) hardener solution is applied by means of a soft brush or with a broad paint brush, and is then distributed uniformly over the surface of the screen with a clean piece of cloth, taking care to remove any excess hardener solution. During this operation the screen is best placed on the degreasing jack. The hardener being a cross linking agent, has to be cured at two stages. First the screen is dried at 650C for 3 hours (overnight). And then cured at 1400C for one houror at 1500C for 30 min.

The whole procedure can be summarised as follows:

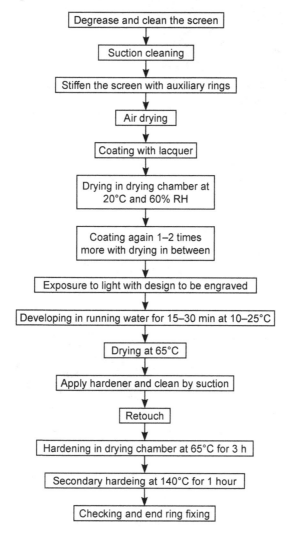

If the defects are found before curing these can be corrected using hardener solution. If the removal of the lacquer coating is necessary in certain areas, this can be done by an electrically operated high pressure water jet. Retouching can be done by hardenerliquidand as per procedure it can be cured.If the clear areas are having small blocks it may also be cleared using water jet of if still nor cleared use fine needles without damaging the screens and then by applying water jet.

Automatic exposing machine

Inspection and retouching stand

If there is correction in the hardened area the lacquer can be removed with concentrated formic acid. Then clean the area with high pressure water jet and after drying retouching can be done with correcting lacquer which doesnot need further curing. Usual problem of pin holes can be directly corrected with correcting lacquer.

Finally, the screens are fitted with the end rings suitable for the screen size used which enables the screen to be used on the printing machine. Mounting by adhesion is done with the help of a cutting jack, support blocks, marking sleeves, gluing device, heating unit, flanges and a pair of screen tongs.

Endring fixing machine

Diffferent types of endrings

3.3.9   Design transfer on rotary screens by etching

Patterns can be transferred to rotary screens with a continuous surface in a manner similar to the method employed in the photogravure of printing rollers, for instance, by the direct half tone or process engraving method. The necessary perforation of the full surface nickel tubes, according to the requirements of the pattern or the colour separation is done by etching, for e.g., with ferric chloride.

Etching is done either chemically or by electrolysis. Easier working according to both methods is enabled by means of etching machines. Care should be taken to see that the coating is completely dry so that it does not become detached permanently while etching. The duration of the etching depends, among other things, on the wall thickness of the nickel rotary screen, and on the distance or the size of the screen dots. They should not be too narrow or thin, to prevent their breaking during printing. Etching can be better controlled in an electrolytic process. In any case it will be difficult to attain uniform etching over the entire length of the screen. This will depend on the quality of the screen (wall thickness, structure and density of the metal) tubes, printing films (transfer film should be absolutely uniform in respect of the sharpness of screen lines and opacity in all areas which are impervious to light), and of course the skill of the operator. After this the screens may be cleaned and end rings fixed and the screen is ready for use.

The whole procedure can be summarised as follows:

Procedure

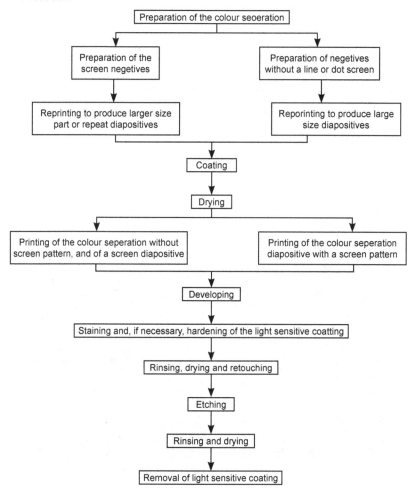

3.3.10 Production of half tone screens

After preparing the colour separation by hand or by photographic reproduction half tone diapositives are made. Screen dot negatives are now made of colour separations. If it is done by means of a process camera, the picture is separated into dots during the taking either by means of a crossed or rotatable line screen, or by a gravure contact screen. It is, however, also possible to separate the picture into screen dots by reprinting, using a contact screen. Depending on the colour intensity from area to area the size of the screen dot varies, while the number of dots per cm^2 remains the same. The network of the lines produced by the screen must be uniformly sharp, which means that the print must be of good quality and the diapositive impervious to light. After developing, fixation, and drying of the screen negatives, they are examined for faults and fitting accuracy, and retouched, if necessary. Diapositives are then prepared which are required for transferring the pattern on to the screen itself. For patterning, the carefully degreased and cleaned full surface rotary screen is fastened to a supporting roller in such a way that its position remains fixed and that no liquid can penetrate sideways. The rotary screen is then coated with a suitable printing solution, preferably on a coating machine using the circular coating technique. The light-sensitive coating is then dried in a drying chamber which is protected from light, or in a dust free room, if necessary, in subdued yellow or orange coloured light. Transfer of the pattern to the light sensitive coating can be done either with a film or large size diapositive which covers the entire surface of the screen, or step and repeat method. After the large diapositive is rounded around the screen and exposure to light developing is preferably done in developing machines, according to the kind of light sensitive coating used.

An etched screen   Lacquered screen

The developed design is controlled by staining it with a colouring solution which contains hardening chemicals. After drying or hardening at an elevated temperature (burning in) the light sensitive coating must be sufficiently resistant to acid. If necessary, small defective areas of the light sensitive coating must be covered before etching, using an acid resistant (asphalt) lacquer. The surface of the screen should not be touched during painting because traces of fat from the fingers will affect etching.

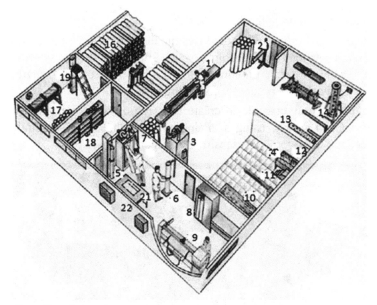

Plan of an Engraving Section. 1 - Unpacking Trough, 2-Inspection Stand, 3-Polymeriser, 4 - Degreasing Tank, 5 - Automatic Coating Machine, 6 - Manual Coating, 7 - Climatizer, 8 - Conditioning Chamber, 9 - Exposing Machine, 10 - Screen Devloping Tank, 11 - Spray Gun, 12 – Shower, 13 - Retouching Stand 14 - Endring Gluing (Vertical), 15 - Endring Gluing Unit (Horizontal), 16 - Screen Storage, 17 Stripping Tank (Horizontal), 18 - Chemical Store, 19 - Stripping Tank (Vertical), 20 - Screen Carrying Trolley, 21 – Table, 22 - AC Units.

3.4   Modern methods of design development and screen making

3.4.1   Computer aided design

By far the most important developments in textile printing of recent years have been in the field of "computer –aided design" (CAD).Associated with these advances has come laser engraving, and INK-JET printing. All these depend upon the successful digitisation of the design that means the conversion of design information into binary code in a form that can be stored in a memory of a computer.

Customers generally placing orders after strike-offs have been submitted. (A strike-off consists of a few meters of the correct fabric printed with each colourway of the design.)These are normally produced on a sample table or even on the production machine. Unfortunately, many of the designs that are engraved, and strike-offs are made are never printed in bulk, and so the expense of screen engraving has been wasted.

The use of a CAD substantially reduces the time taken to produce repeat sketches, colour separations and colour ways. When the design system is linked to an Ink-Jet printer for proofing, this may allow the decision to engrave screens to be delayed until orders are forthcoming.

Although it is now possible to create a design on the colour graphics monitor of a CAD system using a "Paintbox" or "Photoshop" or similar software package with a pressure-sensitive stylus or a mouse.

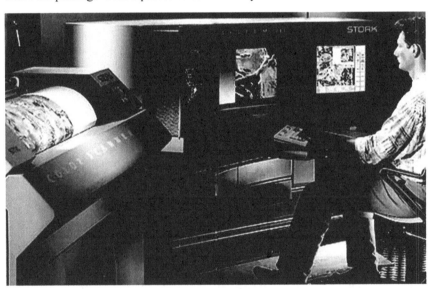

3.4.2   Scanning

The original artwork is then digitised by a scanner. The design information is stored in the computer memory, one pixel (picture element) at a time. Most design-creations are still produced in the traditional manner, with paint brushes or airbrushes on to paper or card.

3.4.3   Drum scanner

A scanner analyses the design one line (that is, one row of pixels) at a time, converting analogue colour information into digital form.

3.4.4   Colour reduction

The normal procedure is to scan a design at low resolution, and then display it on a colour monitor. At this stage large numbers of colour will show. The next step is colour reduction, whereby the design is simplified into a manageable number of colours.

3.4.5   Typical CAD software

A typical CAD design for textile printing purpose should have following Facilities
- colour reduction
- putting the design into repeat
- scaling the design to fit screen
- drawing and brushing
- mirror, inverse and replicatefunctions
- overlap
- fall-ons
- capturing, copying, moving andscaling motifs

- zooming in on small areas in order to edit them
- changing colours
- shading and stippling
- colour separation
- dot removal (deleting stray dots in a colour separation)

When the design is satisfied, design information can be downloaded to a disc, to a film plotter, laser engraver, etc.

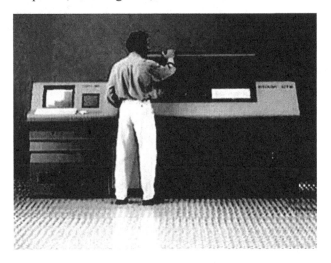

3.4.6   Scanner and Film plotter

Film plotter

3.4.7   Film plotter (the digital manufacture of films)

A scanner scans the design and stores this digital information. For further processing of the design the data are transferred to the design manipulating station. This process may include (see before) the data resulting from this processing stage which are stored on the optical disk and are fed to the film plotter. Thanks to modern electronics the required films are exposed at high speed by a laser diode at extreme accuracy. The plotting resolution is 3-80 points/mm, at a speed of 900 revolution / minute. Films up to 2m² can be processed. With this system a "long" film can be produced, so a step-and-repeat machine to copy the design to the screen is not needed.

3.4.8   Colour manipulation system

The colour manipulation system can be used to colour designs by combining colour separations for those designs. One can select the colours from colour catalogues, like Pantone or via colour measurement with the aid of a spectrophotometer.The colouring system is used to colour designs in a number of colour ways and to make a prediction on the appearance of the print result. This prediction is made on the basis of the dyestuff that are available in the colour kitchen or by the dyes which can be used based on the Colour Matching Software

On the basis of the approved colour recipes, one will know beforehand that the selected colours can actually be produced. The colourways could then be printed on a Ink-Jet printer to get the "go ahead "from the customer.

3.4.9   Laser engraving

The use of high-powered laser for engraving screens is a development from their use for engraving rubber-covered flexographic printing rollers.

A British company ZED and the Austrian company STK (Stork STK) came up in the mid-eighties with the first screen laser engraving machines which revolutionised the rotary screen making process. With laser engraving, the digital information that contains the design for a specific screen is directly transmitted from the CAD system to the engraving machine. The pattern is then applied to pre-lacquered screen, i.e., the lacquer is removed with the aid of a high-energy laser beam and clears the perforations as per the design. Direct conversation of digital design data into an engraved without the intermediate stage of films offers a number of important advantages:

1. Speed: A screen produced on the laser can be ready for use within 30 min. Ascreen produced by the standard method could take up to 2 hours minimum.

Flat screen laser engraver

Rotary screen lster engraver

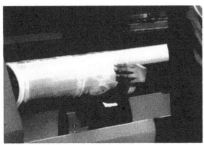
Inside of a rotary screen laser engraver

Details of the laser had of a rotary loser engraver

2. Accuracy and reproducibility: Digital info is direct converted into a design. Consequently, joining problems as they could occur with film are excluded. The laser engraving process is reproducible even after years and the information are stored in digital form.

3. Environment-friendliness: The laser engraving process neither involves development nor does it produce wastewater. The lacquer particles that are removed are sucked away and collected in filters, so no pollution.

4. Efficiency: Thanks to the high speed and high production capacity, this process saves cost in comparison with conventional engraving.

5. Not depending on skill: As the whole process is computer controlled, it does not need human skill.

6. No need for huge screen storage for repeat orders.

7. A damaged screen can be replaced in a short time to continue production a printing order, which otherwise has to wait for a day or so to replace the screen in the conventional method.

3.4.10   Mask exposure machine

The Maxis an engraving system where a wax mask is applied to the light-sensitive lacquer on a screen with the aid of a jet print system. The jet print system is directly actuated by digital design information. Once the negative part of the design has been coated with this mask, the screen is exposed to light in this machine. After exposure, the lacquer on the screen must be developed and then polymerised, in precisely the same way as it would happen with the conventional engraving method. Advantages of the MAX are the speed with which a screen can be engraved, the high quality that is achieved and the fact that much time is saved by the direct digital processing of the digitised design, without the use of films. Joining problems are thus a thing of the past.

3.4.11   Ink-jet engraving

Ink-jet engraving or Wax – jet engraving technique is a method of producing economic and environmentally conscious, high quality rotary and flat screens. With this method, rotary screens or flat screens are engraved without the transparencies and hence CAD system is directly used for producing the exposed screen. The screen to be engraved is coated with the photosensitive lacquer as in the traditional method.

The coated screen is printed or coated with finest wax droplets with the aid of the ink-jet technique. In the case of the rotary screen, this is done with a cylinder rotating about its longitudinal axis, along which the ink-jet head is guided (Fig.). Flat screens on the other hand are inserted in a vertical machine frame, and the ink-jet head "removes" the screen areas to be engraved line by line (Fig.). The graphics file, which is retrieved directly from the CAD machine via a LAN or by an integral computer, serves as the control data file. The spray technique used is the so-called "Drop-On-Demand" process (as explained in the inkjet printer section.) the negative print which adhere to the screen is impervious to light. The printing head does not spray aqueous dyestuff, for example, but wax. This material is thermoplastic. To prepare it for spraying, it is heated in the printing head. Even though the viscosity of the drop is less and it may flow in undefined form over the screen surface on impact it is avoided since the wax droplet changes its viscosity in the flight stage as it begins to cool, finally hardening in contact with the screen surface. Actinic light absorbent additives, which reliably prevent irradiation of the wax particle, can be introduced into wax. The wax develops natural adhesion to all known photoelectric layers, without "sticking" too hard. In this way, it adheres securely to the screen photoelectric layer on the one hand, and on the other can be washed off easily enough in the developing process. Further development of the screen is done in the traditional way except for the removal of wax which is very easy. This new technology has many advantages.

A   B
Ink jet wax printer for engraving (A) Rotary (B) Flat screens

Advantages of Ink-jet engraving

1. The technology is very simple and except transferring the engraving transparency, though by a new method. All other processing stages, i.e., coating, exposure, development, thermal or chemical hardening if necessary, through to retouching, remain unchanged.

2. The ink-jet technique imposes minimal requirements on screen positional accuracy, because, in contrast to the laser technique, no focal range has to be adhered to. The screen is very simply pneumatically stabilised with a minimum over pressure of 0.02 bar. In contrast, the flat screen is brought into the plane position by means of a partial vacuum. Screen positional tolerance has very little effect on picture detail representation.

3. The inkjet technique speed can be increased by increasing the number of jets, whereas in case of laser increasing heads is restricted due to cost.

4. The ink-jet technique produces no seam, as is familiar from conventional film engraving, and is dreaded with certain patterns.

5. The ink-jet has a special advantage of checking the print on a paper before actually engraving. For example,suppose the printer has a doubt about the resultant print of a critical half-tone, it can be printed out on paper which has been laid on the screen by way of a proof. (It has been found that an inkjet engraving can produce on the screen an image (graphic) almost identical with the subsequent print with this screen. Therefore, when in doubt about the resultant).

6. In inkjet method the thickness of the coating does not pose a problem, whereas in case of laser engravers it may be a problem in some cases.

3.4.12   Laser exposure machine

Stork introduces a completely new technique, suitable for both the engraving of screens and for applying designs to galvano mandrels:

3.4.13   The laser exposure technique

Laser exposure combines both the advantage of light-sensitive material and of the laser technique. It gives the sharpness of film exposure without the use of film, without joins, without cutting and without side lighting effects. Laser exposure requires a special lacquer that is sensitive to laser light that has a wavelength of 488 nm.

In laser engraving machines, which is quite common, a laser beam that can be modulated is made to travel in a spiral across a classical nickel screen which had been completely blocked-off with lacquer. The CO_2 laser would be sufficiently powerful to burn the lacquer off in microseconds leaving the specified voids in the screen open. Theoretically, a laser flash such a this would last from 5–7 µs. Up to 27 000 spots could be picked off per second. A device has therefore been produced for this purpose and can apply the design to the screen directly at high speed and with high precision. The edges of the burnt out holes are sharply contoured and are sealed by the singeing and melting process. There is no need for a photosensitive lacquer for laser engraving and hence be selected from an existing range of the most durable lacquers (even self-crosslinking). This gives the advantages like (a) lacquered nickel screens can be stored for long time, (b) in case of some self-crosslinking lacquers drying alone may be sufficient and no curing is required which can save cost (c) after laser engraving no after treatment is necessary (c) Seldom defects are found in engraving as all the data are supplied by the computer and it controls a laser beam and nothing mechanical, (d) the engraving speed depends on the mesh of the screen as this determines the line density of the advance movement, (e) the time used is shortened by the fact that the film no longer needs to be exposed, nor does the exposedscreen need to be developed (washed out), dried, hardened or indeed re-lacquered with extraction or subjected to heat treatment a second time.

The screen is fixed on to the laser machine on which bearings are open on one side and the screen can be changed so that different repeat sizes from 640–1017 mm can be used. The nickel cylinder is pneumatically set under excess pressure which gives rise to the slight stretching which is enough to hold the precisely fitting screen in the exact position. The cylinder is now allowed to run in the two precision bearings which are centred exactly. From a mark on thescreen, its exact circumference is calculated by a computer and, using this, the number of pulses from the laser impinging on the surface during a single revolution is established. With high resolutions screens can be engraved with patterns involving the highest detail. In regard to adjustment, avoidance of "seams" or having to work with adhesive stripes, this is better and simpler than the copying or exposure process using a film. Apart from this, the difficulties encountered in mounting the films on the circumferences which are not quite exact, in spite of high precision adopted in making the screens using electrode position, are excluded. Engraving can be done with the endring on, for an immediate use or without the end ring on the screen.

Laser engraving suitable for flat screens are also available principle of engraving is same, but designed for a flat surface instead of rotary surface. The laser engraving can be used for stripping the screen also (stripping cost may be higher).

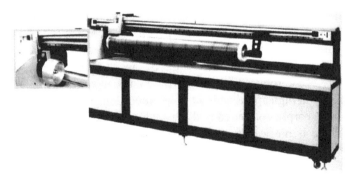

The water-cooled Argon+ laser is mounted on the exposer head and moveswith this head along the screen or mandrel this guarantees a uniform exposure across the complete surface.

3.4.14   Inkjet exposing machine

Repeat gear sets for various repeats
64 cms to 145 cms repeat

Printing paste additives

4.1   Thickeners

Printing is a process whereby the dye particles as per design requirement. In the first step in printing the dye in the dissolved or undissolved or disperse form is forced through a screen or any other media on the surface of the fabric in paste form so that the dyes adhere at the desired place on the fabric till its fixation is complete. A thickener imparts stickiness and plasticity for the print paste so that the above described requirements are accomplished. The viscosity of the printing paste should be sufficiently high to prevent rapid diffusion or flushing of the colour through the fabric which will result on poor definition of the print. The thickener should give a stable paste viscosity, which would allow an even quantity of the paste during the run so that no uneven printing happens. The viscosity should be stable during the storage times like week or even months sometimes. The viscosity should not be affected by the addition of dyes and chemicals including electrolytes sometimes. At the same time, I should not precipitate by the action of dyes or chemicals including alkali and acids which are necessary in the case of some classes of dyes. In the case of rotary screen and flat screen printing the paste undergoes shearing action during the squeegee movements. The thickener film should not become brittle and break off during the drying and subsequent handling. Breaking off the film and the particles of the paste film falls on the other colours or white background can cause colour specks and get fixed during steaming.

The solid content of the printing paste is important for producing good under prints under different conditions in limited period especially while printing lightweight materials. In such cases high solid content will safeguard against the bleeding and to retain the sharpness of the print. Since the thickener is not a part of the design it has to be washed off after the fixation process. Thus, the thickener has to easily removable in the after wash without staining the background. If it is not washed off it will impart harsh handle, since it is an adhesiveproperties. The adhesive property should be such that it should not resist the transfer of paste from the print transfer medium (screen, block, etc.) to fabric surface.

Print pastes may be thickened by any of the following methods:

(a) a relatively low concentration of a long-chain thickening agent

(b) a relatively high concentration of a shorter-chain thickener or one having a highly branched structure

(c) an emulsion

(d) a finely dispersed solid such as bentonite (derived from clay)

The first two methods, particularly the first, are most frequently used today; combinations of these methods are also possible.

From the above explanation one can understand that the material to be used as thickener it should have some basic qualities or requirements:

1. The thickener should have some physical and chemical characteristics such as viscosity, flow properties, wetting and adhesive. The adhesive property should be such that it should not resist the transfer of paste from the print transfer medium (screen, block, etc.) to fabric surface.

2. It should have good swelling properties in cold water.

3. It should be stable to various print paste ingredients like acid, alkali, oxidising agents, reducing agents, etc.

4. It should be fast and easily dryable so that any spreading of the colour beyond the design boundaries is prevented.

5. It should not be any barrier for the condensation of water from steam and the transfer of the dye molecules from the print film layer to the fibre to allow the fixation but at the same time preventing the spreading of the print.

6. The dye should not have any affinity or react with the thickener preventing its movement to the fibre and reducing the colour value in any manner.

7. The thickener should not cause flushing.

8. The thickener should be easily washed off after fixation.

9. Since large quantities are used in print paste it should be easily procurable and cheap so that it should not increase the paste cost.

10. It should be easily soakable/soluble so that the preparation of the thickener solution and washing off is not time consuming.

11. The properties of the thickener, explained above and other requirements, should be able to retain for a long period or in other words the properties should not deteriorate on storage probably for weeks and months.

12. The thickener should have good solid content.

13. It should be compatible with dyes, chemicals and auxiliaries in the paste.
14. High degree of purity and conformity to standard.
15. It should be non-dusting.
16. It should be biodegradable and non-toxic.
17. Preferably manufactured from replenishable raw materials.

4.1.1   Theory of viscosity in relation to thickener

A *thickener* is a thick viscous mass which imparts stickiness and plasticity to the printing colour, so that it may be applied. To the fibrous surface without bleeding and be capable of maintaining the design outlines, even at the application of a high pressure. The thickeners are prepared from *gums* which upon dissolving or swelling in water give thick, adhesive and viscous masses.

In the event if printing by a screen the squeegee presses the paste between the blade and the screen developing a hydrodynamic pressure. Even though this pressure is relaxed when the paste has passed through the holes but it also has to help to push the paste between and into the fibre aided by the capillary forces. This pressure also should ensure the volume paste planned by the printer to be transferred on to the fabric. Penetration of the paste into the fabric depends upon the pressure and the kinetic energy of the paste leaving the screen holes. Before or after the actual printing paste transfer there should not be any paste passing through the screen.

The hydrodynamic pressure developed is a function of the squeegee blade angle, the paste viscosity, the screen hole size and the screen speed, rather than the pressure applied to the squeegee. The Poisseuille equation gives the flow rate of liquid through cylindrical holes, even though it may not represent a screen hole (rotary):

$$Q = \frac{Pa^4}{8l\eta}$$

Where Q is the volumetric flow rate of the liquid transferred, l is the length of the hole, a is the hole radius, P is the constant pressure drop across the hole, and is the liquid viscosity. It can be seen if P increased the hydrodynamic pressure is also increased, but is inversely proportional to the viscosity (η) and proportional to the hole radius to a power greater than two.

Thus, one can see that the volume of paste passing through a screen hole more dependent on the size (radius) than viscosity.

But then, where the viscosity of the paste is important because we all know that the print paste is a viscous liquid. The viscosity plays an important part after the paste [pressed through the screen holes and the scree in removed (flat screen printing) or moved away (rotary screen printing). As the screen separates after the paste transfer the pass of the paste splits into two portion – one part stays on the surface of the fabric and another part stays with the scree surface. We are mainly concerned about the paste left on the fabric surface. As the paste is formed on the fabric it will be in the form of small mounds of colour on the film surface leading to a spotty or 'mealy' appearance. At this stage some flow is desirable at this point to give a level colour.

Viscosity is a measure of the resistance of a fluid to flow when subjected to a force. It is the ratio of the shearing stress to the rate of shear. (See fig. below)

$$\text{Viscosity} = \frac{\text{Shear stress}}{\text{Shear rate}}$$

Where, shear stress is the resistance of the liquid to flow under the influence of an applied force.

$$\text{Shear stress} = \frac{\text{Force}}{\text{Area sheared}}$$

$$\text{Shear rate} = \frac{\text{Velocity}}{\text{Clearance}}$$

Considering a fluid on a solid surface the resistance to flow is maximum at the solid surface, as we go up each layer the resistance is decreased. The fig. represents the flow of liquid below a rigid plate of area. A when a force f

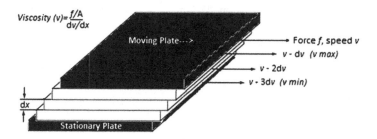

moves it at a velocity v. Because of the liquid's viscosity, the lower layers of liquid move with decreasing values of v. Higher the viscosity he greater the force required to make it flow at a given velocity and vice versa. The shear stress is f/A, and the rate of shear is the velocity gradient in the direction perpendicular to the plate, dv/dx. The viscosity is therefore:

$$\text{Viscosity } (v) = \frac{f/A}{dv/dx}$$

Viscosity can be determined by measuring the flow of the liquid through a capillary tube but this is not suitable for print pastes. A cone and plate viscometer are used. A film of paste is wedged between two plates. The lower one is flat and the upper a flat cone tapering from the centre outwards at a very small angle. The diameter of the cone selected depends on the viscosity of the paste. Driving the cone at a given velocity against the viscous drag of the liquid requires a measured force. The shear stress and the rate of shear can be calculated from the cone dimensions, the rate of rotation and the measured torque required maintaining it. A graph of the rate of shear as a function of the shear stress allows calculation of the viscosity. The units of viscosity are thus Pa s (Pascal seconds). This unit is ten times the older unit 'poise', after Poisseuille.

4.2   Thickeners for printing

Thickeners used in textile printing are high molecular weight compounds giving viscous pastes in water. These impart stickiness and plasticity to the printing paste so that it can be applied to a fabric surface without spreading and be capable of maintaining the design outlines even under high pressure. Their main function is to hold or adhere the dye particles in the desired place on the fabric until the transfer of the dye into the fabric and its fixation are complete. As the printing paste is printed either by roller or screen by the squeegee pressure, the viscosity of the printing paste should be sufficiently high to prevent rapid diffusion or flushing of the colour through the fabric, which would result in poor print definition or mark. Further, the thickener should give a stable paste viscosity, which would allow an even and measured flow through the screen. If the viscosity changes during the run, the shade (depth) changes in the printed cloth. The viscosity stability must not only be durable in terms of the time, during which the printed cloth on the machine, but it must also hold during storage times in terms of weeks/months. Generally, it is necessary to have a large volume of stock paste of the thickener of such viscosity (higher than that of the printing paste) that on dilution with the proper amount of the dye, chemicals and water, that will bring it to the desired printing viscosity. The viscosity should not change extra-ordinarily by the addition of dyes and chemicals. Thus, the thickeners should perform its function over a wide range of physical and chemical conditions.

A printing colour or paste differs from a dye bath in that it has a spatial internal structure which confers to it a shape-retaining capacity; owing to this property, it is possible to obtain a pattern on the fabric. This structure of the printing colour is usually obtained by mixing a dye solution with various thickeners and auxiliary substances. The properties and the internal structure of printing colours are determined mainly by the properties and the internal structure of the thickeners taken. However, in many cases the structure of the printing colour may be strongly modified under the influence of various additions. Particularly, it is necessary to take into account the ability of some dyes and auxiliary substances incorporated in the printing colour to form a certain structure, especially in colloidal disperse condition.

Thickeners have properties which may be conventionally classified into spatial and surface properties. Spatial properties are predominant, the most important of them being mechanical and due to deformation, as well as the dye-retention capacity. The mechanical and deformation properties of thickeners are determined by their internal structures. Systems similar to thickeners manifest two kinds of deformation: (1) in the zone of non-destroyed internal structure (structural and mechanical properties of thickeners), (2) upon the decomposition of the remain of the initial structure in flow (rheological properties). The deformation in the zone of non-destroyed structure in various critical conditions of deformation gives a most comprehensive representation of the whole complex of mechanical properties. At present, a method for representing the mechanical properties of non-destroyed internal structures of such systems by means of mechanical models is coming into use. The essence of this method consists in the selection of mechanical models which represent sufficiently clearly the principal features and particularities of the mechanical behaviour of the thickener. The mechanical models reflect general and characteristic features of the internal structure of thickeners which determine their mechanical properties and are of greatest importance for describing their deformation properties. Particularly, by means of such models it is possible to give a characteristic of the resilience, elasticity and plasticity of various thickeners in the zone of non-destroyed internal structures.

The investigation by this method of the mechanical 'properties of aqueous solutions of a whole range of high-molecular thickeners (starch, tragacanth, fruit glue, sodium alginate, dextrin, and carboxymethyl ester of cellulose) in the zone of non-destroyed internal structures has demonstrated their resilience, elastic and plastic properties, as well as structures of two kinds: a condensation and a coagulation. Condensation structures are formed upon cooling of hot solutions of thickeners in the absence of mixing, while coagulation structures are obtained upon mechanical mixing of preliminarily cooled solutions. The properties of the two types of structures may be quite different. The main difference is in the degree of their thixotropy, i.e., coagulation structures are fully thixotropic, while condensation structures are only partially thixotropic.

The degree of thixotropy determines the dynamical equilibrium of the processes of destruction and restoration of internal contacts in the system at a predetermined routine of the mechanical action to which it is subjected. The fuller and quicker the process of contact restoration the less is the virtual degree of system destruction and the greater 'is its homogeneity during the whole operational process (as for instance of the printing colour during work on the printing machine). If the aforesaid equilibrium cannot be achieved for the given substance, the substance should not be used as a thickener. Many thickening agents are of the shear thinning type, the apparent viscosity progressively decreasing as the shear rate is increased which is reversible, viscosity returning to its original level as soon as the shear is removed. Thixotropic fluids show time-dependent effects in that apparent viscosity depends on both the rate and duration of shear, the return to original viscosity being delayed. On the other hand there are thickeners which behaves opposite to the above, in other words shear thickening, called dilatant behaviour and is not suitable for textile printing.

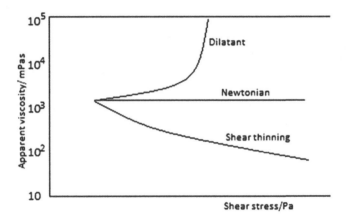

Natural starch is such a substance whose gels in the absence of special additions are unstable non-thixotropic systems. However, undesirable properties of starch may be eliminated by its hydrolytic destruction in the presence of acids as well as by the introduction of various plasticisers. A peculiar feature of starch is that at a certain stage of hydrolytic destruction the consistency of its gels considerably increases. Thus, hydrolytic destruction of starch improves its printing properties and is not followed by a sharp reduction of its thickening ability.

One of the characteristic features of thickeners which is manifested atcritical loads is toughness which may be considered to be the result of the superposition of at least two different processes: the destruction of the internal structure and the viscous-plastic flow of the system. The printing colour behaviour during the work on the printing machine is clearly demonstrated by the rheological flow curves showing the dependence of the deformation speed gradient *(dv/dx)* and viscosity on the shearing stress (p). These curves are usually used for comparing the deformation of various disperse systems upon the destruction of their initial structure remains in the flow. Full rheological curves of various disperse systems are described by very complicated equations. However, as the investigation of thickeners and printing colours has shown, their rheological curves in large ranges of shearing stresses may be described by very simple equations written as follows:

$$\rho = \eta_0 \frac{dv}{dx}$$

$$(\rho - \theta) = \eta_{pl} \frac{dv}{dx}$$

Where θ = maximum shearing stress (yield point); 0 = general viscosity; η_{pl} = plastic viscosity; η = are constants. The equation (2) comprising the yield point θ fully describes the rheological properties of thickeners. In this equation θ characterises the mobility of the thickener and η shows its plasticity. When $\eta = 1$, equation (1) becomes a usual Newton equation describing the flow of an ideal viscous fluid (for instance water), while equation (2) becomes a Bingham equation describing the irreversible deformation of an ideal plastic body. Thus, in the case of equation (2) $\eta = 1$ corresponds to ideal plasticity. Both equations represent the dependence of *dx/dv* on p. These relationships are schematically illustrated in Fig. below, where curve 1 corresponds to equation (1), curve 2 to equation (2), curve 3 to the Newton equation and

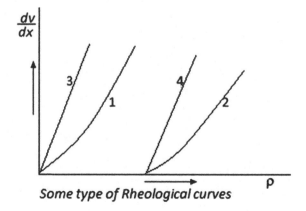

Some type of Rheological curves

curve 4 to the Bingham equation for body with ideal plasticity. In addition to the curves illustrated in fig. below, curves of the effective viscosity η versus the shearing stress ρ are sometimes also drawn by substituting the values of $dv/dx = \rho n$ obtained from the usual Newton equation into the afore-mentioned rheological equations.

As a result, two charts are obtained: $dv/dx = f(p)$ andη $= f(p)$, establishing a connection between the change of three different values: dv/dx, ηand ρ. In case of true structure less liquids such as water, the viscosity is constant and does not depend on the shearing stress, while in structural systems, such as thickeners, the viscosity is reduced when the shearing stress increases, which is explained by the destruction of the internal structure of the system. It was shown on the example of equation (1), that in case of thickeners, both charts $dv/dx = f(p)$ and $h = b(\rho)$ may be combined into one common chart called a "complex diagram". This diagram establishing a relationship between all the three factors dv/dx, ηand ρ is given in Fig. below. Along the abscissa and ordinate axes of this diagram we have log ρ and log dv/dx, and along the third sloping axis the values of log η. Instead of the logarithms of these three values, the values themselves may be taken, but in a logarithmic scale. With the help of this diagram, it is possible to determine simultaneously all the three values for any point of the rheological curve by drawing perpendicular

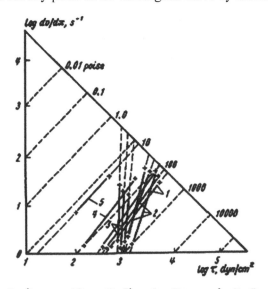

Composite diagram: Viscosity-Shearing Stress-velocity Gradient
1 - 12% Sodium alginate duration of boiling 2, 4, 6 h;2 -10% solvitose made cold; duration of boiling 1-2 h; 3-12% starch, 4 - 8% alginate 5 - 4% alginate

lines from this point to the corresponding axes of the diagram. The tangent of the angle of inclination of experimental curves for the flow of various thickeners drawn on this diagram (see fig. above) is equal to the exponent in equation (1) as well, and may be considered as characteristic for the structure of the thickener andits deviation from the ideal viscous fluid.

As we understood, the principal function of thickening agents is toincrease the viscosity of print pastes or pad liquors, but it should have following properties also to be considered as suitable to be considered as an effective thickener:

1. Stability and rheology of the print paste
2. Adhesion and brittleness of the dried thickener film
3. The effect on colour yield and penetration
4. Ease of preparation andremoval
5. Economical

The viscosity of print pastes, i.e. the physical behaviour of the fluid while it is in shear, may depend on four parameters:

– physical/chemical quality of the substance

– temperature

– shear gradient

– time

These parameters result in the fact that the viscosity is not a constant, but varies.

Different thickener solutions response differently to an applied force in an aqueous media. They follow different types of flow:

1. In an aqueous solution the viscosity is not dependent upon the time or shear rate but dependent on concentration and temperature. This type is called Newtonian Flow

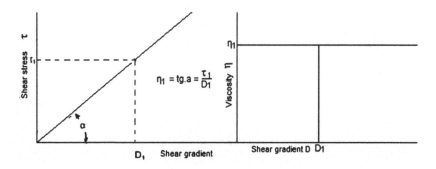

2. This type of flow is called dilatant flow and is characterised by the increase in viscosity with the increase in the shear rate. Starch is an example, and some materials in dispersions having about 50% solids. η depends on D, and where η grows with increasing D = dilatant behaviour where η is viscosity and D is Shear rate (gradient)

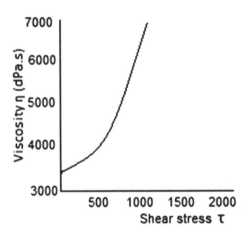

3. Property of certain liquids to change their viscosity reversibly under mechanical strain and to convert from a thick gel-like consistency into a thin liquid solution state. Here

η decreases with increasing t = thixotropic behaviour.

Thixotropic or rheopexic liquids require a regeneration time in order to develop their typical viscosity, i.e,. their internal forces, against a load (see fig. below).

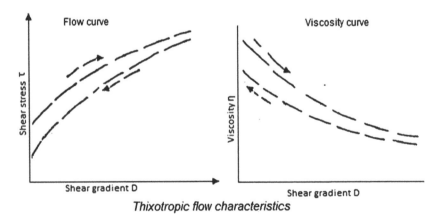

Thixotropic flow characteristics

Thixotropic flow properties never exist alone. It is a superimposition of the viscosity time relationship upon either Newtonian, dilatant or pseudo plastic flow (See below). Thixotropic behaviour of a print paste may be desired with rapid movements (particularly in roller printing). The thinner and lower   its viscosity, the quicker the break contours run into a homogenous, smooth surface and the easier the penetration of the fabric surface will be.

4. A fourth kind of flow is called Pseudo plastic flow. The rheological behaviour of metastable foams and most polymer solutions of moderate concentration is characterised by their pseudoplastic flow, whereby apparent thixotropy can also occur. Materials, which degrade their viscosity due to increasing shear rates, are called structurally viscous or pseudoplastic.

   η decreases with increasing D = structural viscosity (pseudoplastic) behaviour.

In rotary screen printing especially in carpet printing where relatively low viscosities are used to get good penetration in the fibre substances, a pseudo-plastic printing paste is particularly effective. Due to this, the viscosity drops as the shearing action of the squeegee is applied and a rise in viscosity after the shearing stress is released takes place.

Before the fixation of the dye by steaming, the thickener filmdeposited on the fabric surface should not be dried quickly withtoo much heat as this causes excessive surface colour mark off.One of the functions of the thickener is to prevent the migration ofthis colour to the surface through viscosity and the adhesive natureof the thickener solid present. The thickener film must not be brittle or flake off during drying and subsequent handling; otherwise, scattered flakes of the thickeners film cause coloured specks all over the cloth during steaming. In addition, it must not react or precipitate by the action of various dyes and chemicals which are added to the printing paste. The total solids content of the printing paste is important for producing good prints under different conditions in the limited period of time, especially while printing lightweight materials. In such cases, thickeners of high solids content should be used to safeguard against bleeding and to retain the sharpness of the print. The thickener must maintain the print sharpness (prevent spreading of the colour beyond the boundaries of the design till steaming or curing is completed to effect dye fixation). During steaming, the adhesive nature of the thickener holds the dye particles, while the fabric becomes saturated with steam and the chemical reactions take place, giving the dye its fastness properties, while the whole system is exposed to the conditions of high temperature, moisture and time. The presence of sufficient amount of the thickener is a necessity, if good mark or definition of a pattern is to be maintained throughout the steaming operation, especially in fine line patterns in deep shades. Another requirement of the thickener is the ease of removal after steaming by an afterwash. The residual dye and products of decomposition of dyes and chemicals have to be removed from the fabric without staining the ground or areas printed with other colours.

For a printing thickener to be effective, it should have the following characteristics:

1. Stability to keeping (physico-chemical stability) should be good. Aqueous dispersion/pastes of high molecular weight thickeners (carbohydrates) may undergo fractional crystallisation on standing, upsetting the flow property and concentration of the dye in the printing paste, leading to the interference with the regular distribution of the dye.

2. It should have certain physical and chemical properties such as viscosity, flow property, ability to wet and adhere to the internal surface of etchings of the engraved roller. The flow property must be such that the print paste should remain into the engravings for a short time. It should not adhere to these engravings too fast, so as to allow transfer of the paste of the fabric.

3. It must be compatible with the other ingredients of the printing paste, e.g., compounds such as oxidizing and reducing agents, electrolytes, dispersing and wetting agents, solvents, hygroscopic substances, acids and alkalies.

4. The thickener film should dry properly on the fabric to prevent spreading of the colour by capillary action beyond the boundaries of the design (to ensure print sharpness).

5. Proper extraction of water from steam during steaming should be ensured to provide free space for the dye molecules to move towards the fabric and the free water to carry it.

6. The thickener should not have affinity for the dye and should not keep the dye from the fabric.

7. The thickener molecule should have a control over the free water pickup and should not carry the dye beyond the boundaries of the impression (flushing).

8. Once the dye is transferred from the thickener film (during steaming), the removal of the exhausted thickener film without fetching water-soluble dye should be easy.

9. The thickener should be cheap and available in abundance.

The qualities of thickener used for synthetic fabrics is little different than for cotton and other natural fibre fabrics.

Thickeners for Polyester

The thickener film should adhere well to the fabric, should produce an elastic film to prevent cracking, splintering and dusting off. Level prints with sharp outlines should be produced. The thickener must be easily removable in the after-treatment. thickenings with a high solids content bring less water on to the hydrophobic fabric and produce sharper and more level prints than those with low solids content. The dry film is less elastic, and splinters more easily. Thickenings with low solids content behave better with respect to splintering and in thermosol process better colour yield is obtained with such thickenings, but the sharpness of prints is not as good as with high solids content thickenings. In roller printing the sensitivity of the thickener film to crushing is important. Thickeners with high solids content are less sensitive to crushing.

Natural gums, starch ethers, and modified locust bean gums, Gum Indalca are the commonly used thickeners. Sodium alginate is less suitable, since it impairs the brightness of the print with disperse dyes, especially in the thermosol process. None of these is ideal for polyester printing, when used alone. Hence probably a combination of thickeners should be used. With natural thickenerssharp prints are obtained and the dried thickener film is relativelyelastic. If blotches are printed using these thickeners, levelness isno satisfactory. Prints with CMC are normally level and sharp andhave good yield, clarity and brightness. If carboxymethylation ofcellulose (from which CMC is made) is not proper, some insolubleparticles may remain, which clog the screen.Starch ethers have been most successfully used in polyesterprinting, but their wash off properties are not good. Therefore, starchethers are mixed with locust bean gum, which produce dull printswith poor colour yield when used alone.

Emulsion thickenings are helpful since mineral spirits (used in emulsion thickenings) are excellent wetting agent for hydrophobic fibres. These thickenings increase levelness and penetration of the dye in the fibres. They minimise the colour transfer in roller printing and reduce screen frame marks in screen printing. With hot air fixation, full emulsions (without conventional types of thickening agents) are of advantage, since complete fixation of the dye can be achieved.

Thickenings of reducing nature (British gum) produce dull prints. Addition of mild oxidising agents like Ludigol (1 to 2%) to the print paste prevents shadechanges during fixation. Addition of suitable carriers to the printpaste can drastically increase the colour yield, but different dyesbehave differently towards carriers. P-Phenyl phenol is the mostsuitable carrier in ageing processes. In thermosol fixation, special thickeners like LuprintanATP (BASF) or other brands may be used.

Cold-water-swelling gums are the preferred thickeners. In flash ageing, a thickeners, which coagulates in contact with alkali, is used to preserve print sharpness and to prevent flushing. In thermosoling, only those dyes having high sublimation fastness should be used. Staining of the cotton component can be removed (by reduction clear) during the fixing of the vat dye if the disperse dye used belongs to the azo class but not to anthraquinonoid class. Selection of vat dyes for the cellulosic component is much simpler than that of disperse dyes.

Thickeners for Nylon
Thickeners used in polyamide printing are numerous. Crystal gums produce prints of good colour value and excellent definition and may be easily washed off. These thickeners have relatively little effect on solvents and fixing agents as far as flocculation, coagulation and gelling are concerned. However, they have the drawback that blotches dried in a hot air chamber, become brittle. They may also cause fibre damage if over-dried. This tendency may be overcome by adding low viscosity sodium alginate, starch ethers, etherified locust bean gums, etc. Alginate thickeners produce soft prints and the thickeners are easy to wash off. Some acid azo dyes give lower colour value with alginate but metal-complex dyes produce prints of full colour value. A few metal-complex dyes coagulate alginate thickening, but because a large number of dyes are available for printing on polyamide fibres, these dyes may be omitted. In general, low viscosity alginates print best on hard fabrics like taffetas, whereas low or high viscosity alginates can be successfully used on spun and knitted materials.

Modified locust bean gums print well and do not become brittle when dried. On some weaves, the colour is likely to spread, which may be overcome by adding another thickener. British gums produce prints of good colour value and the pastes are not coagulated by solvents and fixing agents. The prints can become quite crisp on drying and even after washing, they may be harsh. Emulsion thickenings produce prints of good sharpness, which wash easily, but often give poor colour value, since penetration is usually very good. The print is soft even before washing thus eliminating ovendrying. Care should be taken in the preparation of the printing paste to ensure stability, electrolytes, used as fixing agents.

The chemistry of many types of thickening agents is extremely large. Only a brief overview is given here:

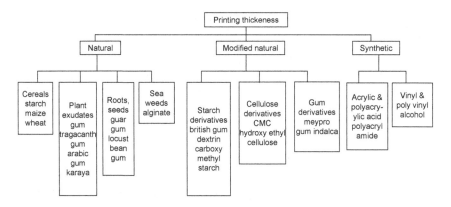

Different printing thickeness

4.2.1   Natural Thickeners

4.2.1.1   Starch

Native starch is a naturally occurring polymer and abundant in many plant products, was the first thickener examined in detail by the textile printer. Because of the requirement of boiling and the amount required to achieve viscosity they are little used unless they have further chemical treatment. The derivatives are still in use.

Starch is a homopolymer of glucose (D-glucopyrnosyl) units, with most of the units joined α-D-(1→4) linkages. Most starches contain 20 – 30% linear chain polymer known as amylase together with an irregularly branched material called amylopectin. Amylopectins have α-D-(1→6) branch for each 15 – 30 glucose units. The so called waxy starches contain little or no amylase separation can be done by fractional solution or by precipitation.

Commercial starches are available from cereals and tubers, etc., like barley, wheat, rice, maize, potato, sago, etc. All starches possess the property of swelling when heated with water giving a viscous paste. They are first wetted out with cold water into a thin suspension which is strained to remove gritty matter if any. The suspension is then heated with constant stirring. On heating for about fifteen minutes at or near about 70°C the starch swells, the paste first becomes thick and opaque and then at once it thins down becoming transparent and smooth in appearance. Every starch has a different gelatinising temperature.

Starch type	Gelatinising Temp 0C	Starch type	Gelatinising Temp 0C
Potato	65	Barley	62.5
Wheat	80	Sago	70
Rice	80	Tapioca (Cassava)	68.7
Maize	75	Arrowroot	70

Further heating should be avoided. It is then allowed to cool. While cooling, the paste must be constantly stirred otherwise it will form a jelly which will be quite useless for printing. A mechanical device is employed by the printer for preparing starch paste. The apparatus is known as " Colour Boiling Pans". It consists of double cased copper vessel provided with an agitator, swing arm water tap, steam taps to inlet and outlet for steam, between the casing; also water taps to inlet and outlet between the casing for cooling the paste or colours, quickly without formation of a skim on the surface. The pan revolves on pivots and is tilted by the wheel with a handle so as to be more easily emptied and cleaned. The pan temperature is gradually raised to the boil and kept boiling for half an hour, stirring all the time in order to produce a smooth paste. The liquor gradually thickens on heating and forms a consistent mass. At the gelatinising point, it at once thins a bit giving a transparent appearance instead of the opaque at the start. Steam is now stopped and water tap is opened. The water is admitted at the bottom, which gradually rises up the pan cooling the paste in the pan. The hot water gets out at the outlet at the top and cold water enters at the bottom. Stirring is continued during the period of cooling too, to avoid formation of lumps or crust. The cooled starch paste is then removed and stored for use.

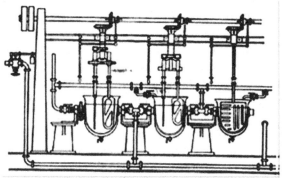

Cooking pans for preparing thickeners

Another important factor is of straining the colour paste before taking it to the printing machine. The ready starch paste must be first strained and then added to other ingredients of a printing paste. In all such mixings it is a good practice to add the bigger lot to the smaller one with constant stirring so that the ingredients get equally distributed throughout the mass of the paste. When all ingredients are added the colour must be strained before use. Straining is a mechanical operation the object of which is not only to remove any lumps or grit from the colour, but also to effect a more thorough mixing, and to give more homogeneity and softness to the paste. Straining of bulk paste for machine printing is best carried out by the vacuum filtering unit. A basic vacuum strainer consists of three parts.

1. A funnel cup at the top.
2. Cylindrical chamber at the bottom.
3. An electric motor with suction fan on the side. The open door of the bottom cylinder shows the can which receives the strained paste.

Working:The door is closed, the can occupies central position below the funnel cup. The paste is poured into the conical cup, motor started which exhausts air from the bottom cylinder, a vacuum is produced inside the cylinder, this sucks the paste from the conical cup, through a filtering gauze of the finest mesh (up to 140,000 meshes to the square inch). Colours contaminated with a great deal of lint are filtered without choking, since the gauze can be fitted as a moving band which is back washed automatically to clear away clogging fibre-simple but valuable practical feature. Many automated vacuum strainers are available in the market now.

4.2.1.2   Wheat starch

Wheat starch is the common starch used by the printer for preparing thickenings, especially for roller printing. Good quality of wheat starch should not contain more than 15% of moisture nor more than ½ % of mineral matter nor any grit. The admixture with potato or maize can be detected under a microscope. Each type of starch has a peculiar shape (a) Potato-elliptical with concentric rings. (b) Ricerectilinear, polygonal and flat. (c) Maize-rectilinear, polygonal with striations. (d)'Wheat-circular outline. (e) Sago-bag shaped. Wheat starch alone gives a stiff feel to the printed cloth. This drawback is overcome by mixing gum tragacanth with wheat starch. This mixture is a common thickening paste used by the machine printer. Wheat flour contains gluten in addition to starch. It is a powerful thickener imparting a harsh feel but has good binding quality. Maize starch is used with alkaline pastes as it can stand alkali far better than wheat. Its common use is in printing sulphide, vat, naphthols, sodium aluminate and alkaline discharges.

Rice starch is never used in printing at all. It is employed in finishing printed goods for giving a stiff finish in old times. Potato and tapioca are also used in finishing but not in printing. Starch is also available in a soluble form known as "Soluble starch." This is very useful for spray printing which requires the thinnest paste in calico printing. In fact, simple solution of colour in water can be printed without any difficulty with spray.

As thickeners the starches show many disadvantages. They exhibit syneresis (a spontaneous reversal of the gelation which occurs during preparation of the paste) and so are not stable over long periods. They are highly anomalous and tend to give unlevel prints. They have poor wash off properties. Starch thickenings can tolerate only a limited amount of alkali, and if caustic alkali is to be used it is advisable to use maizestarch, which is more resistant to alkali than is wheat starch. In case of vat prints to improve the alkali resistance containing starch, and at the same time improve their working properties and penetration into the fabric, additions of more expensive thickenings such as gum tragacanth and British gums are made. Increases inpenetration and smoothness are invariably obtained at the expense of yield; hence a judicious balance between the amount of starch and the amount of other thickening agent is necessary.

Some of the many important advantages they possess are the high colour yield obtained from starch thickeners, because it gives surface printing without penetration and not thinned down with alkali as with acids and their cheapness and availability almost anywhere in the world. Since starch thickeners have a low solids content (up to 10%), they offer little protection against attack by atmospheric oxygen to the sodium sulphoxylate formaldehyde present in prints waiting to be steamed, but do permit rapid diffusion of the leuco vat dye on to the fibre during steaming. Because of these and other problems, research has been done to make starch derivatives which rather cover the disadvantages of the starch and give a better thickener. Thus today starch derivatives are used in preference to starch.

Starch undergoes hydrolysis with the breakdown of starch molecules and loses viscosity when it is boiled with dilute mineral acids (hydrochloric or sulphuric acid). But organic acids do not have much effect on viscosity. But it is not compatible with certain organic acids like citric acid like citric, tartaric or oxalic acids in the presence of sodium bisulphate and hence cannot be used for printing pastes involving these chemicals. Sodium hydroxide causes a transparent jelly formation. Viscosity of starch decreases with storage. The rate of change of viscosity with storage depends on the pH. For starches like maize or corn starch the viscosity can be maintained at pH range of 4.5 – 6.5.

Since it is available abundantly it is one of the most widely used starch. It is not affected by dyes and mordants. It gives a good workable paste which can be used alone or in combination with gum tragacanth, etc. It gives good stable viscosity for print paste requirement at 10-12% concentration and gives good colour yield.

In the actual preparation of a 10% stock paste, 100 g of starch is stirred up with cold water into a smooth slurry and further 900 ml of water can be added (4 ml castor oil is sometimes added to make the paste softer. It is slowly heated to boil with stirring to get a thick paste. Boiling is continued till this paste this down to a smooth paste. It is further cooled to room temperature while stirring is continued, if not a jelly may form which is not useful for print paste preparation.

4.2.1.3   Gums of plant origin

There are many gums of plant origin which are used in textile printing. Except Alginates other thickeners may not be of much importance, except in special cases, due to various reasons which will be explained below:

Source	Name of the Gum	Trade Name
Trees and Shrubs	Gum Arabic, Gum Senegal, Gum Tragacanth, Gum Karaya	
Plant or Tree seeds	Locust Bean Gum	Gum Gatto and Cesalpiniagum (Cesalpina, Italy)

	Guar Gum and starches	
Seaweed	Sodium alginate	Manutex, Lamitex, etc.
	Carrageenan	Irish Moss (Blandly, UK)

Similar to cellulose, galactomannans are another source of natural thickeners. They are polysaccharides composed of main-chain mannose and side-chain galactose. (See below)

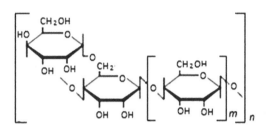

D-Galactomannoglycan

units, though structurally related to starches. Examples are Locust bean gum (m = 3, n = 375) and guar gum (m = 1, n = 440). Some of the natural gums and the distribution of galactose units are shown below:

$$--M-M-M-M-M-M-M-M-M-M-M--$$
Polymannose

$$--M-M-M-M-M-M-M-M-M-M-M--$$
$$\quad G \qquad\qquad G \qquad\qquad G$$
Cassia gum from Cassia tora/obtusifolia seeds Galactomannan-1,5 (m = 4)

$$--M-M-M-M-M-M-M-M-M-M-M--$$
$$\qquad\;\; G \qquad\quad G \qquad\quad G$$
Carob gum from Ceratonia siliqua seeds - Galactomannan-1,4 (m = 3)

$$--M-M-M-M-M-M-M-M-M-M-M--$$
$$\;\; G \qquad G \qquad G \qquad G$$
Tara gum from Cesalpinia spinosa seedsGalactomannan-1,3 (m = 2)

$$--M-M-M-M-M-M-M-M-M-M-M--$$
$$\; G \quad G \quad G \quad G \quad G \quad G$$
Guar gum from Cyamopsis tetragonoloba seedsGalactomannan-1,2 (m = 1)

Preparation of Natural gums is lengthy and tedious process involving cooking boiling, straining, etc. To make the preparation easier and make them more suitable for a particular process many modified natural gums are available in the market. They are chemically or physically or combines modified to improve the products result. These are sold under several trademarks by the manufacturers, some of them are given in the below table. The various brands available represent differing degrees of modification and make the product suitable for different requirements. Cellulose, which is normally very insoluble in water, will, by chemical modification, produce valuable thickeners.

Product	Source
British gums (various suppliers)	Starch
Nafka Crystal Gum (Scholtens, Holland)	Gum Karaya
Indaica series (Cesalpina, Milan)	Locust Bean gum
Polyprint series (Polygal, W. Germany)	Probably Locust Bean derivatives
Meyprogum series (Meyhall Chemical, Switzerland)	Probably Guar gum derivatives
Solvitex and Solvitose series (Scholtens, Holland)	Starch-derivatives such as ethers

Some manufacturers do not reveal the compositions fully. The quantity of thickening powder required to make a print paste of the correct viscosity varies from product to product. This largely depends on the chemical constitution and the chain length and degree of chain branching in the thickening agent constituent molecules. Where relatively large weights of dry thickener are required, say 8% and upwards, the resulting thickening is described as being of medium or high solids content. Conversely, thickenings made with less dry agent are of tow solids content. Thickening agents must possess satisfactory mechanical properties under printing conditions, otherwise an even print will not result. In addition, they must be fully compatible with the other components in the print paste. The viscosity of a print paste is most important, but an added complication is that with the majority of thickenings available, the viscosity falls at increasing rates of shear. The shear rate is determined by the mechanical stresses to which the paste is subjected at the printing stage. In effect it is a measure of the rate at which adjacent layers of thickening are being forced along relative to each other.

(a) Gum Arabic (Gum Senegal)

Gum Arabic, also known as acacia gum is obtained from a tree known botanically as 'Acacia Arabica'. The tree is a native of India, Africa and Arabia. Producers harvest the gum commercially from wild trees, mostly in Sudan (80%) and throughout the Sahel, from Senegal to Somalia—though it is historically cultivated in Arabia and West Asia. Gum Arabic is a complex mixture of glycoproteins and polysaccharides. It is the original source of the sugars arabinose and ribose, both of which were first discovered and isolated from it, and are named after it.

The gum has a colour from a light yellow to dark brown. It is available in the form of lumps. It is slightly acidic and requires neutralisation if it is to be used with a neutral or alkaline printing paste. 1: 1 gives a very thick solution for use in printing. There is lot of woody matter and sand in the commercial product. These impurities are removed by soaking the gum in cold water for several hours. The woody portion floats on the top which is easily removed and the sand settles down at the bottom. It is also removed by filtering through a cloth. Gums give transparent and uniform shades in printing. They are easily removed from the cloth by mere washing with water cold or hot. Starch is not easily removed from the printed cloth. It requires the use of an enzyme, like diastase, for converting it into a soluble form. It does not clog the pores of the screen gauze, like synthetic thickeners or starch pastes. Gum is the most common thickening agent used by block and screen printers. It has a moisture content of about 12-16%. They dissolve in water to give colloidal solutions possessing nearly Newtonian flow properties. The advantage of this gum is that paste viscosity does not change appreciably with increasing load so that at any squeegee pressure the paste viscosity same maintaining the integrity of the print. Other advantages of this thickener is that after the fixation process it can be easily washed off and it doesnot clog the screen. But addition of electrolytes reduced the viscosity.

High solids content is necessary to obtain sufficient viscosity; the thickeners have excellent levelling properties and give good definition, but, of course, only a low colour yield. Such thickeners would only be used when levelness of print is of paramount importance and is used specially for silk and nylon, but mainly used for printing light shades. It is an excellent substitute for British gums in printing vat dyes, but is costlier.it may be used in printing direct and azoic colours on cellulosic fibres and disperse dyes on cellulose acetates and polyester. It can be used for printing acid dyes on nylon and useful for printing level light shades.

A 50-60% paste is normally used.

Recipe

Quantity	Unit	Additions
600	Parts	Gum Arabic lumps are stirred into cold
500	Parts	Water. And the mass is bulked to
1000	parts	

Any floating bits of chips and woody matters are skimmed off and then brought to the boil and maintained thus with constant stirring for about 3h. It is then cooled, bulked to 1000 parts and strained. After dissolution, it is ladled into deep casks and allowed to stand for several days and finally strained. It can be used alone or mixed with starch or gum tragacanth.

Gum Arabic is neutral or slightly acidic (calcium, magnesium and potassium) salt of a complex polysaccharide. On acidifying it gives Arabic acid.

Gum arabic

Gum Arabic has a branched chainstructure with L-rhamnopyranose or L-aranbinose, galacto pyranose and glucuronic acid units, having a molecular weight of 300,000 to 1,000,000.

Commercial Gum Arabic has a moisture content of 12-16%. The paste viscosity of this gum does not change appreciablywith increasing load. Addition of electrolytes to 11% gum solutions decreases the viscosity. The print paste prepared using Gum Arabic gives uniform shades and sharp prints and can be easily washed off after fixation of the dyes. It doesnot cause screen blocks and can be used for outlines as well as level blotches.

(b) Guar Gum

The most popular of the Galctomannan product used as thickener.The guar bean is a seed of the legume plant Cyamopsis tetragonolobais, principally grown in India, Pakistan, U.S., Australia and Africa. It is primarily the ground endosperm of guar beans. Guar Gum Powder is extracted from the Guar Seed after a multistage industrial process. The most important property of Guar Gum is its ability to hydrate rapidly in cold water to attain uniform and very high viscosity at relatively low concentrations. Its colloidal nature gives excellent thickening to the solution.

Guar Gum, either modified or unmodified is a very versatile and efficient natural polymer covering a number of applications in various industries like food, beverages, pharmaceuticals, cosmetics, paper, textile, construction, oil & gas well drilling, mining etc., due to its cost effective emulsifying and thickening properties. It is typically produced as a free-flowing, off-white powder. Chemically, guar gum is a polysaccharide composed of the sugars galactose and mannose. The backbone is a linear chain of β 1,4-linked mannose residues to which galactose residues are 1,6-linked at every second mannose, forming short side-branches.

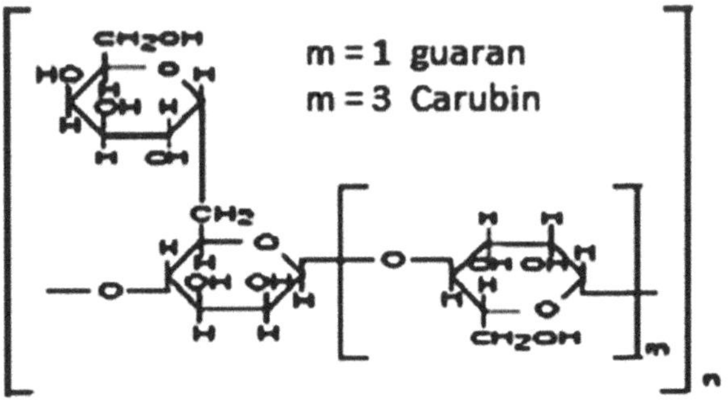

m = 1 guaran
m = 3 Carubin

D-galacto - mannoglycans

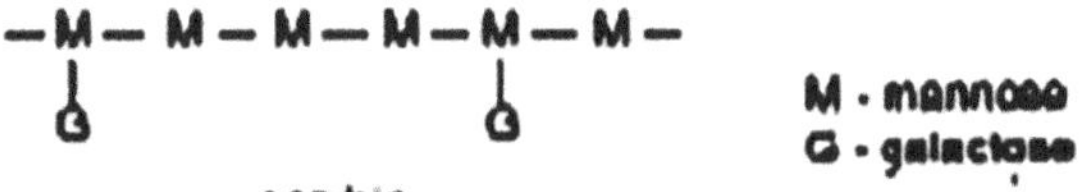

M . mannose
G . galactose

carubin

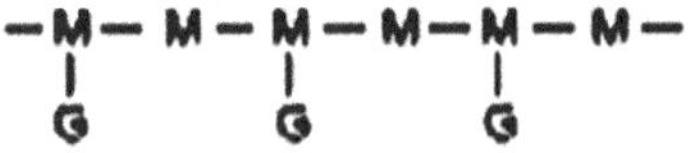

guaran

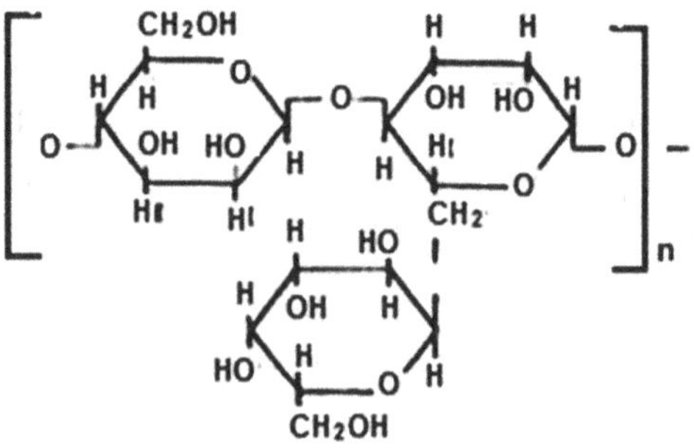

Guar Gum

Guar gum is more soluble than locust bean gum, as it has more galactose branch points. Unlike locust bean gum, it is not self-gelling. However, either borax or calcium can cross-link guar gum, causing it to gel. It remains stable in solution over pH range 5-7. Strong acids cause hydrolysis and loss of viscosity, and alkalis in strong concentration also tend to reduce viscosity. The lower the temperature lower the rate at which viscosity increases and the lower the final viscosity. It can be used for printing disperse dyes on nylon, etc.

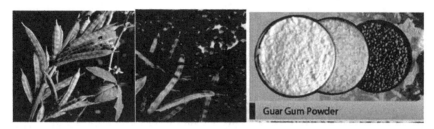

Although the cold water dispersibility is a major advantage, attention must be paid to the technique used for used for dispersion to avoid the formation of lumps. The hydration of powder should not be faster than the rate of surface wetting.

It is industrially produced involving several steps like hull removal, germ removal, endosperm grinding, etc. The seed is first treated with 55% Sulphuric acid and the acid is washed off. The hull is charred and loosened by washing and flame treatment. Outer layers are removed by grinding and sifting. Differential grinding is done to remove the germ. The endosperm only

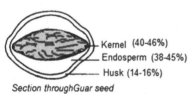

Section through Guar seed

Guar gum seed and powder

is removed after the above two process and ground to a fine particle size and marketed as guar gum. It is a fraction of the endosperm ground from the seeds (diameter approx. 5 mm) of the guar plant, a grey-white powder comprising 65% mannose and 35% galactose (Galactomannan), aqueous-colloidal neutral solution, has an approximately 8-fold starch thickening power is used as thickener in print paste for many textile materials.

Preparation of paste

Hydrophilic guar gum is difficult to disperse (supported by pre-mixing with alcohol or glycerine, as well as through more intensively-acting homogenisers), (whereas easily dispersible guar gum derivatives in vigorously stirred water are considerably easier to handle). The guar gum is sprinkled into water with vigorous stirring when the powder gets dispersed. Stirring is continued till it become viscous. Further the gum is allowed to hydrate or soak and thicken for 30 min.-2 hours. It is kept for 24 hours to become homogeneously soaked. The viscosity depends on the soaking time, temperature, compensation of the powder, pH, mixing, etc. Viscosity of guar gum changes according to temperature but the maximum viscosity is obtained at 20-400C which is very much suited for textile printing. Water temperature has a great influence on the hydration time: the colder it is, the slower the rise in viscosity; however, at low temperatures, higher end viscosity can be achieved. For this reason, control/regulate the water temperature as precisely as possible. Once lumps have occurred, they are insoluble, even when high shear forces and longer hydration times are used. Lumps can be caused by poor quality of guargum, incorrect procedures, extreme water temperatures, pH too low. Under normal mixing conditions, with guar gum 95% of the maximum achievable viscosity is reached in 30 minutes.

Storage more than 24 hours can reduce the viscosity of guar gum due to fermentation and enzymatic hydrolysis. Which can be prevented to some extent by addition of preservatives like formaldehyde, chlorophenols (restricted), etc.

Guar gum produces high viscosity at lower concentration compared all other products. (See comparison below). It is stable over a wide range of pH. Since the gum is non-ionic it can give constant viscosity from 1 to 10.5, above this pH the hydration of the gum is decreased and the viscosity is negatively affected. The best hydration occurs at pH 7.5-9.0. Since it is non-ionic it is seldom affected by the presence of electrolytes. At low concentration like 2-3 % it loses fluidity and becomes a gel. Borate ions act as cross linking agent for hydrated gum to form cohesive gels. In the alkaline range, guar gum can gel in the presence of borate ions and antimony ions, whilst oxides of lead, cobalt, antimony and chromium ions encourage gelling in the acid range. Oxalic acid in the presence of iron very quickly lowers the viscosity through acid hydrolysis. The oxidative breakdown of the galactomannan (and other water-soluble polymers) is catalysed by iron ions and tin ions, accelerated by sequestering agents (EDTA); for this reason, polyphosphates are more suitable than EDTA in handling hardness if necessary. Some of the advantages of guar gum usage in textile printing are (1) it is very cheap, it is even cheaper than alginate) (2) best viscosity at lower concentrations (a 2% solution can give an consistency required for textile printing) (3) simple preparation process (4) since it is non-ionic in is compatible with any dyes and print paste ingredients.

The guar gum can be modified to convert it to an anionic gum by carboxymethylation. For example, it can be hydroxyl methylated by the action of ethylene chlorohydrin.

Guar Gum - OH + NaOH ⟶ Guar Gum -ONa + Cl - CH$_2$COOna ⟶ Guar Gum - O - CH$_2$ - COONa
Guar gum anionic   Anionic

Comparison of various thickener pastes at 250C after 24 hours after their preparation

Thickener	Gum paste %	Viscosity cP
Gum arabic	20%	50
Locust bean gum	1	100
Methyl Cellulose	1	150
Gum tragacanth	1	200
High viscosity CMC	1	1,200
Gum karaya	1	1,500
Sodium alginate	1	2,000
Guar gum	1	3,300

(c) Gum Tragacanth or Gum Dragon

This gum is also from the vegetable source like Gum Arabic. Tragacanth is a natural gum obtained from the dried sap of several species of Middle Eastern legumes of the genus Astragalus, including *A. adscendens, A. gummifer, A. brachycalyx* and *A. tragacanthus.* Some of these species are known collectively under the common names "goat's thorn". The name derives from the Greek words tragos (meaning "goat") and akantha ("thorn").

It is marketed in the form of dry horny scales or leaves varying in colours in different colours like dark yellow to white.

Iran is the biggest producer of the best quality of this gum. It is slightly soluble in water. Six per cent solution gives a thick paste useful for printing. The flakes are allowed to soak in water for 24 hours. It dissolves completely in water giving a thick smooth paste. After that period, water is heated to boiling and boiled for several hours till it thins down and smoothens further. The boiling period can be shortened by adding some dispersing agent.

Gum tragacanth gives well penetrated shades and can be mixed with starch to give solid shades of good penetration. This gives a highly anomalous thickening at low solids content. The thickening is of the "gel" type, but the particles are smaller than those in starch pastes and it has better levelling and penetration properties. It is stable to alkali carbonates, but becomes unstable with caustic alkalis. No other gum works well with starch. In fact, the starch paste is completely spoiled by the addition of any other natural gum to it. A mixture of starch and gum tragacanth is the most common thickening used by the machine printer. It is one of the best thickeners, where the dyestuff is required to penetrate to the reverse side of the cloth. It also gives smooth and sharp prints.

Astragalus gummifer tree and sap extraction

The flow properties of gum tragacanth thickeners are critically dependent on the method of preparation. Prolonged boiling reduces both the viscosity and the degree of anomaly owing to excessive gel$\rightarrow$ sol conversion, which is not completely arrested by the cessation of boiling and so leads to a gradual thinning during storage. Gum tragacanth is most widely used in admixture with starch as the familiar starch-tragacanth thickeners.

Tragacanth is a mixture of a neutral watersoluble polysaccharide and a complex alkali soluble polysaccharide. Hydrolysis with dilute mineral acid gives D-galactose, L-fructose, D-xylose, L-arabinose, D-galacturonic acid.

Preparation of tragacanth stock paste
70g of gum tragacanth is mixed with 930 ml of cold water and allowed to stand for 2-3 days with occasional stirring. The mixture is then boiled with continuous stirring until the gum is fully dissolved and a homogeneous paste is obtained. It is further cooled to room temperature with stirring and strained.

Preparation of Starch-tragacanth paste

Mix 140g wheat starch with 400 ml water and 600 g 7% gum tragacanth paste. The mixture is further boiled with constant stirring and kept at boiling for 30-40 min. The paste is then cooled and made up to 1 kg and strained.

(d) Locust Bean gum

Amongst other things, this distribution has an influence on ease of dispersibility. For example, warm water is required to effect complete dispersion of locust bean gum (the 1,4 form) but guar gum (the 1,2 form) disperses readily in cold water because of decreased molecular association arising from the greater frequency of side-chain substitution. As with the starches, modified gums can be obtained. In particular, etherification improves the cold water dispersibility of locust bean gum. In Table 10.37, various derivatives of galactomannans are listed together with their main applications.

Locust bean gum (LBG, also known as Gum gatto, gum tragon, Tragasol, carob gum, carob bean gum, carobin) is also vegetable origin. It is obtained from the seeds of certain variety of locust trees. It forms a colloidal solution and can be used as a substitute of gum tragacanth. It forms a film in the presence of an alkali. It is mixed with warm water and then heated to boil and boiled for 15 minutes to get a solution. 3% gives a thick paste for working. It is susceptible to bacteria and a germicide must be added if the paste is to be preserved to working condition for a longer time.

Locust bean gum

Some variety with resistance to bacteria is being supplied by manufacturers like "Cesaltex." It resists fermentation and the paste can be kept for several days in good working condition. This product, however, is easily affected by alkalis. It cannot, therefore, be used with alkaline pastes. This drawback has also been overcome by the efforts of the same firm. The manufacturers claim that this product gives shades of exceptional purity and uniformity, with sharp outlines and deep penetration of the dyes. It has a very high thickening power and its solutions are homogeneous, transparent and free flowing. It can be easily washed off from the printed cloth.

It is supposed to be a molecular combination or mixture of galactan and mannose. Like 29.18% Galactan, 58.42% Mannan, 2.75% Pentosan, 5.29% Albuminoids, 3.6% Cellular tissues, 0.82% Mineral matter, traces of Laevulan.

A solution of the gum is likely to be precipitated by high concentrations of electrolytes but can be prevented by the addition of glucose, sugar, gelatin, glycerine, etc., before adding the electrolytes. Addition of acid decreases the viscosity whereas addition of alkali increases the viscosity. Borax is the best ingredient to increase the viscosity. Paste is prepared by sprinkling the powder over cold water and stirring continuously while the gum swells up. The mixture can be boiled till we get a smooth homogenous paste. It can be further strained for use. In another method, gum powder is sprinkled into a solution containing about 0.5% orax, while stirring. It is then heated to 80-900C to get a smooth paste. Phenol may be added as preservative and strained.

Probable structure of the complex of locust bean gum with borate ions

Locust bean gum forms an interesting and unusual crosslinked complex by association of cis-dihydroxy groups in the mannose chains with borate ions, diagrammatically represented in structure (see below). This complex forms a gel, which has been made use of in printing with vat dyes in a two-stage fixation process. The crosslinks are relatively weak, being in a state of dynamic equilibrium, and are ruptured in the presence of hydrotropes such as glycerol.

(e) Gum Karaya

It is an acetylated polysaccharide extracted from plants. It is acidic in nature and relatively insoluble in water. It has a molecular weight of about 9.500,000 and constituted of D-galactose, L-rhamnose, and D-galatcuronic acid units in the proportion of about 14:15:43. This is the only commercial gum not soluble in water. When soaked in water it readily absorbs water and swells. In can be made soluble in water by the addition of sodium peroxide, persulphate or per carbonate and boiling. It can also be dissolved by boiling under pressure.

Normally, around 20-25% paste is used in textile printing, and 0.5-0.6% sodium peroxide is added which also acts as a bleaching agent for the gum, to dissolve the same. The viscosity of the paste decreases with the addition of electrolytes like sodium, calcium or aluminium chlorides and sulphates. It can maintain the viscosity for several days and further extended by addition of preservatives like Formaldehyde, benzoic acid, salicylic acid, chlorinated phenols.

It has good adhesive properties at 20-25% solution which can be further enhanced by the addition of a mild alkali. The gum can be easily washed off from fabric. Modified gum karaya is a good textile printing thickener for printing cotton blotches and sharp outlines and sharp prints at the expense of colour yields.

Paste preparation

Quantity	Unit	Additions
200	g	Gum karaya, fine grounded, is mixed with
500	ml	Cold water, then
0.6	g	Sodium peroxide dissolved in
125	ml	Cold water is slowly added with stirring for 1 hr.
0.65	g	Sodium peroxide dissolved in
75	ml	of Cold water is further added and stirred for 15 min

The mixture is gradually heated to boil and boiling is continued for a further 45-60 min. and cooled. Some alkali will be formed from sodium peroxide which has to be neutralised and the paste is made up to 1 kg and strained.

(f) Derivatives of vegetable gums

Derivatives of Locust Bean gum and guar gums are more important because of their suitability for printing synthetic fibres. Some of the main chemical modification is carboxymethylation, carboxyethylation, hydroxyethylaton, hydroxypopylation, etc.

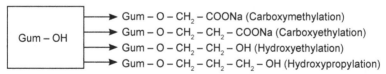

Even a very low degree of substitution can make significant difference in the performance of the thickeners in printing. Guar gum derivatives can be easier to handle and dispersed since they have good emulsifying properties, compared to guar gum which is difficult to disperse with higher chances of lump formation. Guar gum takes 30 minutes minimum to achieve 95% of the maximum achievable viscosity, whereas guar gum derivatives hydrate even more quickly. They are stable to alkalis and acids. They soluble in boiling water and have excellent flow property. It can be mixed with any other thickener and can be easily washed off after fixation of the prints. Indalca is another brand of modified guar gum which is widely used for printing in any machine like roller, flat bed or rotary. They don't have any reducing action on dyes and unaffected by water hardness.

Derivatives of guar gum is made by carboxymethylation, carboxyethylation, hydroxyethylation (See below fig. a) hydroxypropylation (See below fig. b), etc. the first two modifications give ionic polymers, while the other two produce non-ionic polymers. Both are extensively used in printing as thickeners, especially in polyester and nylon printing.

(a) Production of Hydroxy ethylated Guar gum

(b) Production of Hydroxy propylated Guar Gum

These derivatives can be chemically modified to make it suitable for various precise purposes. Given below the modifications and their uses in textile processing, mainly printing:

Parent Galactomannans	Chemical modifications	Used for printing of
Galactomannan-1,2	As such, without modification	Acid, Metal-complex dyes on carpet
E.g. Guar gum	Depolymerised	Acid, Metal-complex dyes on carpet
		Vat, Direct and azoic dyes on cotton and viscose
		Disperse dyes on Polyester
		Acid, Metal-complex dyes on Nylon
		Cationic dyes on Acrylic
	Hydroxy methylated	African prints on cotton with azoic dyes
		Disperse dyes on Polyester
		Acid, Metal-complex dyes on Nylon
		Cationic dyes on Acrylic
		Acid, Metal-complex dyes on carpet
	Hydroxypropylated	Acid, Metal-complex dyes on carpet

		Carboxymethylated	Acid, Metal-complex dyes on carpet
			Vat, selected reactive dyes on cotton
Galactomannan-1,4	Hydroxyethylated		Acid, Metal-complex dyes on carpet
E.g., Carob gum			African prints on cotton with azoic dyes
			Disperse dyes on Polyester
			Acid, Metal-complex dyes on Nylon
			Cationic dyes on Acrylic
			Acid and metal complex dyes on wool and silk
	Carboxy methylated		Vat dyes on cotton
Galactomannan-1,5			Acid dyes on wool and silk
E.g., Cassia gum	Depolymerised		Vat dyes on cotton
			Acid dyes on wool and silk
	Carboxy methylated		Acid, Metal-complex dyes on carpet

Preparation of stock paste

The powder id added to cold water under vigorous stirring, using high speed stirrer for 15 – 30 min. Keep for overnight and is ready for use. It can be dissolved in warm water also much faster.

Alternatively, the powder is mixed in cold water with high speed stirring and then boiled and simmers for 15-30 min. without stirring to avoid the formation of bubbles. On preparation the paste will be slightly alkaline and has to be acidified with dilute acetic acid or hydrochloric acid, if it has to be used for printing acid dyes, disperse dyes, basic dyes, metal complex dyes on silk, wool, nylon, polyester which are printed in the acid pH. But in case polyester where fixation has to be done by pressure steaming, high temperature steaming or thermofixation, the acid used for neutralisation and acidifying, one may use non-volatile acids like tartaric, citric acid, etc.

Characteristics and suitability of various grades of Indalca in printing

Fibre	Class of dye	Thickener	Conc. To be used	Stability to heavy metals	Compatibility with half emulsion thickener	Comments
Polyester	Disperse	Indalca AG, AG/BV	5.0-8.0	Gum arabic	Suitable	Acidify the thickener
Polyester	Disperse	Indalca PA, PA/MV, PA/30, PA/30/LV	3.5-9.0	Locust bean gum	Suitable	Suitable for rotary
Polyamide	Acid, Metal complex	Indalca PA, PA/MV, PA/30, PA/30/LV	3.5-9.0	Yes	Suitable	Stable print paste

Polyamide	Acid, Metal complex	Indalca AG, AG/BV	5.0-8.0	Yes	Suitable	Neutralise with citric or tartaric acid
Cellulosic	Vat	Indalca U	8	No	Suitable	can be mixed with British gum and starch
Cellulosic	Direct	Indalca PA, PA/MV, PA/30, PA/30/LV	3.5-9.0	Yes	Suitable	Good for blotch and good penetration
Cellulosic	Direct	Indalca U	8	No	Suitable	Good washing off

(g) Nafka Crystal Gum

Nafka crystal gum is a highly purified or modified natural gum. In the purified form it is supposed to have 3-4 time thickening power than the other natural gums. It requires a high solids content (around 25%) to yield a satisfactory thickener. It is moderately anomalous has good penetration and levelling power. Behaviour of the gum is similar to gum Arabic but is easily soluble in cold water and very stable. It can be easily washed off after fixation of the dyes. Its viscosity is not changed by increase in temperature or do not gel or ferment on long standing. The gum can be used for all class of dyes like direct, vat, acid, basic, mordant, solubilised vat, etc. and all types of machines including rotary. It gives sharp out lines and even blotch prints. Preparation of Nafka crystal gum is given elsewhere in this book.

(h) Alginates (Derivatives of Seaweeds) $C_{10}H1_8O_{10}(COONa)_2$

Derivatives of seaweeds are done by extraction of alginic acid and treatment with sodium carbonate to produce sodium alginate. The name alginic came from the fact that it is extracted from algae. The see weeds found near the seashore are two types - bladder wort (Fucus vesiculosis) and laminaria species like horsetail kelp (Laminaria digitata), broad leaf kelp (Laminaria saccharina) The latter species have high lignin content and used for extracting alginate. The cell walls of brown algae contains calcium and magnesium salts of alginic acid.

Manufacturing of alginates

The dried sea weeds were used as manure and cattle fodder. The dry seeds may have following composition:

20-30	%	Ash
6-10	%	Protein
8-12	%	Cellulose

10-15	%	Laminarin
10-12	%	Mannitol
20-25	%	Alginic acid

The seaweed which is grown only near the shore area is harvested by different methods found suitable to that area. There are mainly two processes for manufacturing alginic acid from sea weed.

The first process which is called the Greens cold process or Stanfords process, the seaweed is first leeched repeatedly with dilute hydrochloric acid immediately after harvesting and the leeched water is discarded. Once the salt content is reduced to around 80% it is chopped and milled in a hammer mill. Next, it is treated with aqueous solution of sodium carbonate at pH 10 and temperature of about 100C for 30 min. and the diluted with about 6 times ins volume with soft water. The liquid part is filtered in rotary drum vacuum filters which is sodium alginate solution. The alginate is precipitated as calcium alginate by the addition of 12-15% calcium chloride. The precipitated alginate is washed well and bleached with sodium hypochlorite (1%) and again washed till there is no chloride left in the alginate. It is then converted to alginic acid by the addition of 5% hydrochloric acid. It is filtered out on wire mesh and washed thoroughly to till chloride free. Filtered material is pressed to remove the excess water and converted to salts of sodium and ammonium as per requirement as the alginic acid is not stable.

In the second process which is known as Leglohec-Herter process, the harvested sea weeds are dried and treated with calcium chloride solution to dissolve laminarin, mannitol and most of the inorganic salts. The extract is drained and the residue washed with soft water all dissolved solids is removed. It is then treated with dilute hydrochloric acid and the acid is again washed off with soft water. The desalted weed is then chopped and macerated with twice the volume of sodium carbonate solution (4%) at 400C. The mixture is well beaten throughout the period till it gives a smooth paste. It is then diluted with water 3:7 ratio. Compressed air is forced through the slurry and emulsified by stirring with a high speed stirrer and allowed to stand for 8 hours. Undissolved cellulose material comes to the surface floated by air bubbles and is skimmed off.

The clear solution underneath is sodium alginate is coloured due the presence of pigments and it removed by absorbing on a suitable jelly (prepared by co-precipitating a mixture of Alumina, gelatinous silicic acid, and aluminium alginate) and the jelly may be removed by centrifuge. The decolourised sodium alginate is converted to alginic acid by adding hydrochloric acid to a pH 3. Precipitated alginic acid is separated and washed in dilute hydrochloric acid for a short time, then in water and then with alcohol. It is then dried at low temperature of 760C. The purified alginic acid is converted to sodium or ammonium alginate as required. The main advantage of this process in the avoidance of filtration to remove cellulosic and other impurities, as it is difficult to filter sodium alginate is very difficult due to its high viscosity even at low concentration.

They are important for reactive dye printing because of the absence of free primary hydroxyl and the repulsion of dye anions by the ionised carboxyl group under alkaline conditions. They are available in a range of molecular weights which produce grades ranging from 'low viscosity' (high solid) to 'high viscosity' (low solid). It is also used as sizing agent in some cases.

Sodium alginate have become very important for print paste thickening because of their ready solubility, even after high temperature fixation treatments. They are used for reactive dye printing, vat dye printing by the two stage method (flash ageing) in admixture with starch ether and as a component of many polyester thickeners. The alginate macromolecules is a linear co-polymer of β-D(1-4) linked mannopyranosyluronic acid units (Man pA) and

Man pA   Gul pA   Aginic acid

of α-L(1-4) linked gulopyranosyluronic acid units (Gul pA) and Alginate thickenings are suitable for use with many dyes and are particularly recommended with the following types:

Reactive Dyes	Pigment Dyes
Disperse Dyes	Direct Dyes
Vat Dyes	Stabilised Azoic Dyes
Acid Dyes	

Alginate thickenings are also recommended in emulsion, discharge, flock and metal powder printing. It is compatible with Anionic & non-ionic wetting agents. Sodium formaldehyde sulphoxylate, Starches, Vegetable gums, alkali metal & magnesium salts in conc. up to about 15% and mixtures with pH between approx.3.5 and 11.

It is not compatible with Salts of heavy metals. E.g., iron, zinc, chromium, copper, aluminium, calcium, cationic wetting agents and other cationic auxiliaries and mixtures with pH below pprox.. 3.5 or above 11.

Usually, Alginates are made in different grades for textile purpose:

1. High viscosity alginates for use at low concentrations for low solids thickenings.
2. Medium and low viscosity alginates for use at medium and high concentrations-for medium and high solids thickenings.
3. Special grades for special applications.

The thickeners with long preparation procedures are not very common these days unless it becomes necessary for particular print. The requirement of n thickener for a modern printer is their ability to be dissolved in cold water with the aid of a high-speed stirrer and if it can be used directly it more acceptable. Natural gums of the exudate type, sold in lump form, require soaking in cold water for up to 24 h. This is followed by boiling the swollen lumps in a double-jacketed steam heated pan (see under starch thickener), with constant stirring. The boiling and cooling routine, albeit without preliminary soaking, is necessary for powdered gums such as locust bean gum and starches.

Use of alginates in high temperature processes

Alginate thickenings are also particularly suitable for fixation processes depending on High Temperature Steaming (HT Steaming) or dry heat (Baking or Thermosol Process). The film of the alginate thickening on the fabric is easily removed in the subsequent washing off process. Many other thickenings tend to form a hard film which is difficult to wash off resulting in harsh handle of the printed fabric.

Alginates, even with added sodium metaphosphate is gelled by strong caustic alkalis. This gelling can however be prevented by the inclusion of 0.5% triethanol amine or 0.3% basic ammonium phosphate or disodium phosphate (based on the total weight of the paste) in the mixture. This applies particularly to the printing of vat and stabilised azoic dyes. There are however special grades of alginates which are already stabilised to withstand caustic alkaline conditions. The property of some alginates grades of coagulating under strongly alkaline conditions is used to advantage in the printing of vat dyes by a two-phase process.

Alginate thickenings in common with all other grades of sodium alginate thickenings, are liable to produce a slight gel if used in print pastes with a pH below 4.0 especially in areas where hard water is used. Any tendency to gel due to the above conditions can be prevented by the extra addition of sodium metaphosphate (Calgon) to the print paste. Stable print pastes down to a pH of 3.5 are obtained even in water of 40° Hardness (400 p.p.m. CaCO3) by the extra addition of 0.5 to 1.0% sodium metaphosphate (Calgon) to the print paste. This is particularly important for the printing of some acid dyes and for the printing of modified basic dyes on acrylic.

Different alginates thickeners

Thickeners vary in vary in both solid content and, what is most important to a textile printer, behaviour whilst actually being used in printing process (see figs below) – flow characteristics.

1. Low solid grades, high viscosity (Manutex RS Type). More pseudoplastic more sensitive to velocity gradients.
2. High Solid grades Low viscosity(Manutex F type).

Low sensitivity to Velocity gradients (D) η

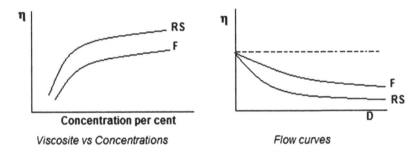

Viscosite vs Concentrations   Flow curves

Manutex RS – For hydrophilic fibres (Low Solid high Viscosity)

Manutex F – For hydrophobic fibres, less shear sensitive than Manutex RS (High solid low Viscosity)

It was usual to refer the thickenings as being either short or long. When a rod dipped in a thickening and lifted out again and the adhering paste allowed to fall back into the container if it flows readily forming long streamers it is called long and if it stays on the rod or falls off in lumps, leaving no streamers it is called short. The properties of long and short thickeners are given in the below table.

Properties of high and low solid type thickeners

Characteristic Property	Type of thickening	
	Low Solids	High Solids
Flow	Short	Long
Effect of increasing shear on viscosity	Viscosity falls off rapidly	Little or no falling off of viscosity

Colour yield	Higher	Lower than with low solids thickeners
Levelness of print	Poorer levelness of prints	Levelness of prints better
Viscosity	High	Medium-low
Handle of the printed fabric	Moderate to good handle	Film tends to be hard, brittle and may crack off fabric
Nature of polysaccharide chains	Very long and straight with little or no side branching	Shorter than low solid types. Heavily blanched, coiled in solution
Price of dry thickener to make required thickener paste	Cheaper	More expensive
Chemical resistance	Chemically resistant to the same extent	Chemically resistant to the same extent

In selecting a print paste thickening the printer will naturally endeavour to work at the lowest cost which is compatible with producing the required quality of print. If the cost of thickening were the sole consideration a low concentration of a high viscosity Alginate grade or in some cases an emulsion with a high viscosity grade of alginate as film forming agent would be the choice. However, a more detailed study of costs of production may show that a more expensive thickening leads to a more efficient use of dye, a higher rate of printing and more efficient washing off with resultant savings.

High Viscosity grades are the most economical in use and are suitable for a very wide variety of printing conditions. However, under more exacting printing conditions such as definition of fine lines (contours), "fall-on". "wet-on-wet" level blotches and the printing of hydrophobic fibres (synthetic polymer fibres) it may be necessary to use one of the higher solids. less shear sensitive thickenings prepared from grades of Medium Viscosity. As the choice of grade to use will depend very largely on the type and construction of the fabric, the actual design being printed, together with other variables such as method of printing, type of dye and fixation process.

Used and of course the result desired. it is not possible to forecast exactly which will be the best grade to use for any particular case The following suggestions based on works practice, are given for general guidance:

Cellulosic fibres(Cotton. viscose. Linen, etc.)
Simple designs   – absorbent fabric   – High Viscosity grades
Simple designs   – less absorbent fabric – High Viscosity grades or possibly
Medium Viscosity grades

Fine line definition – any fabric   – Medium Viscosity grades
Large blotch area   – any fabric   – Medium Viscosity grades

Synthetic fibres
As these fabrics are less absorbent than those made from the cellulosic fibres and the definition of the printed mark is more difficult to achieve the use of thickenings prepared from the high solids less shear sensitive grades are recommended. As there are however many synthetic fabrics being printed satisfactorily from low solids grades, on economical preliminary tests should be made with other grades also.

Concentration of alginate used
Depending on the actual grade of Alginate being used, a concentration of between 1 ½ % to 6% of alginate in the final print paste will generally give suitable viscosity and flow properties. This corresponds to concentrations in the stock thickenings in the range of 3% to 12%. It should be noted that the high viscosity grades require concentrations between 3% to 5% in the stock thickenings, which are classed as low solids thickenings. Medium to low viscosity grades of Alginate require concentrations in the stock thickening of between 6 and 12% which are considered as medium to high solids thickening.

Use of sequestering agents
The flow characteristics of alginate grades which give "short" flow when made into a paste with water can be varied over a wide range by using sodium metaphosphate (Calgon) in proportions varying from 5% to 25% of the weight of alginate. The flow of the paste is made longer by increasing the concentration of the sodium met phosphate (up to about 25% of the alginate in soft water). The flow can also be adjusted by using other calcium sequestering agents (e.g., the salts of ethylene diamante tetra etic acid-EDTA). However, EDTA is not recommended for use with dyes containing complexes metals. Medium to low viscosity grades can be used without the addition of a calcium sequestering agent giving pastes with intermediate flow. The addition of a calcium sequestering agent will lengthen the flow of most grades in this group, but the change in flow properties will be much less than with high viscosity grades. It should be noted that the addition of some calcium sequestering agent may be necessary to all grades (even though they are sequestered) to counteract the presence of calcium or other metallic salts in the water supply or in other ingredients of the print paste. This is particularly important with the printing of reactive dyes, when Colon is the recommended sequestering agent. All the above additions are when soft water is used (hardness below 25). If hard water is used it is necessary add calcium sequestrant to prevent

the gelation and precipitation due to the formation of Calcium Alginate. As a general guide for every 1 ppm of Calcium carbonate present in the water, 6ppm of sodium metaphosphate (Colon) is required to sequester the calcium present and thus soften the water.

Preparation (dissolving) of Alginate stock paste
Method 1. Using High Speed Stirrer (all viscosity grades)

Measure the correct quantity of cold water into the container. Agitate with a high speed stirrer (if necessary set at an angle to the vertical so as to create a vortex). If a sequestering agent is to be used, gradually shake powdered sodium metaphosphate (Calgon) into this vortex. It will dissolve almost at once. If the sodium metaphosphate is available only in the plate form which dissolves slowly. it should be made up into a 25% stock solution in advance. Other sequestering agents are best added as solutions. Then gradually shake in the alginate at such a rate that no unwetted streams of particles persist on the surface but fast enough so that all the powder has been added before the viscosity has risen appreciably. Continue stirring until the alginate particles have swollen and formed a thick suspension (5 to 10 minutes). Then allow to stand until a uniformpaste is obtained (about 1 ½ hours depending on the type of stirrer used. Stir again for few minutes before use.

Method 2. Hand Stirring or Using Slow Paddle Stirrer (High grades only)

First dissolve the Calgon or other sequestering agent in the cold water and then whilst stirring the water add the alginate powder in a steady stream. Continue stirring until the alginate particles have swollen and formed a thick suspension. Then allow to stand until a uniform paste is obtained (preferably overnight). Stir again a few minutes before use. The use of a dispersing funnel avoids lump formation when only a slow speed paddle stirrer is available for preparing the alginate solution. Dispersion of the alginate particles is also improved by forming a paste ofalginate powder in alcohol, glycerine or oil before dissolving in water as described above. If the alginate paste has to be retained for several days it can be preserved by adding formalin (40% formaldehyde), sodium pentachlorphenoxide, sodium dichlorphenoxide, esters of p- hydroxyl-benzoic acid, mixtures of benzyl cresols and mixtures of chloro-phenyl phenols, and 1 :2-benzisothiazolone. However, preservatives has to be selected as per the local regulations as some of these restricted usage regulations.

Carrageenan
This thickener is also similar to alginate, extracted from see weeds called 'Irish Moss'. Mainly used in food industry, this is sometimes added in the

emulsion thickeners to achieve better stability of the emulsion. They are basically linear sulphated polysaccharides and it is called Irish Moss as it is common all around the shores of Ireland (can also be found along the coast of Europe including Iceland). It is a small sea algae, reaching up to a little over than 20 cm in length. It grows from a discoid holdfast and branches four or five times in a dichotomous, fan-like manner. There are three main varieties of carrageenan, which differ in their degree of sulfation. Kappa-carrageenan has one sulphate group per disaccharide, iota-carrageenan has two, and lambda-carrageenan has three. All carrageenans are high-molecular-weight

Irish moss or carrageen moss, chondrus crispus

polysaccharides made up of repeating galactose units and 3,6 anhydrogalactose (3,6-AG), both sulphated and non-sulphated. The units are joined by alternating α-1,3 and β-1,4 glycosidic linkages.All are soluble in hot water, but in cold water, only the lambda form (and the sodium salts of the other two) are soluble.

4.2.1.4   Carboxy Methyl Cellulose (CMC)

Earlier we have explained the roasting of starch to make it more user friendly for printing. Another method of modifying starch is etherification and esterification. These modification give starch ethers and starch esters and both are used although the ethers, being resistant to hydrolysis in acidic or alkaline media, are much the more important as thickening agents for textile printing.

Etherification carboxy methyl, hydroxyethyl and methyl celluloses and carboxy methyl cellulose (CMC) is the most important in these. The starch may be first partially decomposed before etherification and the degree of etherification itself may be varied to produce these celluloses. It is further reacted with concentrated caustic soda solution, when it is converted to soda cellulose, and then reacted with sodium monochloroacetic acid. The degree of alkylation is said to be low or high depending on whether it is less or greater than 0.3 substituents per glucose (or other) unit; the products are termed modified starches if the degree of substitution is low and starch derivatives if it is high. Crossbonded starches can be obtained by treating, for example, a starch ether of low degree of substitution with bifunctional agents such as ethylene oxide, propylene oxide, epichlorohydrin or phosphates. The corresponding derivatives of cellulose can also be made and used as thickening agents if the chain length is appropriate. The steric hindrance effect of the substituents gives thickening agents of improved all-round properties and certain derivatives have ousted their parent products in terms of commercial importance.

Carboxymethyl cellulose   Hydroxymethyl cellulose   Methyl cellulose

There are different grades available according to viscosities – low, medium and high. CMC forms smooth viscous solutions, but the viscosity reduces on heating (but regains on cooling). Even though some electrolytes (barium nitrate, calcium chloride, magnesium chloride, etc.) do not affect the paste but some electrolytes like aluminium sulphate, barium chloride, ferric chloride, ferrous Sulphate precipitate them. It is used for mixing with other water soluble gums like gum Arabic, gum tragacanth locust bean gum, starch, sodium alginate. It can be easily washed off after fixation. They are available in different brand names like Cepol, Cellpro, etc.

4.2.1.5   Cellulose Ethers: Methyl Cellulose, Hydroxyethyl Cellulose

Cellulose ethers like methyl and ethyl cellulose can be easily made by reaction cellulose with concentrated sodium hydroxide solution and then with methyl and ethyl halide respectively. Hydroxyethyl cellulose can be made the same way reacting with chlorohydrin:

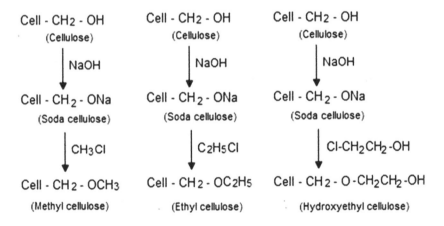

Methyl cellulose

Methyl cellulose is prepared as follows: Methylation should be carried out to 25% methoxy content equivalent to 1.5 OCH_3 groups per glucose unit of cellulose in order to impart water solubility to the final product. If the degree of methylation is high, alkali is needed to dissolve it. Very highly methylated cellulose is insoluble in water, but soluble in organic solvents. In order to protect it from degradation by microorganism, some preservatives may be added. Methyl cellulose is soluble in cold water, but not in hot water (above 50°C). At higher temperature, the solution gels. Addition of thiocyanate and iodides raise the gelling temperature of solutions of methyl cellulose.

Methyl cellulose is not soluble in organic solvents, like acetone, benzene, alcohol, etc., but is compatible with diethylene glycol, glycerine, sorbitol, casein and water, dispersible natural gums. Addition of these gums increases the viscosity and raises and gelling temperature.

Methyl cellulose stock thickener paste is made by dissolving first mixing with 20 to 25% of the required quantity of water and heating to 80-900C with stirring. It is allowed to soak well at this temperature for 5-10 minutes and the balance cold water is added with stirring till we get a homogeneous paste. A 6% paste is sufficient for normal prints.

They give strong but elastic prints which does not affect the brightness of the prints, especially azoic prints. It can be used for finishing making it water resistant with urea formaldehyde. Since they are non-ionic it can be easily mixed up with electrolytes (no problem of salting out) and other anionic or non-ionic ingredients, alkali, reducing and oxidising agents. It is stable under alkaline conditions in the absence of oxygen, but loses its viscosity in the presence of alkali and oxygen. Hence it is used in printing vat, azoic, Rapid Fast, Direct, Rapidogen, etc. Since it is easily soluble off it can be easily washed after fixation.

Solubility in cold water makes methyl cellulose an ideal thickener in textile printing from the point of view of ease of removal from the fabric after the fixation of the dye.

Hydroxy Ethyl Cellulose (also called glycol cellulose)
This product is soluble in water to give translucent solutions, which on drying on textile material, give clear transparent films and find use in sizing, dyeing, printing and finishing. It is compatible with solutions of other ingredients like starches, gelatine, gum Arabic, gum tragacanth, dextrin, CMC, etc. It is partly compatible with methyl cellulose and polyvinyl alcohol. Aqueous solutions of hydroxyethyl cellulose tolerate 105 to 15% of methyl, ethyl and propyl alcohols and acetone and up to 20% methyl acetate. The thickening power (viscosity) of aqueous solutions of hydroxyethyl cellulose is compared with that of some other thickening agents.

Solutions of hydroxyethyl cellulose on drying give films, which are stable to heat and light and are soluble in water, but not in organic solvents. Low molecular weight polyhydroxy compounds and acetic and formic acids dissolve the film with difficulty. The film flexibility increases with moisture content. The film does not become tacky, neither does it dust off and does not build static charges. The films of hydroxyethyl cellulose can be made partially water resistant by adding 6% of 30% aqueous glyoxal solution to it before drying at 105°C. Urea-formaldehyde or melamine-formaldehyde precondensates may be used to make it water-resistant by adding a plasticiser (10 to 50%) such as polyhydric alcohol (ethylene glycol, diethylene glycol, polyethylene glycol, glycerine, etc.) and sulphated castor oil.

Hydroxy ethyl cellulose is used in sizing, dyeing, printing and finishing. It can be mixed with CMC, methylcellulose, dextrin, gum tragacanth starches, gelatin, gum Arabic etc. The viscosity of hydroxyl ethyl cellulose varies as per the grades for example, 5% Low viscosity grade gives around 75-125 cPs, Medium viscosity gives 400-600 cPs and High viscosity gives 900-1100 cPs. This thickener gives printing results better than gum tragacanth and gum Arabic giving better penetration and sharp outlines. In printing it performs better than gum tragacanth or gum Arabic by way of facilitating good penetration and promoting the production of sharply outlined patterns. The printed fabrics acquire a soft hand.

4.2.1.6   Starch Derivatives – Starch Ethers and Esters

These are produced by the reaction of one or other of monochloroacetic acid, ethylene oxide, dimethyl sulphate in the presence of alkalis. They offer a range of flow properties similar to that offered by the British gums, and give a thickening at relatively low solids content. They are more stable than starch pastes, are more soluble than starch or British gum and, in fact, are cold-water- soluble. They do not show a reducing action with

$$CH_2 - O - CH_2COONa$$

Carboxy Methyl Ether of Starch

alkalis. They readily produce cold soluble with good printing properties and colour yield although the handle of the final print can be harsh in under certain conditions especially where H.T. steaming is used or given a bake fixation. Normally, used in admixtures with sodium alginate for vat printing by the two stage method (flash age) or as a component of a blend product for polyester printing (by HT steam fixation) where the starch ether can increase colour yield.

It is sold in different brand names and one of the most common products is Solvitose (C, H and H4). They have many advantages over other starch derivatives like easy solubility in cold water, highly viscous, stable paste, good dispersing power and easily washable because of it solubility in cold water. It is useful in the printing of both acidic and alkaline printing and thus can be used in printing of vat, rapidogen, rapid fast, solubilised vat dyes, etc. Some printers prefer to use a mixture of starch and solvitose.

4.2.1.7   British gums and dextrins

British gum

Natural starch has many disadvantages. The amylose component is substantially crystalline, forming helical structures that uncoil in an aqueous solution. It can also aggregate to give a gel or precipitate, an undesirable phenomenon known as retrogradation. Amylose is completely hydrolysed by the β-amylase enzyme. Amylopectin is substantially amorphous, having a globular structure that can expand considerably in aqueous solution. Its branched chains give rise to a much more stable solution, substantially free from retrogradation, and it is much more resistant to the action of β-amylase. Starches containing little or no amylose are known as 'waxy starches'.

The natural starch can be made more suitable for printing purposes by a dry roasting treatment at 135–190 °C. These also comes under derivatives of starch, but manufactured from starch by roasting it at different levels and other various processes. It was found long back that if starch is roasted it is easily soluble in cold water. In the beginning of 19th century this knowledge was used to make some branded products for printing usage called Dark British Gum by roasting wheat starch to dark brown colour at 200-2100C,accelerated by trace quantities of acid, to give random hydrolysis of the 1,4-links to decrease the chain length but with the formation of 1,6-links (branching). This product is called British Gum which has an increased stability, and stability but has some reducing characteristics, which can affect certain susceptible dyes, are enhanced by formation of more aldehyde end groups. Control of the hydrolysis and branching reactions yields a varied range of products.It a variety of grades according to the degree of dextrinisation. Dark (or thin is also called dextrin. These are partially dextrinised starches, available in

British Gum

boiling) British gums are nearly pure dextrin and give thin solutions showing little anomaly. Light (or thick-boiling) British gums are lightly dextrinised, and give a thickening which is in effect a mixture of dextrin solution and starch paste, and resemble starch in their behaviour. Owing to the presence of dextrin, British gum thickeners containing alkali have a reducing action, which can be sufficient to bring about fixation of some easily reducible vat dyes.

During roasting the degree of polymerisation of starch is reduced (by breaking the molecular chains) and thus the molecular weight is reduced. In case of dark British gum the roasting is continued till the whole starch chain is broken where as in the light British gum the roasting is sopped before complete solubilisation. Thus, different grades of dextrin or British gum are manufactured by different extent of roasting. The portion of the starch which is not affected forms a part of the British gum. Where the starch portion is more either by the extent of roasting, or by physical addition, on boiling it will give more characteristics of starch. Higher the roasting, the gum becomes more the reducing properties as there is an increase in the presence of aldehyde groups at the end of the molecules. Hence, if used with dyes which are susceptible for reduction it is always better to add some resist salt in the paste.

From the grades of British gum available, almost any desired compromise in properties can be selected. The colour yield of British gum thickenings can be increased by an addition of starch, with which they are compatible. The high solids content ensures that sodium sulphoxylate formaldehyde on a dried print is well protected, but causes some delay in diffusion of the leuco vat dye on to the fibre. For this reason, a starchy (lightly dextrinised) British gum, which thickens at lower solids, is sometimes preferred if short steaming times (3-5 min) are employed.

The main advantage of British gum or dextin is that the paste will be smooth, easy solubility, good levelling and easy wash off. They are not coagulated with solvents and give sharper prints. The colour yields given by British gum is not to the level of starch or other low solid gums like gum tragacanth, locust bean gum or guar gum which is due to the capacity of starch and other gums to retain large quantities of water in the interstices of the gel.

Disadvantages associated with. British gum thickenings are a tendency to clog the screen - hence they are not widely used in screen printing and its marking-off difficulties during ageing or subsequent processing. Marking-off difficulties usually occur on the mouthpiece or the rollers of a roller ager; much less trouble is experienced where a fest on steamer is employed, although care has to be taken to ensure that the "take-in" rollers on the mouthpiece do not become soiled.

4.2.1.8   Viscous Emulsions

If immiscible liquids are emulsified the viscosity increases and this can be exploited to prepare thickenings for textile printing. Oil and water do mix, if enough energy is used to break up one component of the mixture into small droplets, dispersed in the other component. Such simple emulsions are not stable, but their stability and ease of preparation can be considerably increased by the incorporation of a surface active emulsifier. The nature of the emulsifier and the nature of the two immiscible liquids determine which liquid will be dispersed (the disperse phase) in the other (the outer continuous phase). The size of the droplets in an emulsion is inversely related to its viscosity, typical diameters ranging from 100 to 7000 nm. Theoretically no more than 75% of oil can be incorporated in an aqueous emulsion, assuming uniformly spherical droplets, but distortion due to packing allows significantly higher proportions of oil phase to be added.

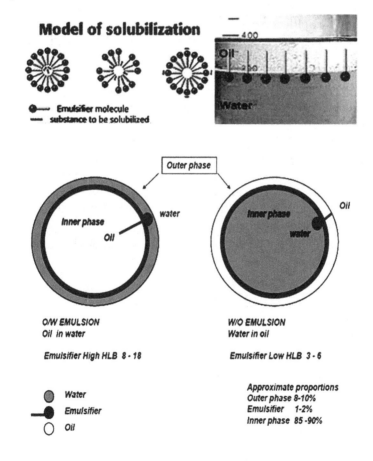

Model of solubilization

●— Emulsifier molecule
━━ substance to be solubilized

Outer phase

Inner phase
Oil
water

Inner phase
water
Oil

O/W EMULSION
Oil in water

Emulsifier High HLB 8 - 18

W/O EMULSION
Water in oil

Emulsifier Low HLB 3 - 6

● Water
●— Emulsifier
○ Oil

Approximate proportions
Outer phase 8-10%
Emulsifier   1-2%
Inner phase 85 -90%

If there is two layers oil and water in a beaker and we add an Emulsifier, the molecule tails in oil and molecule heads will be in water.

The emulsifier forms a film between the two liquids reducing the interfacial tension. Each molecule of the emulsifier contains a region that is hydrophilic and another that is hydrophobic (lipophilic) and it is the balance between the dominance of these regions that determines the relative solubility in oil and in water.

This hydrophilic lipophilic balance (HLB) has been evaluated foe many surface active agents and expressed in a standard manner over the range of 1 – 40. Experience has been shown that with HLB values of 3 – 6 stabilise water-in-oil (W/O) emulsions and those with values 8 – 18 are effective for oil-in-water (O/W) emulsions.

HLB Values (Hydrophilic Lipophilic Balance)

The HLB system (HLB = hydrophilelipophile balance) was introduced by Griffin, as an empirical scale from 1–20, to characterise the balance between the hydrophobic and hydrophilic moieties in surface-active textile auxiliaries (surfactants, emulsifiers) and is independent of their constitution (but highly temperature-dependent). A low HLB value (< 10) signifies a predominance of hydrophobic groups indicating that the surfactant is lipophilic, and a higher HLB value (> 10) that more hydrophilic properties predominate. The HLB describes the ratio of relative intensity between polar and apolar portions of a surfactant molecule. The HLB value is therefore determined by the type and size of the hydrocarbon chain as the apolar representative and the strength of the dipole in the polar portion. Consequently, a long straight-chain hydrocarbon portion has greater intensity than a short branched hydrocarbon chain. In the polar range, an ethylene oxide group is less, and a carboxyl, phosphate, sulphate, sulphonate group (in order of increasing effect) more intensive. The following properties are determined by the HLB value:

(a) surfactant solubility,

(b) emulsifying properties of a surfactant,

(c) soil suspending properties of a surfactant.

In simplified terms, the 1–20 scale of the HLB system has the following significance:

1 – 3	solvent soluble, water insoluble
3 – 7	solvent soluble, water dispersible
7 – 10	solvent soluble, water soluble
10 – 13	water soluble, solvent soluble
13 – 18	water soluble, solvent dispersible

18 – 20   water soluble, solvent insoluble

Once the HLB value of the surface active agents are known we can predict its properties and can be employed for particular purpose, for example:

1.5 – 3   antifoaming agents

3 – 6   dry-cleaning detergents

4 – 6   W/O emulsifiers

7 – 9   wetting agents

8 – 13   O/W emulsifiers

13 – 15   wash-active surfactants

15 – 18   solubilisers

HLB values are very important in the preparation of water in oil or oil in water emulsions. Oil and water do mix, if enough energy is used to break up one component of the mixture into two small droplets, dispersed in the other component. Such simple emulsions are unstable, but their stability and ease of preparation can be considerably increased by the incorporation of a surface active emulsifier. The nature of the emulsifier and the ratio of two immiscible liquids determine which liquid will be dispersed (the disperse phase) in the other (the outer, continuous phase).

W/O emulsion water in oil   O/W emulsion oil in water

Lipophilic Type (Low HLB)   Hydrophilic type (High HLB)

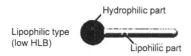

Type of Emulsion

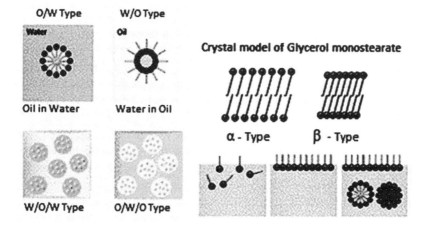

By using the petroleum distillate faction (b.p 150 – 2000C) known as white spirit, water and emulsifier we can prepare (O/W) emulsions suitable for pigment pastes. Now a days these emulsion are not used much due to ecological reasons.

An Emulsion thickener is made by mixing the two liquids with a high speed stirrer. High Speed mixers with a rotor and stator with a small clearance gap between the two, produce tighter and therefore more stable emulsions. Vigorously high speed mixing the two immiscible liquids together with a suitable emulsifier produces an emulsion.

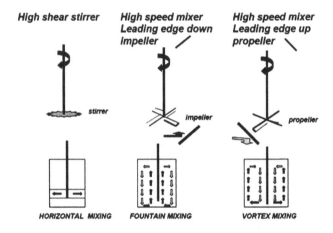

Mixing procedure
- To the outer or continuous phase an emulsifier is added and mixed
- The inner phase is slowly added with vigorous high speed mixing
- Viscosity of the emulsion increases
- Depending on the colorants to be used, fixation chemicals are added

When an emulsion is mixed with water, it is stable and homogenous it is an oil in water emulsion, while water in oil cannot. For an emulsion to be used as a thickener it should stable printing paste ingredients like dyes, salt, acids, alkalis, metal slats solvents, etc. When a cationic emulsifier is used for making the emulsion anionic dyes cannot be added to it and vice versa. An electrolyte has a destructive effect on an emulsion as it may change the distribution of the emulsifier between the two phases or alter the charge of the dispersed phase and it may break up. Emulsion can be made viscosities from 500-5000 by changing the proportion of the two phases. This characteristic is advantages for washing off the paste from, screens, squeegees rollers, etc. as the water is added to the emulsion its viscosity is reduced and easily washed off. (Water in oil emulsion thickens in the presence of water but finally breaks but oil may collect on the surface of printer parts and may cause problems. Actually, oil in water emulsions is easy to make (only a low speed stirrer is sufficient) whereas oil in water emulsion has better lubricating properties.

Emulsion thickeners are mainly used for pigment printing. However, it can be mixed with low concentrations of either natural or synthetic thickeners, especially when applying fibre-substantive dyes; these additions act as film formers, taking the place of the binder used with pigments to increase retention of the dye by the substrate prior to fixation.

Preparation of emulsion thickeners

Quantity	Unit	Additions
100	g	Sodium alginate 5% with seq. agent
10	g	Emulsifier
90	g	Water
800	g	White spirit with stirring
50	g	Urea (optional, helps to avoid surface drying)
1000	g	

4.2.1.9   Synthetic polymer thickeners

Even though synthetic polymers based on acrylic acid have been known since the 1930s, it was not being used as thickener for printing until late 1970s. Of late, more and more use of synthetic thickening agents is being promoted as an alternative to emulsion thickening which was very popular until recently, mainly because of the environmental pollution and hazards in using it. Typical repeat units are shown in these thickeners are shown below:

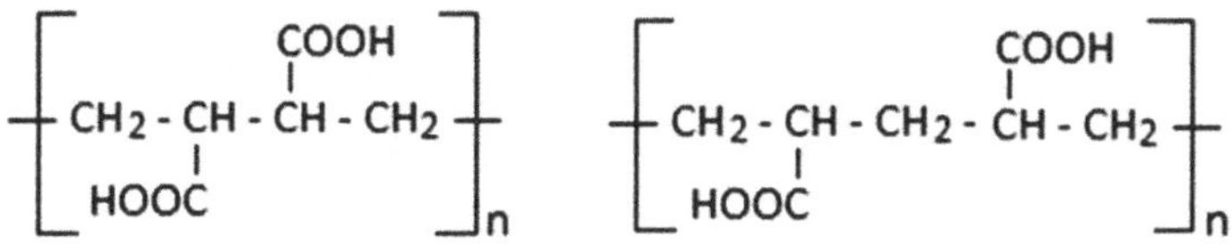

$$\left[\begin{array}{c} COOH \\ | \\ CH_2 - CH - CH - CH_2 \\ | \\ HOOC \end{array} \right]_n \qquad \left[\begin{array}{c} COOH \\ | \\ CH_2 - CH - CH_2 - CH - CH_2 \\ | \\ HOOC \end{array} \right]_n$$

Repeat units of acrylic polymers

Usually the number of repeat unitin first type (see above) may be 50–750, but crosslinked grades of higher relative molecular mass are also available. In the second type products will have a a range of n values from 3200 to 30 000. Thus in general these thickeners are based on co-polymers of acrylic acid (e.g., Methacrylic acid, ethyl acrylate) which in water give low viscosity dispersions (the molecules are randomly coiled). On adding an alkali, the carboxyl acid groups are ionized, which cause the polymer chains to extend by mutual repulsion with consequent increase in viscosity. (See below) Only the longer-chain grades are of significant interest for textile printing in the form of their sodium or ammonium salts. The below fig. gives a schematic interpretation of this change from randomly coiled to chain extension (viscosity increase) with alkali addition. Viscosity develops as water is absorbed, causing swelling and rearrangement of the polymer chains, a process that is assisted by, indeed is critically dependent on, neutralisation. On neutralisation a thickener of comparable to emulsion thickening is achieved. (See fig. below) It is important that the degree of crosslinking is the optimum required to maintain the polymer in this swollen state and prevent it from dissolving, which would result in loss of desirable properties.

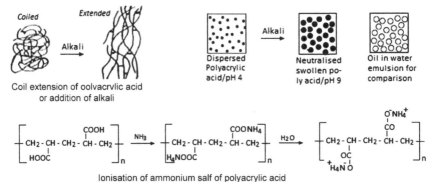

Coil extension of oolvacrvlic acid or addition of alkali

Ionisation of ammonium salf of polyacrylic acid

Advantage of thickeners include rapid make up since they require no waiting for hydration to occur, sharp print boundaries and controlled penetration which usually provide better colour value and levelness.

Binders used in pigment printing and synthetic thickeners both are acrylic based polymers. But the binders are copolymers and contain an integral crosslinking agent which fixes pigments on the substrates. Synthetic thickeners are used in pigment printing as an alternative to emulsion thickeners. One major problem with traditional (PAA poly-acrylic acid) thickeners is the sensitivity to electrolytesalthough this is less of a problem in pigment printing than in printing with dyes. A high pigment concentration may cause a large decrease in viscosity of the required viscosity. Traditional PAA thickeners sometimes exhibit difficulty in wash off after fixation of the print. New innovation and technology in synthetic printers solved most of the above limitations. The sensitivity of poly(acrylic acid) to electrolytes can be reduced by copolymerising with acrylamide. But the problem here is that a higher level ofcopolymerisation affects thickening efficiency. Two important and interrelated parameters for acrylic thickeners are relative molecular mass and degree of crosslinking. Simply increasing the molecular mass of linear poly(acrylic acid) yields thickeners that give stringy pastes unsuitable for use in printing. Hence a degree of crosslinking is necessary to minimise stringiness by decreasing the water solubility and promoting dispersibility. Thus, a balance has to be maintained between molecular mass and cross linking to achieve a quality thickener with good penetration, levelness of ground colours and sharpness of the print.

Generally, PAA are supplied as partially neutralised form, which has to be neutralised by the printer during the print paste preparation to get the full thickening effect. This neutralisation has to be done according to the dyes or pigment used in the print paste. While using along with binder and pigments the neutralisation can be done with ammonia, which in turn will be evaporated during the curing process to liberate the free polyacid, which can catalyse activation of the resin binder. The same neutralisation process is not applicable when the thickener is used for a reactive print. It should be noted that when the free polyacid is liberated on steaming can lower the pH of the print, and there by affecting the fixation negatively. (Moreover, reactive dyes can be deactivated by reaction with ammonia to form their non-reactive amino derivatives). In such cases the neutralisation of PAA may be dine with a non-volatile alkali like NaOH.

Thickeners are manufactured separately for pigments and dyes with ingredients supporting the particular colour. Thus, a synthetic thickener for pigment printing may not be suitable for dyes printing due to many reasons. For pigment systems the thickener may also contain additives (surfactants or polyelectrolyte dispersing assistants), such as the ammonium or sodium salt of linear poly(acrylic acid) (See below) which not only modify the behaviour

$$\left[\begin{array}{c} CH_2-CH \\ | \\ CO\overset{-}{O}\ NH_4^+\ or\ Na^+ \end{array} \right]_n$$

of the acrylic thickener but also assist dispersion of the pigment. In such a mixture, the surfactants can affect the fastness properties and the liberated ammonia from ammonium salt of poly (acrylic acid) can affect the fixation of reactive dyes (as explained above).

Thickeners in general

Shear sensitivity of thickeners are very important. The thickener should not change its characteristics under shear stress which is very important in screen printing especially rotary printing. The solid content and viscosity play an important part in this.

Low Solids High Viscosity High Shear Sensitive Thickener

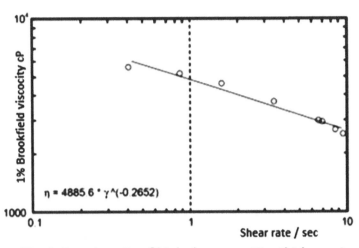

1% solution viscosity of high shear sensitive thickener A as function of shear rate

High Solids Low Viscosity, Low Shear Sensitive Thickener

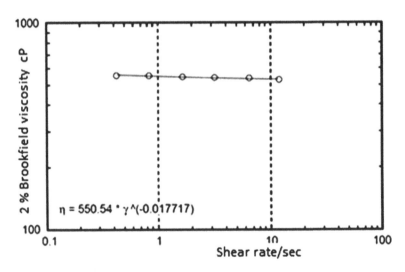

2 % solution viscosity of Low shear sensitive thickener B as a function of shear rate

General use of Low Solids high viscosity High Shear Sensitive Thickeners *with short flow properties:*

1. On medium to heavy weight hydrophilic (absorbent) fabrics with course /medium yarn, low / medium twist, low ends and picks
2. Low quality cotton, rayon and P/C fabrics)
3. Printing Towelling, Flannel, Fleece, Carpets
4. Where fine print detail is NOT required, using course / medium mesh screens with high printing pressures

General use of High Solids low viscosity Low Shear Sensitive Thickeners *with long flow properties*

1. On light weight fabric with fine counts and high twist, high density ends and picks, hydrophobic (poor absorbent) fabrics
2. High quality fabrics
3. Printing Poplin, Silk, Fine Polyester, Discharge and Resist styles
4. Where fine print detail is required using medium / fine screen mesh with low print pressures

Causes of splashing and cracking of print paste (Hand Printing)
Splashing and cracking of print paste easily occur mostly when the drying process is being carried out after printing. The following can be some of the causes:

1. Poor elasticity of thickener film
2. Poor compatibility of thickener with dyestuffs, auxiliaries and chemicals
3. Over-drying on printed areas
4. Unevenness on printed surface Poor or too good penetration on printed areas
5. Treatment after printing
   (a) During removal of printed materials from the table after printing
   (b) During rewinding the cloth before steaming
   (c) During steaming
6. Adhesion due to static electricity
7. Designs such as line-delineation, all over and white ground
8. One of the ways to prevent problems is, to choose thickeners with good film elasticity

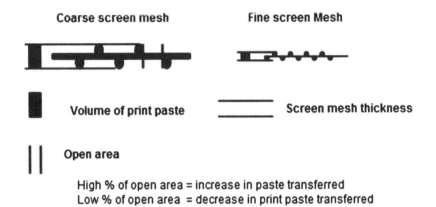

Selection of thickeners

Selection of a thickener depends on many factors: (1) a prime requirement is that the thickener functions satisfactorily by the printing method chosen, e.g. screen blockages should not occur in screen printing or 'sticking-in' (i.e. blocking up the engraving) in roller printing (2) The substrate to be printed e.g., it should be easily washed off from delicate fabrics like silk, it should not hamper the handle of the fabric, etc. (3) it should not react with the dye used for example the reactive dye react with most of the thickeners other than alginate, emulsion and synthetic thickeners. (4) The thickener should be easily washed off in the washing off method suitable for the dye and the substrate. For example, starchy thickenings are difficult to remove during washing-off and hence their use with dyes of poor washing fastness would be unwise. (5) Should be compatible with the ingredients of the print paste. (6) Should be easily washed off from machines, equipments screens, etc. The greater ease of cleaning-down equipment, rollers, screens, printingblankets and containers often results in oil-in-water emulsions being used in preference to water-in-oil systems. (7) Available equipment is often decisive in determining which thickening is chosen from those available. Thus, if gum boiling pans such as the ones that are required for the preparation of British Gum Thickening are not available, an alternative cold-dissolving thickening agent may be used. The cold-dissolving thickenings are also convenient for small scale operation. (8) Stability of the thickener.

Most chemically modified thickeners already contain a preservative but natural gums do not. It is customary to add a preservative in all thickeners. To improve the lubrication properties of thickening agents, especially under commercial conditions, printing oil may be added. Mineral oil Grade SAE 10 at the rate of 10-20 parts per 1 000 parts usually effects an improvement in, for example, starch-gum tragacanth and thickenings containing relatively large amounts of electrolytes.

4.2.1.10 Preparation of various thickeners

Gum Tragacanth

This thickening is widely used, both on its own and in mixtures with starch. It is particularly useful where the goods are not to be washed well and for blotches.

Recipe

Quantity	Unit	Additions
70	Parts	Gum tragacanth are mixed with
1000	Parts	Cold Water. And allowed to stand for 2-3 days with occasional stirring

The mixture is then raised to the boil and maintained thus until the gum is dissolved. A steam or water-jacketed pan should be used and the operation usually takes 8—12 h. Further boiling results in a thinner paste. After cooling, the thickening is bulked to 1000 parts and strained.

Starch-Tragacanth thickener

Starch thickeners cannot be stored for a long time and do not give a sharp outline of the design. Addition of tragacanth imparts to .it softness, plasticity, .and stability in storage. Starch-tragacanth thickener is widely used for printing cotton and viscous fabrics with dark and medium-tone colours.

Quantity	Unit	Additions
140	Parts	Wheat starch is stirred into
400	Parts	Cold Water. Then
600	Parts	Gum Tragacanth is added (7%), made as described earlier are stirred into the paste

The mixture brought to the boil and maintained thus for 30—40 min with constant stirring. It is then cooled, bulked to 1000 parts and strained.

Locust Bean Gum

These gums are coagulated by alkali, and this property is utilised in the Flash-Age process for vat dyestuffs. They are also employed for general purposes where alkaline conditions are not encountered.

Quantity	Unit	Additions
0.5	Parts	Borax dissolved in
1000	Parts	Cold Water. Then gradually stirring
20	Parts	of the gum powder are added

The paste is made slightly acid with acetic acid and is then heated to 80°-90°C. It is finally cooled, bulked to 1000 parts and strained. This thickening has poor keeping properties and only a sufficient quantity for immediate use should be prepared at any one time. An addition of 0-5 parts of' antibacterial agent will extend the storage life to several days.

Preparation of alginate thickeners (e.g., Manutex RS, Lamitex L) with lower solid content

Quantity	Unit	Additions
12.5	Parts	Sequestering agent (Sod. Hexametaphosphate) are dissolved in
137.5	Parts	Water at 600C and
800	Parts	Cold water are added
50	Parts	Thickening are sprinkled on the solution and stirring is continued for 5—10 min to break down any lumps. Finally, the mixture is bulked to
1000	Parts	

After standing overnight the thickening is ready for use; straining is not normally necessary.

Alginate thickeners (e.g., Manutex RS, Lamitex L) with higher solid content

Quantity	Unit	Additions
12.5	Parts	Sequestering agent (Sod. Hexametaphosphate) are dissolved in
137.5	Parts	Water at 600C and
700	Parts	Cold water isadded
150	Parts	Thickening is sprinkled on the solution and stirring is continued for 5—10 min to break down any lumps. Finally, the mixture is bulked to
1000	Parts	

After standing overnight the thickening is ready for use; straining is not normally necessary.

Dextrin thickener

The dextrin thickener is used to prepare alkali colours since dextrin is resistant to alkalis and promotes the reduction of the dye. A dextrin thickener is used to produce resists and discharges on dyed fabrics. This thickener has a good covering capacity and is stable in storage. Its approximate composition is:

Quantity	Unit	Additions
600	g	Dextrin
397	g	Water
40	%	Caustic soda

British Gum D and No.5

These are used for printing vat dyestuffs and discharges. Gum No. 5 is less dextrinised than Gum D and hence the same proportion of gum yields a more viscous paste. By mixing the two a thickening of the required consistency and total solids content can be obtained.

British Gum D	British Gum No. 5	Unit	Additions
500	250	parts	Powder is stirred with
500	750	parts	Water and bulked into
1000	1000	parts	

The paste so obtained is boiled with constant stirring for 20—30 min and cooled. The volume is adjusted to 1000 parts and the thickening is finally strained.

'Indalca' U

'Indalca' C/is a useful general thickening, giving good level prints.

Quantity	Unit	Additions
45	Parts	Indalca is gradually added to
1000	Parts	cold water with constant stirring
1000	Parts	

The mixture is brought to the boil and maintained thus for 30 min. After cooling, it is bulked to 1000 parts. Straining is not normally necessary.

'Nafka' Crystal Gum Supra

This thickening is widely used for printing cellulose acetate, nylon and polyester.

Quantity	Unit	Additions
200	Parts	Nafka' Crystal Gum Supra are stirred quickly into
1000	Parts	Cold water and the mixture is allowed to stand overnight

After straining, the thickening is ready for use. If required for use quickly the suspension of the gum may be heated to the boil, boiled for a few minutes, cooled and strained.

'Meypro' Gum CRX

This is a general-purpose thickening that is stable to acids and alkalis.

Recipe:

Quantity	Unit	Additions
200	Parts	Meypro' Gum CRX are sprinkled slowly into
950	Parts	Cold water and the mass is stirred for 15 min
1000		

It is then boiled for 15-20 min and cooled to 50°C, and bulked to 1000 parts. In order to avoid bubbles the thickening is then allowed to cool by standing without stirring. Straining is not normally required.

'Meypro' Gum AC

This thickening is recommended for printing on synthetic fibres.

Quantity	Unit	Additions
80	Parts	Meypro' Gum AC are sprinkled slowly into
900	Parts	cold water and the mass is stirred for 15 min
1000		

It is then boiled for 5 min and cooled to 50°C (120°F) and bulked with cold water to 1000 parts.

Solid content (in g) per kg of different thickener

Thickener	Solid content
Locust bean flour	20-25
Locust bean derivatives	20-250
Guar derivatives	40-250
Starch ether	100-400
Carboxyl methyl cellulose	40-200
Hydroxymethyl cellulose	15-40

Crystal gum	250-330
Carragheenates	50-100
Polysaccharides	40-200
Alginates	30-120
Tragacanth	65-80
Starch	80-100
Polyvinyl alcohol	100-200
Polyacrylic products	200-250
British gum	500

Preservation of thickeners

All natural thickeners are prone to biological degradation, leading to a loss of thickening efficiency. Hence a bactericide is added during the manufacture of the thickener or during the stock thickener preparation by the printer, the latter being more common. Formaldehyde was one of the product easily available and most commonly used but is restricted presently being carcinogenic.

4.3   Resist salt

All Reactive dyes especially those belongs to azoic group, to a greater or less extent, have been found in practice to be susceptible to reduction, leading to loss of yield and/or to variation in the final printed shade. The effects of reduction are especially noticeable in pale shades (1/2 %depth or less) and when prolonged steaming times are employed. The effect can be very pronounced when a pale mixture shade is printed where one of the components *of* the mixture *is* more sensitive to reduction than the other(s). Reducing conditionsmay be caused by one or both of two factors. First the alkali (sodium bicarbonate or carbonate) - cellulose system present when a reactive print paste is applied to a cellulose substrate has a certain reducing potential. Secondly an accidental contamination of the reactive paste can occur when reactive prints are handled or steamed in apparatus which has been used previously for vat prints or discharge styles prints.

To safeguard reactive prints against both these reducing atmosphere,an addition of 1% Mild reducing agents isadded to reactive print paste is recommended to ensure maintenance of consistent yields. The stability of print paste is in no way affected by the addition of this chemical.

One of the very commonly used mild oxidising agent is sodium meta-nitrobenzene which is commonly called as Resist salt. (also called Ludigol)

Resist salt darkens on exposure to light and the print containing resist salt should be given adequate washing to ensure its complete removal from the fabric. However, the normal reactive print wash is sufficient for the removal of Resist salt. Resist salt is used in alkaline printing pastes. But in the case of acidic paste normally sodium or potassium chlorate is used in its place.

4.4   Urea/Hydrotropic agents

Printing is normally done with highly concentrated pastes that tendto lose moisture under adverse conditions which can affect the colour yield. Using an auxiliary that tends to increase the aqueous solubility of the dye, one can achieve better colour yield. Hydrotropes, come into play an important part in printing, which act as an amphiphilic bridge between the dye solubilisate and the aqueous medium. Even though some surfactants can function as a hydroptrope in dyeing dark shades where the dye quantity is much high, this may not be much useful in print paste due to its powerful wetting properties which can promote bleeding and haloing of the prints. Hence for printing hydrotropes with much less strong surface-active properties are more suitable for use in printing. The commonly used hydrotropes are urea and thio-urea (see below). Even if there are a lot of environmental restriction for using urea

in printing paste but forreasons of print technology and colloidal chemicals, the elimination of urea from the recipes is not easy. There are many functions other than hydrtropic for urea being considered as the most efficient printing paste additive. In a printing process the fabric is dried and invariably the moisture content will be very low and hence the fibres would have shrunk. But in the fixation process the dyes in the printed thickener layer has to be transported to the inner parts of the fibre. This can happen only if the fibre is swollen, which is very important especially for viscose fabrics. It should be noted that the swelling is an exothermic process and hence the, it will be better if fibres as well as the dried out thickener is swollen before steaming at room temperature if possible. Even after swelling the dyes need moisture (water) as a transportation medium except may be in some classes of dyes like pigment printing where the colourant is fixed on the face of the fibre by entirely different fabric. Part of the water quantity required is obtained from steam condensation on the "solid fabric" when the dry fabric enters the steamer. The condensate quantity is determined by the specific heat of the printed fabric and by the temperature difference between the fabric entering the steamer and the boiling point of water under the prevailing steam conditions. The specific heat of cellulosic fibres and natural thickeners is approx. 1.26 kJ/kg.°C (0.3 kcal/kg.°C), that of the printing auxiliary at 0.42 kJ/kg.°C (0.1 kcal/kg.°C). Moisture on the fabric has a specific heat 4.2 kJ/kg.°C (1 kcal/kg.°C). So, the presence of water influences the condensation of steam considerably more than the printed textiles. On the other hand, the steam composition in the steamer possibly varies considerably. Dry steam occurs if there is insufficient steam supply, and saturated steam conditions cannot always be evenly adjusted in the steamer, which is why a constant high temperature of 102°C is demanded by "good" steamers.

In the case of reactive dyes if the selected dyes are not compatible there are chances of variation in the yield hence variation in the same batch of batch to batch as a result of temperature/humidity fluctuations in the steamer. This variation can be reduced by moisture injection (as explained under steamer) which has its own limitation. One of the major contributions of urea in reactive prints is in this respect. Urea reduces the demand on the steam quality.

Circulation steamers with moisture injection also increase the reproducibility of the printing result, in particular in reactive dye printing on cellulosic fibres, and with the printing performance. During fixation process, urea with the condensed steam forms a "melt" on the fabric, which only releases the water again at temperatures above 115°C. The composition of the reactive stock colour pastes as regards quantities of alkali and urea is usually a compromise, which is determined by the reactivity of the dyes and by their sensitivity to hydrolysis compared with moisture (absorbed by the urea in dried printing).

As explained earlier as the swelling of fibre happens the dyes have to be released from the thickener paste which is entangled in the dried thickener. The thickener also has to swell again during the steaming process so that the dyes and chemicals has to be released for diffusion in to the fibre and reactions to take place To what extent the solute cases of a thickener are constructed and decomposed, amongst other things, essentially depends on the chemicals, which promote the thickener being absorbed in water. Urea promotes water absorption during steaming in varying ways depending on polymer thickener for example. The hydration spheres of the urea molecules taken up onto the polymers are the amino groups directed outwards. The hydrophilicity of the polymers is therefore increased by the amino groups; but the take up can also promote the swellability of the polymers without increasing hydration if the cohesion of the macromolecules is reduced.

A third function of urea is as a hydrotropic agent. It increases the solubility of the dyes broken down in the dried print paste, depending on the quantity used. In order to overcome the intermolecular forces (due to the short liquor-to-goods-ratio during printing and steaming) of less soluble dyes, a third, solution-conveying component is required, which introduces dissolving properties together with water. The surface tension of the water should firstly be reduced in the course of this before hydrogen bond interactions can come into force in the ternary system. The hydrophilic amino groups of urea reduce the surface tension of the water and so cause the structure of the water in direct proximity to the solid dye to be dissolved, which promotes the dissolving process. The carbonyl group on the other urea dipole end has (in contrast to the amino groups with their hydrogen donor properties) an affinity for electrons: both carbenium C atoms and semipolar absorbed oxygen appear as a mesomeric restricted form, which is why energy exchange is possible with other organic reaction partners (the chemically diverse dyes). For example, in the printing of wool with acid/metal complex dyes or vinyl sulphone reactive dyes urea is added in the print pastes where it functions as a hydrotropic agent in both with acid and metal-complex dye and a solubiliser in the vinylsulphone dyes solubiliser in the printing paste. It also promotes water absorption and assists as a fibre-swelling agent during steaming.

Action of urea in the discharging pastes is many fold. Hydrotropic auxiliaries like urea, accelerate or make possible the discharge process during dye fixation and also dissolve cleavage products thereby facilitating their removal in subsequent afterwashing. Urea also ensures the presence of a certain quantity of water on the fabric during steaming thereby increasing the efficiency of the reducing agent. Urea, which is the most frequently used hydrotropic auxiliary, performs the following functions:

- increases dye solubility in the discharge print paste or pad liquor
- accelerates dye fixation during steaming
- improves dye colour yield
- prevents the fibre drying out during steaming

The quantity of urea can be reduced by using effective even moisturising the fabric before steaming without forming water drops on the fabric. However due to manifold functions of urea in the complex fixing process it may not be possible to avoid the use of urea.

There are many other hydrotropes suitable for different dye–fibre systems under specific conditions. Some of the hydrotropes which can be used in printing are: triethanolamine, N,N-diethylethanolamine, sodium N-benzyl sulphanilate, sodium N,N-dibenzyl sulphanilate, ethanol, phenol, benzyl alcohol, resorcinol, cyclohexanol, ethylene glycol, glycolic acid, 2-ethoxyethanol, diethylene glycol, 2-ethoxyethoxyethanol, 2-butoxyethoxyethanol, thiodiethylene glycol and glycerol. Glycerol is especially useful with practically all classes of dyes.

Various hydrotropic agents used in printing and dyeing

It is not necessary that a hydroterope has to be a surfactant, but cause the surface tension of water to be reduced which is necessary to act as a solubilising agent. This is enabled by the proton donating groups (such as hydroxy, amino and amido groups) and proton-accepting atoms (such as the nitrogen atom of a tertiary amine) in the hydrotropes and further supported by hydrogen bonds, together with weaker dipolar and van der Waals forces, etc. which contribute to this interaction. Thus, hydrotropes are also used in the diluent system in the manufacture of certain dyes, particularly liquid brands, in which they not only increase the apparent solubility of the dye but also help to prevent the formation of surface skin. Dicyanamide or 1-cyanoguanidine is one of the common hydrotrope used for this purpose.

Dicyanamide   *Recorcinol*

Mixture (40/60) of urea and formamide is a hydrotropic agent (which increases the solubility of dyes in water) which helps in reducing steaming time at the same time increasing the colour value by migration of dye from thickener layer to textile substrates. When it is present in a printing paste containing chrome mordant dyes it disperses them and assist in their migration in the printing paste so that they become fixed in a textile material more quickly.

4.5   Hygroscoping agents

Hygroscopic agents are water-absorbing products used in printing pastes to balance the air and fibre moisture. For example, in printing with acid and metal-complex dyes, the print paste contains hygroscopic agents (e.g., thiourea, urea, thiodiethylene glycol). Sodium chlorate can be used as a strong oxidising agent at the same time hygroscopic agent. Hygroscopic agents also help in absorbing the moisture from surroundings, etc., from steam in a steamer. If the hygroscopic agent is not deliquescent it can be used in print paste to avid spreading of the print when higher amount of moisture is absorbed thereby maintaining the clarity of the print. Generally, a hygroscopic agent functions as humidity controller and fibre swelling. Zinc chloride is used as a hygroscopic agent in textile printing pastes for white resists on bromine indigo, etc. In wool printing with acid dyes glycerol is used as a hygroscopic agent (20–30 g/kg).

Main hygroscopic agents used in print pastes are urea and glycerine. Hygroscopic agents are required to absorb the condensed steam (water) during the steaming process otherwise it may spread the colour and affect the clarity of the print. It is essential to use the optimum quantities of the hygroscopic agent. When lower quantities of hygroscopic agents are used all the dye present in the thickener may not get enough water to dissolve ending up in poor colour yield. The larger amount of hygroscopic agent than optimum amount, due to more water supplies to thickener film may cause spreading of the print as mentioned earlier.

4.6   Anti-foamers and defoamers

Wetting agents are sometime added to the print paste for various reasons like penetration of the print, etc. or other surface active agents or dissolution agents, causes foam generation due to the stirring, agitation, squeegee action, etc. and this can cause faulty prints. To avoid this, defoaming agents should be incorporated in the print paste. Usually silicone defoamers, readily emulsifiable hydrocarbons, sulphated oils etc., may be used for this purpose. Another common defoamer used in printing is Pine Oil which is readily miscible with water to give a milky dispersion.

A processor should know the difference between 'defoamers' which are intended to destroy already-existing foam, and 'antifoams' which are designed to prevent the formation of foam Compared to dyeing or finishing the printing, especially print paste preparation, there are more chances of foaming. Foaming can play havoc in printing results. Wherever required antifoams are added considering the dyes involved. Silicone based or silicone free antifoams like straight or branched-chain and alicyclic alcohols (with 5–18 carbon atoms in the molecule hydrophobe such as iso-octanol), pine oil (chief constituent terpineol), insoluble alkyl esters of phosphoric acid e.g., Tributyl phosphate (TBP) can be used antifoam agent. The main requirement for an effective defoamer is that the agent should be insoluble in the foaming system in order that, at low concentrations, it can be retained at the phase interface. The entry coefficient must be positive, and the agent should form quasi-gaseous boundary layers.

4.7   Alkalis

Various alkalis are used in printing taking into consideration the pH requirements of the dye fibre reaction, the ability of the fibre involved to withstand the alkalinity, etc. The main strong alkalis used are sodium hydroxide, potassium hydroxide. As milder alkalis sodium carbonate, potassium carbonate, sodium bicarbonate, sodium silicate, disodium hydrogen phosphate, trisodium phosphate, sodium acetate, triethanolamine, ammonium hydroxide, etc. are used. Milder alkalis are used in the case of fibres like wool, silk, etc., where strong alkali cannot be used.

When a paste has to be acidic while printing and needs alkalinity while fixation Sodium acetate can be suitably used. Sodium acetate being a salt of a weak acid and strong alkali, after printing with an excess of acetic acid remains as acidic preventing the dissociation into acetic acid and sodium hydroxide, while steaming at higher temperature it dissociates and acetic acid being volatile escapes and the medium becomes alkaline.

4.8   Oxidising agents

Oxidising agents are required to develop final colour in steaming or after treatments in case of certain dyes like solubilised vat dyes, aniline black, etc. In most cases it should not be active in the print paste as it may cause premature development of the dye preventing the subsequent fixation of the dyes. Potassium chlorate, sodium chlorate are common oxidising agents used in printing. It is not active in neutral and alkaline medium and hence can be used a volatile base like ammonia. When the paste contains acid liberating agent like ammonium chloride, during steaming at 1000C or higher ammonia gets volatise, liberating ammonia, when it escapes alkalinity reduces. After this the ammonium chloride dissociates into hydrochloric acid and ammonia, again ammonia escapes and the print becomes acidic. At this point chlorates become active (at higher temperatures) and development of the dye takes place. For dark shades and dyes whichare sensitive to reduction in the steamer, an addition of sodium chlorate (10–15 g/kg) is often made to the print pastes, will neutralise the reducing action. Sodium nitrite or sodium chromate is also used as mild oxidising agent. In acidic pH it produces nitrous acid which is an oxidising agent.

$$NaNO_2 + HCl \longrightarrow NaCl + HNO_2$$

$$2HNO_2 \longrightarrow NO + H_2O + NO_2 \text{ (Oxidising agent)}$$

Sodium chromate itself is not an oxidising agent but in acidic medium it produces sodium bichromate, which is a strong oxidising agent. This bichromate which is yellow in colour can change the shade if it is not washed off well.

$$2Na_2CrO_4 + H_2SO_4 \longrightarrow Na_2SO_4 + H_2O + Na_2Cr_2O_7 \text{ (Oxidising agent)}$$

Both these oxidising agents can be added in the print paste, after printing and drying or steaming when the cloth is treated with an acid whereby the oxidising agent is liberated and the colour is developed.

In some cases where the dye used is susceptible for reduction, a mild oxidising agent like resist salt. But it is active in alkaline medium in a reducing atmosphere only. If a reducing action has to be prevented in an acid medium sodium chlorate may be used.

4.9   Reducing and discharging agents

Reducing agents or discharging agents are used in various occasions in printing pastes. A major requirement is in discharge printing where the already dyed ground has to be destroyed and made white and leaves it as white or another colour can be fixed at the discharged area. For example, on a ground dyed with a dischargeable azo dye has to be discharged by printing, a reducing agent is incorporated into the print paste and printed so that the white ground of the fabric is developed on steaming.Two cleavage products are formed from splitting the dyestuffs, both of which are amines and each one of which has a nitrogen atom from the original azo group. The cleavage of dyestuffs is often the basis of discharge and resists printing. If a discharge-resistant anthraquinone dye is added to the discharge paste, then a coloured pattern is developed instead. If the dyes printed on are particularly luminous, then they are termed illuminating dyestuffs in coloured discharge printing. The alkaline discharging agents like sodium hydrosulphite can also discharge vinyl sulphone dyes. It breaks the bond between the vinyl sulphone dye and the –OH group of the cotton. (See fig. below)

| 1. Neutral Discharge of azo group | 2. Alkaline discharge of Dye - Fibre bond |

$$R1 - N \neq N - R2 - SO_2 - CH_2 - CH_2 \mid O - Cell$$

Cleavage products of reaction 1	Cleavage products of reaction 1&2
R1-NH$_2$ H$_2$N-R2-SO$_2$-CH$_2$-CH$_2$-O-Cell	R1 - NH$_2$ H$_2$N-R -SO$_2$-CH$_2$-CH$_2$-OH Cell-OH

In case of disperse dyes alkaline cleavage of the ester groups takes place in discharging

Polyester Dyeable   Water soluble

The main discharging (reductive) agents used in printing are:
• sodium formaldehyde sulphoxylate
• zinc formaldehyde sulphoxylate
• calcium formaldehyde sulphoxylate
• thiourea dioxide
• tin (II) chloride

Eventhough, discharge print brings the reducing agents in the mind of a processor, it is not necessary that discharging is always done by reducing agents. For example, oxidative discharges like chromate coloured discharges on indigo grounds use, sodium bichromate and alkali in the print paste. The nature of the fibre itself also plays an important role in dischargeability. Since the common reducing agents are all hydrophilic and, in some cases, readily water-soluble, it is obvious that water-soluble dyes on hydrophilic fibres (e.g., wool or cotton, are more easily discharged than hydrophobic dyes on hydrophobic fibres (e.g., disperse dyes on polyester fibres).

Due to its strong reducing power to destroy the dyes and easy solubility, etc., sodium hydrosulphite or sodium dithionate or popularly called as hydros ($Na_2S_2O_4$) is the commonly used reducing agent in printing. It can dissolve in 5 times its weight of water at room temperature but is not stable in the presence of oxygen (atmosphere) because of its high reducing power. In aqueous solution following reaction takes place:

$$Na_2S_2O_4 + 4H_2O \longrightarrow 2NaHSO_3 + 6H$$

The stability can be improved in the presence of a small amount of alkali like sodium hydroxide. In the presence of alkali, the bisulphite formed is converted to sulphite:

$$Na_2S_2O_4 + 2NaOH \longrightarrow 2Na_2SO_3 + 2H$$

Hydros is manufactured as a white powder which is stable in sealed container in cool and dry surroundings. Once the container is opened and exposed to air it starts looses its strength immediately to around 75% at which it remains stable for a while and again drops (may be due to the formation of sodium sulphite and its stabilising effect). Sodium hydrosulphite (3 g/l) with NaOH (1.5 g/l) is a powerful reducing agent with reduction potential around -700 mV. In printing the main area where discharging agents are used is in the discharge prints. As dischargeable grounds, either normal plain dyed fabric or, more frequently, pre-padded grounds are employed. The discharge prints open flexibilityand great variety of design possibilities even though the process is more complicated and costly production process (high energy consumption and a high water consumption) which generally involves the following stages: dyeing, intermediate drying, application of discharge print pastes, intermediate drying, steaming (reaction medium), washing-off and drying.

As explained earlier the main disadvantage of Hydros is the stability of the product especially in printing. If it is used in printing the paste is applied as a thin film over the fabric will get oxidised by the atmospheric oxygen faster and won't be available for the discharging of the dye. Hence, it was necessary to stabilise the hydros so that it is active only during steaming or high temperature whereby it can easily act upon the dye.

When formaldehyde is reacted with hydrosulphite you get sodium sulphoxylate formaldehyde (NaHSO2CH2O) (CI Reducing Agent 2)and sodium bisulphite formaldehyde (NaHSO3CH2O).

$$Na_2S_2O_4 + 2HCHO + H_2O \longrightarrow NaHSO_2.CH_2O + NaHSO_3.CH_2O$$

(Hydros)   (Formaldehyde)   (Sod. sulphoxylate   Sod. bisulphite
                                              formaldehyde)   formaldehyde)

Sodium sulphoxylate formaldehyde and sodium bisulphite formaldehyde are stable. Sodium sulphoxylate formaldehyde is available as 100% branded products named Rongolite C, Formasil, Hydrosulphite NF, etc. from various manufacturers. It is soluble in water insensitive to alkali but decomposed below pH3.

When it is used in print paste and steamed (during fixation – 1000C or higher) the sodium sulphoxylate formaldehyde dissociates into sodium sulphoxylate and formaldehyde, of which sodium sulphoxylate is a strong reducing agent, which is used for discharging:

$$NaHSO_2.CH_2O \ 2H2O \longrightarrow NaHSO_2 + CH_2O + 2 H_2O$$

(Sod. sulphoxylate
 formaldehyde)

$$NaHSO_2 + 2 H_2O \longrightarrow NaHSO_4 + 4H$$

Care should be taken to dissolve the sodium (calcium or zinc) formaldehyde on cold water as it decomposes at higher temperatures. Thus, cold solutions, is added to the print paste which remains stable for prolonged periods, even after printing and drying as moisture is not available. However, after printing it is not advisable to store for more time, as the moisture in the atmosphere can be absorbed.

Different stabilised Discharge agents based on sodium hydro sulphite.

Sodium formaldehyde sulphoxylate ($NaHSO_2$-$CH_2O.2H2O$) C.I. Reduction agent 2

This produces the best white effects compared to the other stabilised products. It has extremely high redox potential. But it has relatively high danger of degradation before steaming, e.g., when drying takes place on the table (humidity), it is therefore more stable stabilised reducing agents (Rongolit H – Calcium sulphoxylate formaldehyde) especially for flat screen prints. It is also used for coloured discharges and reducing agents for vat dyes.

Zinc formaldehyde sulphoxylate [$Zn(HS_2O$-$CH_2O)$] C.I. Reduction agent 6

The soluble salt zinc formaldehyde sulphoxylate (e.g., Decrolin- BASF) often produces only weak white effects. Its discharging action is in the acid region, sometimes exhibits rather marked darkening on exposure light. It is mainly used for coloured discharges. It has some restrictions due to the ecological problems due to the zinc salts contained in the effluent.

Calcium Formaldehyde sulphoxylate [Ca(SO$_2$CH$_2$O)] C.I.Reduction agent 12

It usually marketed as an insoluble product in fine dispersion and it has lower redox potential compared to sodium formaldehyde sulphoxylate. The white ness of discharges will depend on the dischargeability of the dyes used in dyeing. Before using it as a reducing agent, one should check the compatibility with the other ingredients of the paste. It does not degrade even in poor drying conditions like sodium formaldehyde sulphoxylate, and there are no problems of contaminating effluents.

Another area where the discharging agents are used is in discharge resist print pastes. Here in addition to discharging the dyes it also prevents the fixation of the dye (Resist). Resist printing is preferred in polyester printing as it is difficult to discharge the already dyed polyester.

Other discharging agents used in print pastes are thiourea dioxide (formamidine sulphinic acid) (CI Reducing Agent 11) and tin (II) chloride. The latter is unsuitable for wool since brown tin sulphate can be formed with the reduced cysteine groups in the fibre resulting in unsightly effects. Selection of these reducing agents is based on the type of fibre to be discharged, and reducing condition applicable, severity of discharging effect required, solubility, presence of other additives in the print paste, etc. The selected product must not cause damage to the fibre nor influence the colour fastness of the dyes used. A prerequisite for the reaction is a sufficiently high reactivity of the discharging agent. This is characterised by the redox potential, i.e., the more negative the redox potential, the more reactive the discharging agent (dependent on temperature, pH and concentration).Tin (II) chloride is applied under strongly acidic, and zinc formaldehyde sulphoxylate under mildly acidic conditions. Thiourea is used under both mildly acidic as well as alkaline conditions. Calcium formaldehyde sulphoxylate (CI Reducing Agent 12) and thiourea dioxide are only sparingly soluble in water. Strongly alkaline sulphoxylate discharges are not particularly suitable to produce discharge effects on alkali-sensitive fibres, e.g., wool, silk, acetate. Since thiourea dioxide is effective under neutral or mildly acidic conditions, it does not cause any damage to alkali-sensitive fibres.

Thiourea dioxide, when heated with alkali and water, a rearrangement of the molecule takes place to formamidine sulphinic acid which is irreversible. Decomposition of this compound releases sulphoxylic acid which acts as a reducing agent:

irreversible rearrangement takes place with the

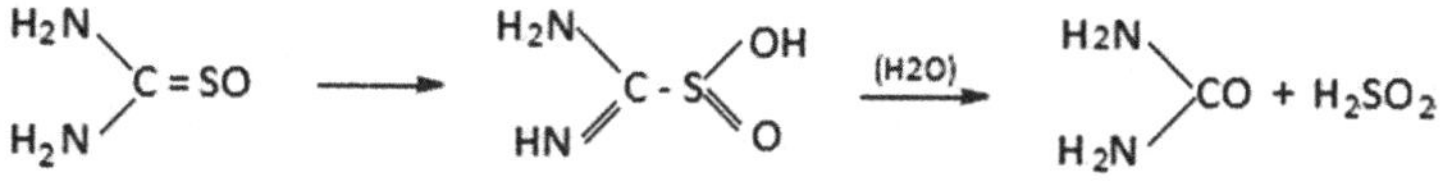

| Thio-urea Dioxide | Formamidine sulphinic acid | Urea | Sulphinic acid |

Another reducing agent, which has been used since the earliest times, is tin(II) chloride. It is a readily soluble compound which reacts with an azo dye as shown below. It is important that tin (II) chloride solutions are used quickly since hydrolysis, which gives a turbid solution, occurs on standing. The hydrochloric acid produced will attack unprotected metal, the steaming equipment is particularly. It is one of the reducing agents with lowest redox potential among the reducing agents for discharges. It is not used for white discharges as there are only limited dyes which can be fully dischargeable with it. But it is highly suitable for coloured discharges as there is no danger of over-reduction. It is very corrosive to printing machines and steamers.

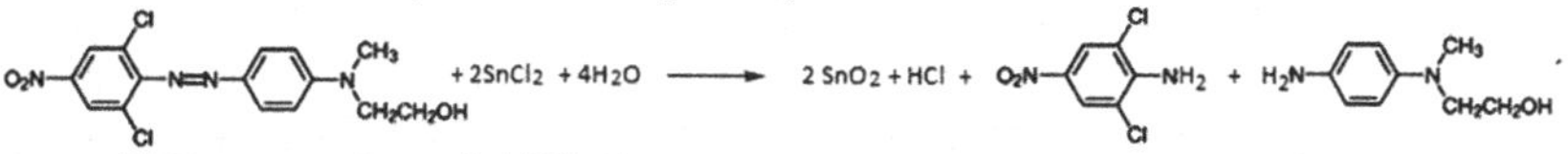

| C. I. Disperse orange 5 | Tin (II) Chloride | | Amines |

Reducing action of tin(II) Chloride

Selection of a reducing agent for a printing primarily depends on the fibre to be printed and of course, based on the dyes involved. Water soluble sulphoxylates can give haloing problems on the synthetic fibres, due to the movement of solution along the yarns. This problem can be overcome by using the insoluble formaldehyde sulphoxylates or thiourea dioxide. This latter product has had a considerable success in the discharge printing of acetate and triacetate, due to its low tendency to haloing and also because it is effective under acid conditions, which do not saponify the fibres as an alkaline reducing system can do.

4.10   Solvents, Dissolution aids, Dispersing agents

In some print pastes solvents are used for the easier diffusion of dyes into fibres. Examples are diethylene glycol diethyl ether (diethyl "Carbitol") and diethylene glycol dibutyl ether (dibutyl "Carbitol") used for cationic dyes, also pasting auxiliaries for vat dyes, print pastes which helps in the dissolution of leuco esters dyes. Glycerol is used in printing pastes as a hygroscopic agent and solvent. In polyamide printing with acid dyes urea, *inter alia*, is used as a dye solvent. Benzyl alcohol and ethanol also are used a solvent in print pastes.

Another purpose of dispersing agents and solvents are preventing aggregation of dyestuffs in the printing pastes. It is more important where dye concentration is higher.

Examples of the solvents, dissolving aids, dispersing agents used in print pastes are:

(1) Acetates of glycerine

$$CH_2 - OOC - CH_3$$
$$|$$
$$CHOH$$
$$|$$
$$CH_2 - OH$$

Mono acetate of Glycerin

$$CH_2 - OOC - CH_3$$
$$|$$
$$CH - OOC - CH_3$$
$$|$$
$$CH_2 - OH$$

Diacetate of Glycerin

$$CH_2 - OOC - CH_3$$
$$|$$
$$CH - OOC - CH_3$$
$$|$$
$$CH_2 - OOC - CH_3$$

Triacetate of glycerin

(All made by reacting glycerine with acetic acid or acetic ahydride)

The solubility of these in water varies. Monoacetate is completely miscible with water, whereas diacetate is less soluble and triacetate is still less soluble but soluble in alcohol and ether. They are used as solvents for basic dyes, rapidogen dyes, etc. Acetin is a mixture of mono, di and tri acetate of glycerine which is also used as a solvent in printing. These additives gives better penetration and higher colour yields in printing.

There are branded products used as solvents like Soledon Developer GE (ICI) or Cellosove which is Glycol monoethyl ether (See below) which is used as a solvent in vat and azoic colours. Another frequently used solvents are triethanolamine and tetraethanol amine ammonium hydroxide.

$$CH_2 - OH$$
$$|$$
$$CH_2 - OC_2H_5$$

Glycol
monoethylether
(Soledondeveloper GE)

$$N \begin{cases} CH_2 - CH_2 - OH \\ CH_2 - CH_2 - OH \\ CH_2 - CH_2 - OH \end{cases}$$

Triethanolamine

$$OH - N \begin{cases} CH_2 - CH_2 - OH \\ CH_2 - CH_2 - OH \\ CH_2 - CH_2 - OH \\ CH_2 - CH_2 - OH \end{cases}$$

Tetraethanol amine amm-
onium hydroxide

Both chemicals are used as solvents and especially used in screen printing using rapidogen dyes for better colour yields.

$$S \begin{cases} CH_2 - CH_2 - OH \\ CH_2 - CH_2 - OH \end{cases}$$

Thiodiethylene Glycol

(Glycine A   Glydote B)

Dissolving (or Solution) salt B

Some other branded products like Glyecine A (BASF) and Glydote B (ICI) are based on Thiodiethylene glycol (See above). They are good dissolving assistant for basic, direct, acid and disperse dyes. They are neutral solvents miscible with water, hygroscopic, solvent for dyes mentioned above and also for vat dyes and their leuco compounds and especially solubilised vat forms. During fixation process even if it is oxidised to (See below) sulphoxide derivatives but still useful as solvent and in increasing the colour yield

Thiodiethylene Glycol
(Glycine A   Glydote B)

Sulphoxide derivative of
Thiodiethylene Glycol

One of the very efficient solvent for vat dyes is monobenzyl sulphanilate which is sold in the name of Solution salt B/SV, Dissolvong salt B by some manufacturers. While manufacturing along with monobenzyl dibenzyl sulphanilate is also formed. Since latter product is also acts as a solution aid usually it is not separated, and the mixture is marketed as solvent (see below)

Sodium sulphanilate   Benzyl chloride   Monobenzyl sulphanilate   Dibenzyl sulphanilate

In addition to the property as dissolving assistant, it has hygroscopic property also which further helps accelerating the dispersion and migration of dyes and improving the yield. It is mainly used in vat printing paste of the Ronolite-Potash or Rongolite soada ash method.

Sodium salt tetrahydronaphthalene -2- sulphonic is an easily soluble powder, is used as an effective dispersing agent used in print pastes which can improve the production of deeper coloured prints.

Tetrahydrnaphthalene - sod. sulphonate

$CH_3 - CH (OH) COONa$

Sodium Lactate

Urea is also used as dissolving assistant, but urea is a very important ingredient in the print paste and it is functions in various purposes and hence dealt separately. Sodium lactate (see above) is a non-volatile steam stable compound. It is useful as a humectant, and is used in print pastes, where the prints are aged or steamed for fixation.

4.11   Whitening agents

In discharge print pastes, often auxiliaries are added which enhances the discharged grounds. Examples are optical whitening agents, white pigments like zinc oxide, titanium dioxide. These auxiliaries should be resistant to the discharging agents in the print paste.

4.12   Wetting agents

Wetting agents are used, especially in discharge prints, to reduce the surface tension between water and air and enable the aqueous medium to penetrate into the air-filled capillary spaces in the fibre easily.

Wetting agents have two purposes in the print paste – to wet out the fibre and helping in dissolving the dyes. Water is the main solvent used in the paste preparation. Because of the surface tension of water (73 dynes/cm) it has a resistance to wet out the hydrophobic surface of the fibre or dyes. There are substances which is soluble in water and reduces the surface tension of water and thus helping in the wetting out the surfaces which are non-wettable (e.g., synthetic fibres, insoluble dyes like vat dyes) or difficulty wettable (e.g., natural fibres, some soluble dyes) – such chemicals are called wetting agents. Wetting agents comes in the group of surface active chemicals.

As far as dyes are concerned, nowadays some manufacturers are providing soluble dyes which can be easily dissolved by sprinkling in warm or even cold water. Wherever the dyes are not easily soluble it has to be helped with a wetting agent to avoid any lump formation, which will hinder the dissolution. In such cases the dye may be pasted with little water and a wetting agent so that dissolution takes place without any problem. But this is not the case always, especially in the case of insoluble dyes like naphthols and vat dyes. They have to be helped in wetting out before dissolution. It is usually done by pasting the dyes with a wetting agent. While dissolving (e.g., vat dyes with alkali and a reducing agent) wetting agent works in two ways. It helps to wet out the outer layer of the dye particles enabling the chemicals in the dissolving action and secondly easily leaching out the dissolved dyes into water and exposing the next layer for the action of the chemicals by wetting again.

Common wetting agents are sodium salts of sulphated vegetable and animal oils and other completely synthetic wetting agents. Sodium salt of sulphated castor oil (Turkey Red Oil) is one of the most common wetting agents used in dissolving insoluble dyes. This has many advantages like they are reasonably stable, easily dispersed in water, give clear solutions and not affected by hard water (calcium, magnesium). Such wetting agents can also be made by sulphation of oleic acid, ricinoleic acid. In these cases, it is first methylated and then sulphated. Chemical reaction of manufacturing Turkey Red Oil from castor oil is as follows:

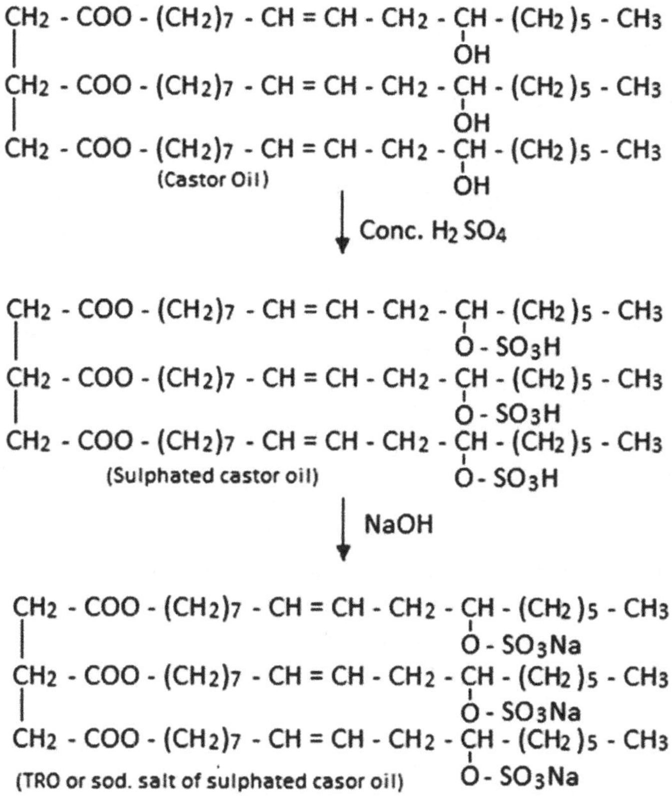

$$CH_2 - COO - (CH_2)_7 - CH = CH - CH_2 - \underset{\underset{OH}{|}}{CH} - (CH_2)_5 - CH_3$$

$$CH_2 - COO - (CH_2)_7 - CH = CH - CH_2 - \underset{\underset{OH}{|}}{CH} - (CH_2)_5 - CH_3$$

$$CH_2 - COO - (CH_2)_7 - CH = CH - CH_2 - \underset{\underset{OH}{|}}{CH} - (CH_2)_5 - CH_3$$

(Castor Oil)

$$\downarrow \text{Conc. } H_2SO_4$$

$$CH_2 - COO - (CH_2)_7 - CH = CH - CH_2 - \underset{\underset{O-SO_3H}{|}}{CH} - (CH_2)_5 - CH_3$$

$$CH_2 - COO - (CH_2)_7 - CH = CH - CH_2 - \underset{\underset{O-SO_3H}{|}}{CH} - (CH_2)_5 - CH_3$$

$$CH_2 - COO - (CH_2)_7 - CH = CH - CH_2 - \underset{\underset{O-SO_3H}{|}}{CH} - (CH_2)_5 - CH_3$$

(Sulphated castor oil)

$$\downarrow \text{NaOH}$$

$$CH_2 - COO - (CH_2)_7 - CH = CH - CH_2 - \underset{\underset{O-SO_3Na}{|}}{CH} - (CH_2)_5 - CH_3$$

$$CH_2 - COO - (CH_2)_7 - CH = CH - CH_2 - \underset{\underset{O-SO_3Na}{|}}{CH} - (CH_2)_5 - CH_3$$

$$CH_2 - COO - (CH_2)_7 - CH = CH - CH_2 - \underset{\underset{O-SO_3Na}{|}}{CH} - (CH_2)_5 - CH_3$$

(TRO or sod. salt of sulphated casor oil)

Reactions in the manufacture of Turkey red oil

These days there are stronger synthetic wetting agents are easily available and the importance of TRO has been reduced.

4.13   Acids and acid donors

Acids are used for maintaining the pH of the pastes and for providing acidic atmosphere in the fixation period. In the former case, normally non-volatile acids are used so that pH changes should not happen during drying or steaming due to the volatility of acids like acetic acid. For example, since several disperse dyes are sensitive to alkalis, the pH of the print pastes must be adjusted on the weakly acidic side by the addition of a non-volatile acid such as tartaric, citric, lactic or glycolic acid. Acid medium is required in the

$$CH_2 - COOH$$
$$|$$
$$CH(OH)COOH$$
$$|$$
$$CH_2 - COOH$$
(Citric acid)

$$CH_3$$
$$|$$
$$CH(OH)$$
$$|$$
$$COOH$$
(Lactic acid)

$$CH(OH)COOH$$
$$|$$
$$CH(OH)COOH$$
(Tartaric acid)

$$CH_2 - OH$$
$$|$$
$$COOH$$
(Glycolic acid)

fixation of several dyes like solubilised vat dyes. In pigment printing Binding agents and crosslinking agents require an acid pH value for their fixation. Additional acid donors such as diammonium phosphate are included in the printing paste to guarantee optimum fixation even if the conditions are unfavourable. In certain cases, acid pH is required to activate oxidising agents like sodium chlorate which is included in the printing paste.

Some of the acid medium is used for printing polyamide fibres with acid dyes are ammonium Sulphate and ammonium tartrate. In the printing of wool with acid or metal complex dyes acidic pH has to be maintained. In all these cases acid donors like ammonium sulphate or organic acids like citric acid or tartaric acid are used. Two types of compounds used for this purpose is (a) esters of organic acids or (b) ammonium salts of organic and inorganic acids. In both cases at room temperature they do not react acidic but at steaming conditions it acts acidic by the dissociation – in case of esters into acid and alcohol and in case of salts of ammonia into ammonia (which escapes into the atmosphere and the parent acid. Since ammonium tartrate has chances of forming ammonium salts with solubilised vat dyes which are sparingly soluble in water and hence prevent them getting fixed it is not used.

$$CH(OH)COOC_2H_5$$
$$|$$
$$CH(OH)COOC_2H_5$$
(Diethyl tartarate)

$$\longrightarrow$$

$$CH(OH)COOH$$
$$|$$ $$+ 2C_2H_5OH$$
$$CH(OH)COOH$$
(Tartaric acid)   (Ethyl alcohol)

OSO3Na / OSO3Na
Dye

$$+$$

$$CH(OH)COONH_4$$
$$|$$
$$CH(OH)COONH_4$$
Ammonium Tartarate

OSO3NH4 / OSO3NH4
Ammonium salt of dye

$$+$$

$$CH(OH)COONa$$
$$|$$
$$CH(OH)COONa$$
Sod. Tartarate

Other frequently used acid liberating agents are ammonium Sulphate $[(NH_4)_2SO_4]$, ammonium gluconate $[H_4NOOC(CHOH)_4C_2OH]$, ammonium nitrate (NH_4NO_3), ammonium chloride (NH_4Cl), ammonium sulphocyanide (NH_4CNS), diammonium hydrogen phospahate [$(NH_4)_2HPO_4$].

4.14   Fixation accelerants

In case of fixation of disperse dyes on polyester by the thermosol or HT-steam process, it is often advantageous to add fixation accelerator in the print pastes for the achievement of good colour yields. In the case of half-emulsion thickeners, no addition of fixation accelerator is necessary. But when printing triacetate with disperse dyes fixation accelerator is necessary even if half emulsion is used as thickener.

4.15   Carriers and swelling agents

Since the synthetic fibres, especially polyester, which is difficult for the fibre to penetrate, carriers are used as swelling agents in dyeing to help dyeing of synthetic fibres, at boiling temperature. Under the same principle, carriers are used in printing also for fixing dyes by pressure steaming at 120-1300C. They basically swell the fibre so that the pores in the fibre are enlarged so that the dye molecules can enter the fibre. The suitable swelling agents for this purpose are diphenyl (polyester) and phenol, resorcinol, benzoic acid, thiourea, diethylene glycol acetate (DEGDA), Polyethylene glycol [$HO(CH_2CH_2O)nH$], ammonium sulphocyanide (NH4CNS), ethyl lactate [$CH3CH(OH)COOC_2H_5$] (cellulose triacetate). Phenol, Resorcinol acts as swelling agents for polyamide fibres and at the same time dye solvents for acid and metal complex dyes.

Diphenyl   Resorcinol   Phenol

4.16   Binders

Binder component in print paste is used only for Pigment printing. Pigments are fixed on the fabric by means of binder. During fixation (curing) binder creates a film, which surrounds the colour pigment and is responsible for fixing to the substrate. Pigments are of the size 0.2-1 μm but a very good quality pigment size is of 0.2 – 0.5 μm. In a pigment, dyes from manufacturer may contain only 25–45% pigments remaining will be additives like dispersants, water, etc.

As explained the binder acts as an adhesive between the pigment particle and substrate. since the pigment does not have any affinity or reaction with the substrate, and can remain as far as the binder is present, there are many requirements for a binder. Also, it can greatly affect the handle of the substrate and fastness properties of the pigment on the substrate, the quality of dyeing greatly depends on the binder. Hence the binder should have following requirements for resulting in a good quality prints:

It should have minimum or no effect on the fastness characteristics of the pigment.

It should be resistant to acids, alkalis and solvents

It should have swelling resistance

It should be resistance to weather ageing and heat

It should have minimum or no effect on the handle of the fabric.

The binder should be easily removable from the screen and other parts of the machines like conveyor, rubber blankets.

It should form film only at higher temperatures. Film formation at lower temperature gives problems like choking of the screen, etc.

The film formed must be colourless, clear, of uniform thickness, neither too soft nor too hard, i.e. possessing elastic properties.

It must coat and adhere to the pigment well and possess resistance to both mechanical and chemical loads.

The listed requirements of the binder make it clear that the development of the pigment printing or pigment dyeing process followed the development of suitable binders.

Binders are selected high molecular compounds, which are built up from single monomers by polymerisation, polycondensation or polyaddition. In textile printing, primarily weak dispersion binders are used, which are manufactured by emulsion polymerisation.

Common monomers used in binders are:

$$CH_2 = CH - \overset{\overset{\displaystyle O}{\|}}{C} - OR \ (R - alkyl)$$

Different esters of acrylic acid

$$CH_2 = CH - O - \overset{\overset{\displaystyle O}{\|}}{C} - CH_3$$

Vinylacetate

$$CH = CH$$

Ethylene

$$CH_2 - CH - C \equiv N$$

Acrylonitrile

$$CH_2 = CH - CH - CH_2$$

Butadiene

$$CH_2 = CH$$

Styrene

The character of the film formed by the polymers formed by these monomers will depend greatly on the glass transition temperature of the monomers. They can give an idea of the film hardness; predictions can be made regarding the handle quality. Generally, film hardness increases with increasing glass transition temperature of the homopolymers. Since one homopolymers cannot give the required characteristics you require for a practical binder for textile. For example, some homopolymer of high glass transition temperature cannot be used in textile due to its high stiffening effect on substrates.

Glass temperatures of homopolymerisates

Monomer	Glass transition temperature of the polymer 0C
Butadiene	−87
Ethylene	−70 to −77
2-ethyl hexylacrylate	−85
n-Butyl acrylate	−52 to −57
Ethyl acrylate	−22 to −27
Methyl acrylate	5 to 8
Vinyl propionate	7 to 8
Vinyl acetate	30 to 37
Vinyl chloride	80
Styrene	90 to 95
Acrylonitrile (not fully crystalised)	100 to 106
Acryl acid	166

Hence generally a mixture of monomers is used as binder considering the requirements of the print. Thus, for example a soft component (e.g., ethyl acrylate) and hard component (styrene) can be mixed in different ratios can achieve the final required qualities. Even though it is possible to get a satisfactory range of softness etc., it should have a wide temperature range i.e., is the lowest wearing temperature (atmospheric say, 00C) to the highest ironing or drying temperature (say, 2000C). It has been found that a mixture of 60 parts of butyl acrylate and 40 parts of styrene (all by weight) yields a binder that satisfies the main requirements but with some disadvantage of poor resistance to higher temperature and solvents which is usually solved by the inclusion of a small proportion of a reactive compound, which brings about the cross-linking of the binder. In order to get even film formation, the monomer should be homogeneously distributed in the print paste and, if possible, uniformly separated. After the paste deposition on the fabric, it is dried when the polymerisation of the monomers sets in. At this stage the dispersed solids coagulate into a layer of thick spherical packings, as the moisture evaporates and the dispersion is broken. During curing stage, the coagulated particles coalesce and simultaneously deforms into a film surrounding the pigments and further cross-linking reaction that takes place during fixation that an elastic film is created. The cross linking can be a self-crosslinking or assisted cross linking. Examples are:

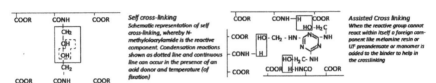

Self cross linking and assisted cross linking

If a thickener like alginate or CMC is used after the crosslinking the dried thickener gets trapped inside the binder film and either difficult to be washed off or if it is swollen in water may break the film which will fail the purpose of the binder. Thus, pigment print was successful only after the incorporation of emulsion thickeners in the print paste where both component of the thickener is evaporated during the curing process and hence no remnants of the thickener to be washed off. The application of pigment printing on synthetic fibres are not very fast due the smooth surface of the fibres, where even after the film formation on the surface of the fibre, it is washed off from the fibre easily. But in case of cotton since the fibre surface is not smooth and ribbon like with twists the binder fixes firmly on the surface with pigment solids trapped inside.

4.17   Resist agents

Resist agents are different from resist salts, which are used for resisting the fixation of a dye in a resist printing method. Discharge method of printing has a limitation of using only fully dischargeable dyes on which discharge is planned. In many cases fully white discharge in not possible where the dye is not fully dischargeable. Resist prints can produce effects of discharge prints especially when a dye is not dischargeable or partly dischargeable in a print. In many dischargeable prints, when a dye has to be fixed in alkaline medium (e.g., Reactive dyes) and acid, acid donor or a dye (white resist) or pigment, which fixes in acid medium, along with the acid or acid donor can be used as a resist (coloured resist) or vice versa. Examples of acid resists are acetic acid, citric acid, lactic acid, tartaric acid, glycolic acid, etc., and acid donors are ammonium chloride, ammonium Sulphate, ammonium citrate, ammonium nitrate, ammonium sulphocyanide etc. These chemicals are used depending on the fixation conditions like steaming, curing, drying etc. where these agents should be able to perform in that condition. For example, a volatile acid like acetic acid cannot work in curing/baking conditions.

A compound with which the dye can react much higher rate than the substrate can act as a resist agent. For example, an amino (-NH2) is more reactive than a hydroxy (-OH) group towards a reactive dye. Thus, an amino carboxylic acid or an amino sulphonic acid can act as a resisting agent for reactive dyes. The amino group is required for quick reaction of the reactive dye at the same time sulphonic or carboxylic group helps in the removal of the reacted compound.

Some of the amino compound used as resisting agents are:

2-amino-4-sulphonic benzoic acid

$$HOOC - \overset{\displaystyle H_2N}{\underset{}{\bigcirc}} - SO_3H$$

4-sulphophenyl hydrazine

$$H_2N - HN - \bigcirc - SO_3H$$

N- methyltaurine $CH_3 - NH - CH_2 - CH_2 - CH - SO_3H$
N-methyl glycine $H_3C - NH - CH_2 - COOH$

There are some special resisting assistants which is used for resisting vinyl sulphones by destroying the reactivity of these dyes by reducing the vinyl group to ethyl group while at the same time are not able to reduce the azo chromophore of the dyes.

$$D - SO_2 - CH_2 - CH_2 - OSO_3Na \xrightarrow{(OH)} D - SO_2 - CH = CH_2 \xrightarrow{(Na_2SO_3)} D - SO_2 - CH_2 - CH_3$$

This resisting agent cannot be used as resisting agent for MCT dyes as they have no reducing action on MCT dyes. But it can be used as resisting agent for mixture of the MCT and VS dyes. Thus, when a fabric is padded with a mixture of VS and MCT dyes are printed with paste containing sodium or potassium sulphite and steamed, both dyes gets fixed on the ground while only MCT dyes get fixed at the printed areas. In a different method if a fabric is blotch printed with a paste containing both MCT and VS dyes and over print with small motif containing these resisting agents and steamed, wherever resisting agents are printed only MCT gets fixed and at the other portions both the dyes getfixed while any white ground is left as such and they remain white.

4.18   Other printing paste components

Mordants, after treatment agents or fixing agents, levelling agents, complexing agents, etc. are also used in printing paste at different occasions. A special complexing agent is used in the discharging of dischargeable vat dyes called Leucotrope W. When a vat dyed ground is printed with discharging agents like

Leucotrope W

sulphoxylate formaldehyde, during steaming the vat dye is converted to its soluble leuco form. It is difficult to wash off the leucoform due to its affinity and will get reoxidised. If it is a white discharge it can show as a tint in the white. In such cases the above complexing agent, when incorporated in the print paste, forms a complex with the soluble leuco form of the vat dye which cannot be reoxidised to the parent vat dye. It is soluble in boiling alkali and hence can be removed by an after treatment with an alkali.

5

Heat transfer printing

The commercial development of transfer printing occurred during the 1960s. It involves the printing or painting of transfer inks onto paper that then adhere to certain fabrics under specified heat-controlled conditions. There are several methods of industrial heat transfer but the most commercially viable is 'sublimation' printing. The sublimation process turns the dye printed on the paper, which is a solid, into a gas, and then returns it to a solid again as it is transferred to the fabric.

While this method was significant in opening up the printing of photographic imagery onto textiles, it is limited in that it can only be used with synthetic fabrics, primarily polyester. With the development of many new and innovative synthetic performance fabrics for active sportswear, heat transfer printing has found a solid niche in the market. Transfer inks can also be painted straight onto paper and then transferred using an iron or a heat press directly onto fabric, allowing designers to take a more spontaneous approach in the studio.

5.1   Transfer printing papers

A customer who wants fabric printed with his own exclusive pattern, or who wants to reproduce weave or knit structure effects, needs no more than a copy of his pattern or a swatch of fabric. In the heat transfer or Sublistatic printing process consists of first printing the required pattern on paper and then transferring the dyes from the paper to a fabric. The whole procedure is remarkably simple, from the choice of a pattern to the method used to put it on a piece of knitted cloth. The paper can be printed using any method, gravure printing, offset printing, screen printing including digital, and this has increased the scope and the range of imagery available to this process.

5.2   Transfer paper printing by screen printing

Screen printing is used for producing transfer papers. There are some disadvantages in for this method, like there are chances of dimensional changes during the printing which may cause dimensional changes in designs and misfitting and even unevenness. Another problem is paper being blown from the conveyer while drying. There are some solutions like using a heavier papers 100 – 110 gm/m2, using dimensionally stable papers and printing pastes using solvents so that drying can be done easily without heavy blowing. Another problem is judging the shade of the print from the transfer printing paper. The actual shade can be seen only after the actual transfer printing takes place. Generally, this method of printing transfer papers are used for printing large sheets which are cut up later for the production of placement designs or logos for garments or garment panels. The paper is fed from large rolls of paper automatically and the machine is run much faster than the normal screen printing on fabrics.

5.3   Transfer paper printing by rotary printing

Printing of transfer printing papers on rotary printing machine

Rotary printing can be used for printing transfer papers the same way as the printing the fabrics. Since paper les less compressible and less absorbent it is possible to print very intricate designs with higher accuracy. However, the paper has to be little more absorbent than that used for gravure or flexographic printing since it is necessary to operate wet-on-wet. It has to have the capability of draining the water from the ink layer, but must be only minimally penetrated by the essential ingredients (dyes and binder). This method has got some advantages over other machines relatively low capital cost and the capability of printing very wide widths. In addition, the ability of screen printing to deliver heavy ink loadings can be useful to produce papers for transfer printing heavy fabrics or thick materials. It can be also used for in house transfer printing paper production for own transfer printing machines.

5.4   Transfer paper production by off-set printing

Off-set is one of the common methods of printing paper products and it can give photographic quality reproduction. Originally, this type of prints were made of particular kinds of stone and was called lithography, but modern offset lithography.is based on modifying the surface properties on flexible aluminium, polyester, mylar or paper printing plates instead of stone tablets. A zinc or aluminium printing plate is coated with a light-sensitive oleophilic (hydrophobic) material. Modern printing plates have a brushed or roughened texture and are covered with a photosensitive emulsion. A photographic negative of the desired image is placed in contact with the emulsion and the plate is exposed to ultraviolet light. After development, the emulsion shows a reverse of the negative image, which is thus a duplicate of the original (positive) image. The image on the plate emulsion can also be created by direct laser imaging in a CTP (Computer-To-Plate) device known as a plate setter. The positive image is the emulsion that remains after imaging. Non-image portions of the emulsion have traditionally been removed by a chemical process, though in recent times plates have come available that do not require such processing. The plate is affixed to a cylinder on a printing press. Dampening rollers apply water, which covers the blank portions of the plate but is repelled by the emulsion of the image area. Hydrophobic ink, which is repelled by the water and only adheres to the emulsion of the image area, is then applied by the inking rollers. If this image were transferred directly to paper, it would create a mirror-type image and the paper would become too wet. Instead, the plate rolls against a cylinder covered with a rubber blanket, which squeezes away the water, picks up the ink and transfers it to the paper with uniform pressure. The paper passes between the blanket cylinder and a counter-pressure or impression cylinder and the image is transferred to the paper. Because the image is first transferred, or offset to the rubber blanket cylinder, this reproduction method is known as offset lithography or offset printing.

Normally this method of printing uses only three primary colours (yellow, magenta and cyan) and black and the print is done in the form of dots (in some cases more colours are used). All the dots have the same colour intensity but depth of colour in an image is managed by the density or size dots and infinite numbers of colours are produced by overlapping or mixing the colour dots. This has important repercussions to produce sublimation transfer printing paper by lithography, because the dot sizes produced on the paper by printing are not those produced on the fabric by transfer due to the lateral diffusion of the dye, however it can be overcome by adjusting the dot size in accordance with the sublimation characteristics of the dyes. The amount of ink delivered to the paper by the plate is small; the ink layer in the dots has a thickness of only 1–2 μm. The colour content of the inks has therefore to be of the order of 50%, leading to very high viscosity compared to the inks used in other systems of printing. The inks are oil-based. Earlier days linseed oil is used for making this oil based inks, so that the printed image was oxidatively cured on standing but today synthetic products are used but following the same principles. Since the printed ink films (or dots) are very thin, the different colours are applied successively without intermediate curing or drying and high printing speeds are possible. This method of production of transfer papers are quite economical, especially when printing longer runs.

5.5   Gravure printing

This method of printing is comparable to roller printing probably in the fact that the design is engraved on copper cylinders. Thus, the design 'screen' used for gravure printing is a highly polished engraved cylinders which may have a solid steel centre with a thinly laid, soft copper coat wrapped around it. By an intaglio process (in-tal-yo means engraved or cut in) in which a negative image is etched into the surface of copper printing cylinder as tiny cells or dots of various size and depths. The tiny recessed chambers really need to be viewed under magnification to fully appreciate how shallow they are, being as they are, only microns deep. These recessed indentations are referred to as 'cells' and it is not unusual for there to be, dependent upon the complexity of the design, tens of thousands per square inch. The copper is chrome plated for durability. Ink is applied to the surface and a flexible metal blade called doctor blade removes the excess colour leaving the surface clean, with ink only in the depressions.

The depth and width of each cell determines how much ink is applied to the paper, and subsequently the strength of colour. The shallower the cell, the lighter the colour; the deeper the cut, the more ink is applied and the stronger the colour. On the print machine the cylinder sits partly submerged in an ink trough, whereupon the engraved cells 'fill-up' with ink as it rotates. Before the cylinder can release this cargo of ink the excess, which is laying on the surface of the cylinder, has to be removed. The acutely angled blade skims the surface of the cylinder scraping the ink that is not stored in the cells, back into the ink tray. The print cylinder is then squeezed against a rubber roller with the paper as its sandwich. During the high-speed rotation of the cylinder the cells release the ink onto the face of the paper. Transfer printing with water based printing pastes on a vertical roller printing machine (Portalrouleaux printing machine Fig.) and dried in an overhead drier prior and batched.

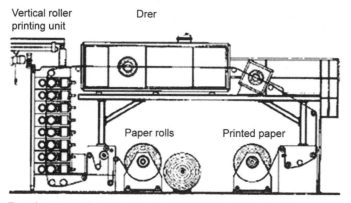

Transfer paper printing on the squeressig roller printing machine

The major benefit of Gravure printing is the ability to print fine tonal work and gradation of colour using a single cylinder. This gradation of colour, governed by the depth and width of each individual cell, can be from a solid colour through to anything as subtle as a 20% tint. Combine this with the fact that a typical Gravure machine may have 8 or more cylinder stations, it is easy to appreciate that the perceived amount of colour achievable is extensive. Using a series of rollers with varying depths of engraving and allowing successive inks to overprint and produce 'fall on' effects, very complex designs with a large tonal range can be produced with high definition and subtlety. Very high printing speeds can be achieved (60–120 m/min but the paper must be dried in between print stations in order to avoid smudging or marking-off. In this respect the gravure printing of paper differs from that of textiles, in which the absorbency of the fabric allows wet-on-wet operation. The need for rapid drying when printing paper by gravure makes it necessary to operate with highly volatile solvents such as toluene or ethanol, and in many countries, this is leading to increasing pressure from environmental protection agencies.

Considering the very high quality of printing and the subtlety of shading in design, gravure printing is one of the main transfer paper printing methods even though it is slightly costlier method.

5.6   Transfer printing papers by flexographic printing

Flexographic printing is one form of relief printing where ink is applied to rubber or polymer plate on which the printing image is raised above the rest of the surface as a 3-D positive mirrored relief. This stereo form is made in the form of roller which are arranged around a large drum (normally there are six, sometimes eight) carrying the paper. Since only the raised portions of the stereo come into contact with the paper there is no need for drying between print stations; as the paper is dried only once, there is no problem of dimensional change on drying and slower drying solvents can be used. If the paper is dried between stations fall-on effects can be produced, but this is unusual in the production of sublimation transfer printing paper production it is frequently necessary to apply heavier than usual ink loadings in order to achieve the required shade depth on the fabric. In practice this means that in the production of transfer paper by flexographic printing output is frequently slower than when gravure printing is used or when flexography is used for normal paper printing.

Normally the image is separated into four colour values cyan/magenta/yellow/black. Individual colours are printed as dots at varying densities Individual dots combine to form additional colours. The number of final colours in a design is limitless.

The major requirements of transfer printing papers are:

1. Printability depends on surface smoothness of transfer printing paper (achievable by means of calendars or coating and/or choice of raw material and paper manufacture).

2. Handling during transfer process requires a certain suppleness which is promoted by low humidity content (4-6%). Even thickness and absence of folds are important.

3. The process of transferring the printing dyes onto the textile substrate must take place uniformly and effectively. The structure of the transfer printing paper has a great impact on this.

Ghosting or double image formation of the transfer printing design on the papers on undue storage is a problem faced by the printers. This happens due to the movement of dye molecules due to abrasion, migration or sublimation.

The abrasion can happen due to rolling and unrolling of the paper which may be due to poor dye adhesion and rough paper surface. The solvents used in the print paste can assist the migration of the dye causing blurred contours and shade shifting of the designs. Storage in higher temperature, poor porosity of the paper and dyes with low sublimation temperature can cause the dye sublimation to from one layer of the fabric to the next layer of the fabric in a batch.

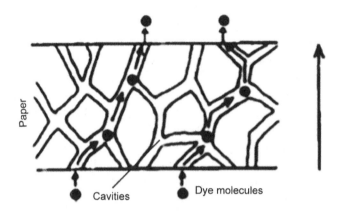

Dye molecule migration due to paper porosity

5.7   Components of printing paste (Ink)

The transfer printing ink or ink paste largely depends on the type of machines used for commercial paper printing machines. The ink system used determines the physical form of the dye which forms the starting material as well as the nature of the ancillary equipment needed to make the ink.

We have explained above the machines used for printing transfer papers. The sophistication both mechanically and automation wise are comparable or better than today's textile printing machines. Most of these machines can run high speeds which will require the printing ink/paste should dry very quickly. Most of the printing machines need drying of the inks within the period of first printing cylinder (engraved roller) to printing by the second cylinder. This is often achieved by having a drying chamber in between each printing station. The system is termed inter-station drying and should have safety features included to remove the inflammable gases produced during the drying operation.

The 'web path' (the route the paper passes on its way through the machine) is much shorter on a flexo machine than on a gravure machine. Since the flexo roller carries its design in rubber, the ink system must not be based on toluene or similar solvents which would cause swelling and distortion of the design. For this reason, a flexo ink is normally based on an alcohol—water mixture. The viscosity characteristics are important too, because the ink has to be accurately furnished to the printing roller via an all-over engraved roller often called an anilox roller. Flexo rollers deposit a much thinner layer of ink than a gravure roller, approximately 0.006 mm (0.00025 in) compared with 0.0254 mm (0.001 in). It follows therefore that to get the same final shade after transfer, the dye concentration in a flexo ink must be proportionately higher than in a gravure ink. This led to a preoccupation with dyestuff strengths for transfer-printing inks both by ink makers and dye makers. A gravure ink has a much lower average viscosity than a screen ink and requires a change of reduction medium at the dilution stage. The value of the screen-printing method for transfer printing on the small scale lies in two main directions. Screen-printing methods, once optimum gauze has been selected (e.g., 6T) give prints of acceptable quality while depositing a relatively thick dye-containing layer. Hence concentrated dye brands are not essential.

The main ingredients of the printing pastes are disperse dye, solvent and the binder. Since the transfer printing is based on sublimation principle, the dyes has to be selected which are sublimable, i.e., normally low molecular weight disperse dyes. The dyes supplied for dyeing which is either powder with dispersing agents other additives or liquid again with dispersing agents in aqueous medium is not suitable for this purpose. Since the dyes are dissolved in the solvent pure dye is only suitable, which is procured from the manufacturers from the dye cakes which are made into fine powder by milling to help in dissolution in the solvent. Solvents used are mainly toluene for gravure printing method, ethanol/methanol for flexographic printing method or drying oil like linseed oil for lithographic or for water based pastes for screen printing. The solvent selected has to be volatile to enable easy drying of the printed paper at the fast speed at which the transfer paper is being printed. Binder is necessary to bind the dye on the paper but at the same time it should release the dye easily during the actual transfer printing. Common binding agents used in solvent based inks are poly vinyl acetate or acrylic polymers, in waterbased inks alkyl cellulose and in emulsion based inks (water/toluene) hydroxyl cellulose. Another factor in selecting the binder is the ability of the binder to release the dye quickly and the dye should diffuse through the binder layer to enable the dyes to get evaporated without any hindrance. Poly(vinylbutyral), and ethyl cellulose has been seen efficient in these respects. The binder should have high resistance for diffusion at room temperature due to its rigid network but the same time at the transfer printing temperature the chain mobility supports the expansion of the network and allows the free diffusion of the dye molecules.

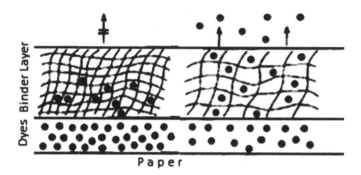

At room temperature the the rigid network resists diffusion of dye.

At tranfer printing temperature the expanded network allows the dye diffusion.

Some firms of ink makers offer ranges of solvent-based screen inks and suitable dilution media (which correspond in function to the reduction thickening in textile printing). The advantage of such inks is that being solvent based, they do not cause paper cockling. Jobbing screen printers in the paper trade have taken readily to using such inks, but they are, of course, used to working with solvent-based inks and the fire precautions (including ventilation requirements) that are necessary. In making inks on a solvent basis for screen printing it is not so essential with highly concentrated powders as it is with gravure and flexo inks. The main reason why ink makers prefer the concentrated brands is that the diluents and dispersing agents present in the textile brands are Unable to interfere with the solvent milling agents. This interference is usually one of the causes of ink concentrate instability.

Example of an aqueous Print Ink for screen printing
Prepare stock emulsion thickening

Quantity	Unit	Additions
8-15	Parts	Dispersing agent or emulsifier dissolved in
192-185	Parts	Water at 60°-70°C, cool
800	Parts	and stirred thoroughly. The mixture is left for up to 15 min to complete dye dispersion, and when cool stirred into
1000	Parts	

Printing ink recipe

Quantity	Unit	Additions
10-100	Parts	Disperse dyes are sprinkled into
490-400	Parts	Water at 30°-40°C, and stirred thoroughly. The mixture is left for up to 15 min to complete dye dispersion, and when cool stirred into
350	Parts	Stock emulsion thickening, in which
150	Parts	Indalca' PA3 9% (or equivalent)thickening have previously been incorporated
1000	Parts	

Notes:

1. Some disperse dyes disperse most readily when sprinkled on to 10–20 times their weight of water. This is not possible in preparing the strongest shades in the above recipe. It has proved practical on a small scale to prepare satisfactory dispersions provided care is taken to avoid lumps.

2. Where a Liquid brand is available this is simply diluted with cold water and added into the thickening.

3. Where a stock emulsion is not available replace it by a mixture of 'Indalca' PAS or equivalent and water sufficient to obtain a suitable printing viscosity without unduly depressing the yield on transfer.

4. The 150 parts of 'Indalca' PAS 9% thickening in the above recipe contain 13.5 parts of dry thickener and 136.5 parts of water. In cases of difficulty in dispersing larger amounts of disperse dye, this amount of water may be added to that already calculated for the recipe above. Once the grains have been dispersed the dry thickener quantity is stirred directly into the cooled grain dispersion and allowed to dissolve.

5. Under no circumstances should stock emulsion thickening be mixed with warm-water dye dispersions as the emulsion is liable to break and the printing viscosity will be lost.

6. It should be noted that the transfer of dye (in transfer printing) is hampered if too much thickening agent is present in the dried down film. The amount of thickening agent present in the film may be reduced by replacing part of it by an oil-in-water emulsion.

5.8   Transfer printing dyes

Transfer printing dyes have to be selected which transfer as evenly as possible. These can be referred to as optimum transfer zones (Fig.); at temperatures which are too low, dyes which sublime too rapidly are frequently the origin of blurred printing edges (so-called bleed-out). Sometimes in classifying dyes for dyeing polyester the terms A, B, C and D are used. In this classification, the A dyes have the lowest heat fastness and the D dyes the highest. Sublimation behaviour is the reverse of heat fastness, so A class disperse dyes sublime most readily and D class dyes feast readily. In practice, only the A and B class dyes (plus an occasional C class dye) are of interest for transfer printing. A slight anomaly is that an occasional disperse dye which is not recommended for polyester dyeing does transfer print satisfactorily.

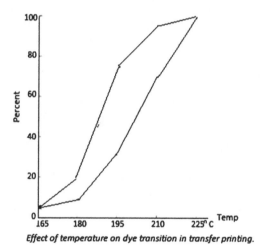

Effect of temperature on dye transition in transfer printing.

The sublimation properties do not depend entirely on the dye molecule but also on the physical form of the dye (commercial form) and the dye concentration used. It should be noted that paper only absorbs a small portion of the print paste quantity as compared with direct printing onto textiles and the dye quantities used in paper print pastes are much higher than in textile printing. The special commercial dye form with high dye content and low electrolyte content has to be adapted to these circumstances. Use of large dye quantities in liquid form is preferable.

Some of the disperse dyes used for Transfer printing:
Originally it relied on available disperse dyes with good sublimation characteristics. New dyes, specifically developed for this process, have

appeared on the market. C.I. Disperse Yellow 54, C.I. Disperse Red 60 are the leading yellow and red dye for the transfer printing of polyester.

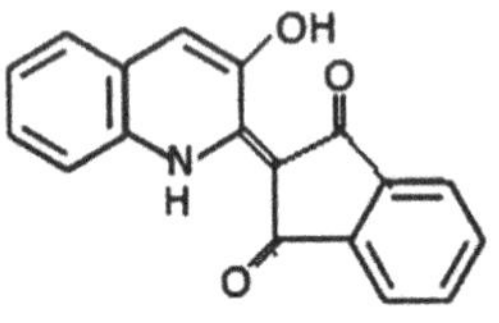

C.I. Disperse Yellow 54 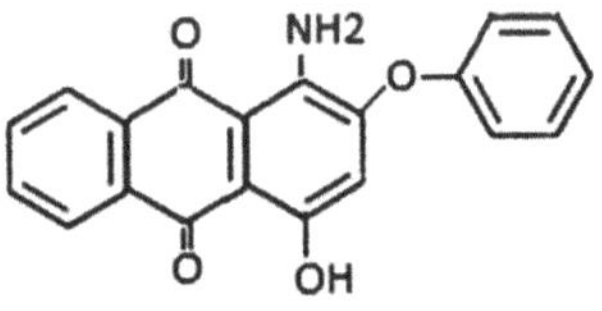   **C.I. Disperse Red 60**

Thermal transfer printing, particularly dye diffusion thermal transfer (D2T2) printing, is still important, especially for colour images. A yellow D2T2 dye is probably the easiest to obtain both in terms of colour and properties. Two of the most widely used dye classes are methine (a) and especially azopyridones (b)

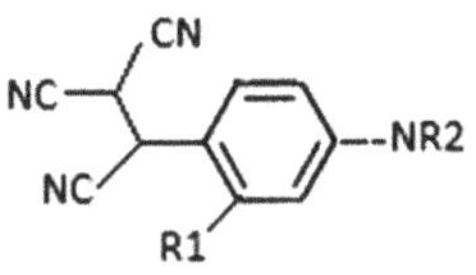

(a) 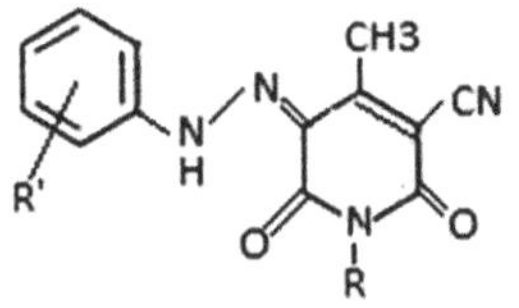   (b)

Magenta dyes are more difficult to produce than a yellow dye, both in terms of colour and properties. Anthraquinones are the leading textile transfer-printing dyes, but these are red rather than magenta and are tinctorially weak (low print optical densities). However, they possess outstanding lightfastness, and dyes such as C.I. Disperse Red 60 (c) are often used in mixtures with other dyes, such as heterocyclic azo dyes. Dyes based on isothiazoles such as (d), had to be designed to produce D2T2 magenta dyes.

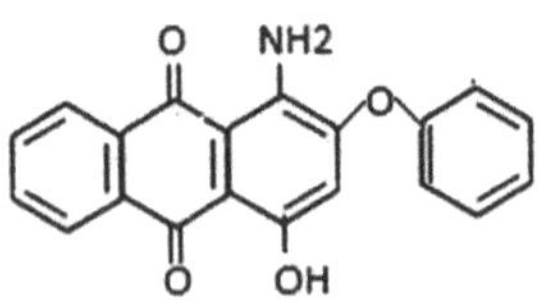

(c) 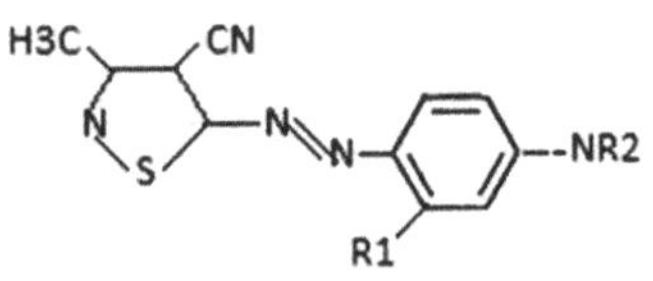   (d)

The best cyan dyes are copper phthalocyanines, but these are unsuitable as D2T2 dyes because of their large molecular size Hence the main cyan dyes used in transfer printing are anthraquinones, with disadvantages like these dyes are blue rather than cyan, low tinctorial value, and of low light fastness. Cyans used in photographic area are indoanilines have the correct colour, but they are also weak and generally exhibit poor lightfastness. Dyes of type (e), however, have overcome this instability problem. These dyes and novel indoaniline dyes (f), which have higher lightfastness than conventional indoaniline dyes, are currently the most prevalent cyan D2T2 dyes.

(e)   (f)

Self shades or a mixture of dyes are possible for a particular shade for transfer printing. Blacks and browns are seldom homogeneous products in the disperse range. Suitable black mixtures are usually possible by using a blue component, shaded with a yellow and adjusted if necessary, with a red. It should be remembered that a black or grey suitable for polyester seldom gives the same shade if printed on nylon 66. It may be advisable to reformulate the ink by altering the proportions of the components when a change of fibre is involved. Suitable browns may make with violets and an orange as a basis and shading bluer or yellower as required.

5.9   Principle of dyes transfer to the substrates

Disperse dyes used for normal dyeing are the ones with higher molecular weight and lower vapour pressure. They give higher fastness and heat fastness. These dyes are sparingly soluble in water and hence require higher temperature and pressure for dyeing. Disperse dyes of low molecular weight and higher vapour pressures were not in demand. But in textiles it is usual that disadvantage in relation to application can be turned to a positive advantage to another. Thus, disperse dyes of relatively high vapour pressure has been put into good use in the sublimation transfer printing. The available disperse dyes are grouped into A, B, C and D groups where A group is of highest volatility and the D group the lowest. Generally,B and C group disperse dyes are used for Transfer printing. A group is so volatile that it is difficult to preserve the integrity of the design during the transfer. Whereas the group D the volatility is insufficient for significant transfer to takes place. There are more and more dyes available in the group D, the range in B and C groups are limited.

Generally, energy is required to make a chemical reaction to happen. This is called activation energy and it is required overcome "energy barrier" before a starting material can be converted into a product. The height of the barrier is directly proportional to the rate of the reaction. In textile dyeing scenario the activation energy depends essentially on the dyeing mechanism. For disperse dyes in transfer printing, a lot of energy has to be put in to convert the dye molecules from their (assumed) form of ideal crystals into the gas phase (see fig. below). This is expressed in the fundamental thermodynamic equation:

$$\Delta G = \Delta H - T \cdot \Delta S$$

Where G = free energy, H = enthalpy, S = entropy. In the case of sublimation (Transfer printing) ΔH is a measure of the strength of intermolecular forces in the crystal. At room temperature almost 15% of the lattice forces are accounted for by the heat $T\Delta S$ that comes from the surroundings, so that only a free sublimation energy ΔG of about 105 kJ/mol is required to destroy the crystal completely.

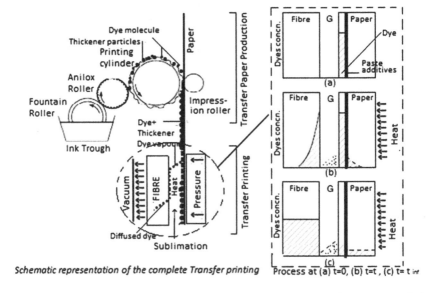

Schematic representation of the complete Transfer printing Process at (a) t=0, (b) t=t, (c) t= t

Once the disperse dye has been converted to vapour by heat energy, it is in contact with polyester fibres in transfer printing process. At this stage the dye vapour, the distribution of the at equilibrium is governed by a partition isotherm, in which the partition coefficient K is defined by

$$K = \frac{C_f}{C_v}$$

Where,

C_f = dye concentration in the fibre phase

C_v = dye the concentration in the vapour phase

Polyester and cellulose acetate can absorb up to 10% of the weight of the material, the value of K is of the order of 105 to 106, which is much higher than that found in aqueous systems.

As shown in the figure above there are dye in three phases – Dyes in the vapour phase, dyes on the fibre surface and the dye in the diffused phase. Even though the equilibrium is reached between the vapour and the fibre surface, but the kinetics of achievement of the ultimate equilibrium are governed by the rate of diffusion of dye from the bulk of the vapour to the surface and by its rate of diffusion from the surface of the fibre into the fibre mass. Hence the dye on the fibre surface is only considered as an intermediate phase. The process can proceed only if the dye diffuses into the fibre to create free sites so that the dyes on the surface can enter the fibre. Considering, the dye vapour possesses ideal behaviour, the concentration in the saturated vapour is given by:

$$C_v = \frac{pM}{RT}$$

Where,

p = the saturation vapour pressure of the dye,

M = the molecular mass of the dye,

R = the gas constant and

T = temperature.

As mentioned above the partition coefficient of polyester and cellulose acetate is very high, it can relates to competition, possibly allosteric in nature, between dye and water for sites on the fibre. Thus, in transfer printing these materials can absorb dye to an extend that it will be 'overdyed' if it is available, which can cause poor fastness to rubbing and washing due to the desorption of dye when in contact with aqueous medium. Vapour pressure of many disperse dyes used for Transfer printing may not be available, Molecular mass can be easily found out.

Once evaporation of the dyes takes place the gap between the paper and the substrate is a factor which affects the migration of dye to the fibre surface. Even though, the paper is pressed against the substrate surface one should consider the fibres inside the yarn and the inter spaces of the construction of the fabric to be away from the fibres touching the paper. It has been seen the air in between the paper and the substrate is no hindrance for the movement of dye molecules to the fibre surface and in the contrary helps in the migration in the desired direction.

Consider a situation where the fabric is pressed against the transfer paper in a hot atmosphere. Even though the paper and fabric is pressed together, when we are talking about molecules and vapour we have to still consider there is a gap G between them. Before starting to heat (figure above (a) t=0) the dye is contained in the ink layer. When the paper is heated on the reverse side (figure above (b) t-t) sublimation of the dye happens (vapour phase) and slowly transferred to the fibre by diffusion. After a brief delay, even though in the beginning there can be a condensation of dye on the fibre surface as the fibre gets heated up a uniform steady state system will be established. The time required for the equilibrium depends on many factors like the size of the gap between paper and fibre, the thermal properties of the paper and the fibre, but it may be assumed to be short relative to the full transfer time. When the temperature stabilisation takes place dye will diffuse from the fibre surface inwards due to concentration gradient within the fibre. This transfer, diffusion and migration continues till a thermodynamic equilibrium is established between the three phases fibre, G and I as shown in figure above (c, $t = t_{inf}$). As the fabric and paper moves out of the calendar, a small part of the dye may remain on the paper surface and there are chances of a little of the dyes diffused into the fibre, even though there is preferential attraction towards fibre. To reduce the quantity remaining on the paper selection of the paper and the binder/thickener selection has to be considered. Thus, the transfer of dye from paper to fibre depends on sublimation of dye from the ink layer into the vapour, movement of dye across the gap between paper and fabric, adsorption at the fibre surface, diffusion into the fibre. Once the temperature of the sublimation is reached first two processes happens rapidly and the third process which is slow will control the printing process speed of the transfer printing machine will depend on this. Parameters helping to quicken this process can improve the situation.

5.10   Transferring the dye to the fabric

5.10.1   Dry heat transfer

Transfer printing may be carried out by sublimation transfer, melt transfer or film release methods. In sublimation (or dry heat) transfer printing, volatile dyes (typically disperse dyes) are pre-printed on to a paper substrate and are heated in contact with the textile material, typically polyester fabric. The dyes sublime and are transferred from the vapour phase into the fabric in this dry-heat transfer printing method. This may be assisted by the application of a vacuum. Melt transfer is principally used on garments whereby a waxy ink is printed on paper and a hot iron applied to its reverse face to melt the wax on to the fabric surface. In the film release method, the print design is held in an ink layer which is transferred completely to the textile from a release paper using heat and pressure. The design is held on to the textile by the strong forces of adhesion between the film and the textile.

To transfer the dye from the paper to the fabric, the one is laid on the other, then both are pressed together for up to 30 seconds at a temperature of about 200°C. The dye on the paper is vapourised and mass transfer of the dye to the fibre is effected by sublimation. The configuration and limits of the printed pattern on the paper are accurately reproduced. No follow-up heat fixation, washing or steam treatments are needed. Once the heat transfer operation has been carried out, the printed fabric is ready for making up or shipment; the fastness properties of the prints meet all normal requirements. The heat transfer operation can be carried out using relatively simple equipment (T shirt etc.) either on the customer's own premises or by a contract printer. However, due to its peculiarities, the heat transfer process does have its limitations. Itcan be used only to apply:

- disperse dyes to
- synthetic fibres.

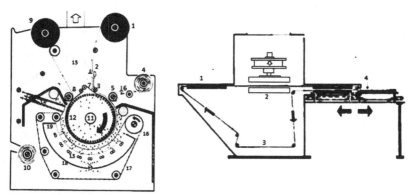

(a) Schematic diagram of Continuous Transfer Printing Station 1. Feed, 2-Fabric cutter, 3-Tension roller, 4- Heat transfer paper, 5-Contact roller, 6-Paper Cutter, 7- colour stop blanket, 8- Exit roller, 9-Printed fabric, 10-Paper after printing, 101- Vacuum Chamber, 12- Perforated drum, 13- Infra Red heater, 14- Infra Red Meter, 15- Heat outlet, 16- Chain, 17- Shield, 18- Reflector, 19- Safety Grid. *(b) Schematic Diagram of Batchwise Transfer Printing Unit* 1- Feed station, 2- Printing station, 3- Conveyor Belt, 4- Piler

In sublimation transfer printing machines, the size of the heated drum decides the speed of the machine. For an 80% coverage print on a modern machine of 2 m dia. at 20s contact time can print at a speed of 15 m/min there are bigger machines with almost double the speed also. The threading of the fabric is such that (see the fig) the face side of the fabric touching the printed side of the paper and back side of the fabric touching the heated cylinder. The fabric pressed against the paper by means of an endless blanket at around 1.4–2.1 × 103 kg /m2. The blanket should be heat resistant as it is endless and repeatedly heated. Usually made od Nomex fabric (an aromatic polyamide fibre – poly (1,3-phenyleneisophthalamide), which has a glass transition temperature of 2750C and a softening temperature above 3500C. Usually the transfer paper slightly wider than the fabric and machines are available which can print fabrics up to 3.5m.

Sublimation Transfer printing machine, the transfer of the dye from paper to fabric happens due to heat application, it is needless to say that if there is any variation happens in the face of the drum it can affect the even ness of the print widthwise or length wise. This is accomplished by making the drum as a shell and the shell carrying the heat transfer liquid only partially filled – usually biphenyl or biphenyl ether. The unfilled portion is evacuated and sealed and the liquid is electrically heated. The vapour from the boiling liquid condenses on the internal surfaces to give uniform heat to the drum throughout. Natural equalisation of the internal pressure ensures uniform heating and, providing the machine is sited sensibly so that there is no accidentally uneven cooling, uniform transfer temperatures are maintained. Other methods of heating, including the use of oil, gas and electricity, have been used with appropriately designed arrangements.

Owing to the relatively high temperature needed to effect dye transfer, the process cannot be used for the printing of synthetic fibres with low melting points, e.g., polyamide 11, certain types of polyamide 6, polypropylene and PVC fibres. Blends can be printed provided that the synthetic fibre component predominates. With 50/50 blends, the shades produced are dull and pale. The synthetic fibre component should make up at least 66 per cent of the blend fabric. Excellent results are obtained' when printing 70% polyester/30 % wool blends.

These units closely resemble the presses ordinarily used to finish knit goods but have no steam lines. The upper buck, and on automatic presses the lower buck as well, are electrically heated to a uniform temperature over their full surface area. During the heat printing operation, paper and fabric are kept under uniform pressure throughout. The pressure used can be varied as required.

5.10.2   Vacuum transfer

There is another version of Transfer printing machine which is based on vacuum-assisted transfer. Even though it is difficult to incorporate in the continuous units nevertheless reliable and successful units are available. Vacuum assisted machines can be designed either by a pressure difference within the machine so as to promote a flow of hot air which assists the migration of the dye molecules, or by producing a reduced pressure in the dye transfer zone. In such machines the transfer paper and fabric are heated to the required temperature as they run over a cylinder heated from within by electric, oil bath or infrared systems. In this method an air flow through the system holds the fabric in contact with the drum, and the paper is held in place without the need for a blanket.

The hot flow of gas directed on to a dye carrier medium (e.g., gas-permeable transfer printing paper) which it penetrates. The sublimed dye is carried along with the hot gas and deposited on the continuous web of fabric which rests on a gas-permeable conveyor belt. The direction of the gas flow passing through the dye carrier medium and the fabric is assisted by suction applied from beneath the conveyor belt. The choice of dye carrier medium, gas flow temperature, pressure difference between the upper and under sides

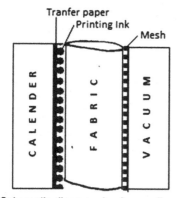

Schematic diagram showing continuous transfer printing under vacuum conditions

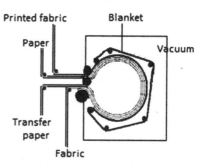

Schematic diagram of Vacuum Transfer Printer

of the fabric and the contact time are decisive parameters for control of the process. The degree of colour penetration achieved on pile fabrics by this process is far superior to that obtained by other transfer printing processes up to now. Since the dye carrier medium is only in light contact with the fabric being printed, no pile deformation occurs. This method also avoids any glaze formation at the face of normal woven fabric. The practical difficulty to make the continuous machine is to make effective vacuum which needs a reliable seal which was difficult with continuous feed of fabric and paper. There were two ways to do this either by wrapping the transfer drum in an evacuable skin or by enclosing the whole machine including the paper and fabric feeding, heating, all in a vacuum chamber. The latter was a direct and easier solution but more elaborate and obviously costlier (see Fig above). In the former method the fabric and paper are held in contact with the heated cylinder surface by a perforated flexible metal band in place of the conventional blanket. Since only a part of the cylinder circumference (78%) is under suction it is quicker and easier to maintain the vacuum. It also allows the drum diameter to be reduced from 2m, even at a practical pressure of 10.7 kPa (80 mmHg) pressures circumference is a continuously maintained vacuum chamber with the entrance and exit seal-operated. Thus, the construction of the machine easier and cheaper and transfer can be effected easily and quickly at a speed of up to 30 m/m.

Enclosing the whole machine in vacuum is an easier solution for vacuum transfer printing design, which was successfully implemented by Stork machines (see fig above). Advantages of these machines are like, Under vacuum the transfer temperature can be reduced by 400C from normal transfer temperature it is safer, chances of glazing is less, more suitable for heat sensitive fabrics like acrylic dyes with less sublimability can be used by higher temperature (but still less than the normal temperature of transfer printing) and achieving better heat fastness.

Vacuum system in flatbed machine can be easily arranged by using a perforated hotplate, while in a continuous machine external heating equipment such as infrared heaters gives the best results. One big advantage from this kind of arrangement in continuous-web printing, apart from more rapid transfer in certain difficult cases, is the absence of compression and consequent glazing of the fabric. Special presses are used for the batch wise printing of piece goods or cut-out apparel parts.

Made-up apparel parts generally are mounted on boards and printed by pressing heated plates. The heritage of heat transfer printing emerged from a new range of images, including those with photographic qualities, which could be printed onto fabrics using more colours than could be achieved with other print methods. The process also increased the number of tones that were achievable, and this led to what became known as illusionary prints. These designs created the illusion of a heavyweight woven textile printed on a finer and lighter fabric.

5.10.3   Melt transfer printing

Sublimation transfer printing methods explained above is suitable only for synthetic fibres, especially, polyester. The success of this method and the transfer of clear and aesthetic designs achieved naturally made the machine manufacturers and technicians to try to print such designs on the natural fibres also. The result was the printing was successful mainly in transferring the designs printed by gravure or flexographic methods, but fixation has to be done by traditional methods like steaming etc. Even though this method is not very popular in continuous printing of textiles but used more in panel/garment printing. The advantage of this process is that it needs much less pressure than the previous systems which will help to retain the integrity of the substrate. This type of transfer printing is suitable for cotton, silk, wool etc.; however, they need the fixation process. Transfer papers available where the coating can melt at 60°C, and at a pressure of 6 kg/cm^2 (59 kpa). The normal transfer printing machine with some changes in the parameters can be used for this purpose. The low temperature for transfer is advantageous to soft fabric like silk. This printing is still a two phase printing process it does not give the simplicity and one phase printing like the transfer printing of dry heat transfer or vacuum transfer of polyester (synthetic fibres) where the transfer and fixation happens in one process. Compared to pigment printing also this method of printing is not simpler.

5.10.3.1   Principle of the melt transfer printing method

Melt Transfer printing is based on all grades of dye being embedded in a fusible resin on the carrier (paper) and the dye is transmitted onto the textile via the phase of the molten resin under the effect of the temperature. The dyes which are neither melt nor sublime can be mixed with a polyglycol or an ether or ester derivative of a polyglycol which act as a 'transfer solvent' for the dye helping to transfer from paper to the substrate. The base paper is coated with an impermeable layer of chemical like nitrocellulose to avoid the ink layer from migrating back into the paper, and the transfer is highly efficient. Gravure printing of the design is done with a print paste after giving another coating of wax along with dispersing agents on the nitrocellulose coated paper. During transfer printing at required temperature and pressure the ink and wax layers melt and are forced into the fabric. During the transfer printing there is no need for the dyes to penetrate into the fibre, it happens during the fixation treatment only. The additives in the print pastes are washed off after fixation.

5.10.3.2   Garment printing

This method has not become very popular due to the reasons given above but the method was adopted for garment or panel printing adopting the process to suit the prints in such a way as it does not need any process further after printing. The principle is based on printing the dye and a resin which melts under the transfer printing temperature and fixes the dyes on the fabric by the resin. Variety of resins were used which can melt and fix at standard transfer press operating at 190–200 °C within 15–20 s. Even though this type of printing was introduced using resins based on poly(butyl methacrylate), acrylonitrile–butadiene–styrene terpolymers, a vinyl isobutyl ether–methyl methacrylate copolymer and similar materials and formulations of waxes mixed with acrylic resins and aminoplast precondensates or other crosslinking agents have also been used successfully, the prints were later used for printing photographic prints on fabrics which was printed on transfer paper using a laser printer and hence modifications were made. Laser printers use 'colour formers', which are essentially pigments coated with a low-melting (80 °C) methacrylate resin. Today the transfer printed papers are available from professional printers or one can use a laser printers and produce their own photographic prints or normal prints and do the transfer on a normal transfer print equipment which is used for hot transfer for disperse dyes for synthetic fibres.

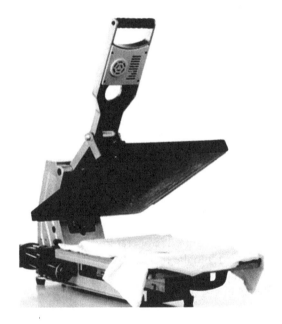

The usage of resins has been adapted to the requirements of prints and as mentioned above variety of resins including acrylic and other systems formulated with modifying agents such as plasticisers or crosslinking agents have been successfully utilised these days achieving all the fastness requirements of the customers.

5.11   Film release transfer printing

Film release print uses a special kind of ink called plastisols which dispersions of plastic powder in plasticisers such as dioctyl phthalate, tricresyl phosphate or chlorinated paraffins, which do not gelatinise the plastic powder when cold, e.g., vinyl resin dispersions plus softener or polyvinyl chloride paste made up of polyvinyl chloride plus softener. The image may be printed on a release paper and then to blotch screen print over the image using a white or grey plastisol backing coat and dried in the oven to set the plastisol layer. The image is placed on a garment/panel and in a normal transfer press or special press unit designed for this purpose, pressed at a temperature of 180-195°C for 15-20s when the plastisol layer softens and enter into the fabric. The garment/panel is allowed to cool and the image is reset and the release paper is removed.

This method can be used for printing metal foil on the fabric. The fabric can be printed with the suitable thermoplastic, colourless special adhesive or the adhesive can be printed on the foil with the adhesive wherever it has to be transferred to the fabric. Printed material and foil (with the matt foil side towards the printed material side) are combined on a transfer calendar or a transfer press and, depending upon the type of adhesive, pressed together at 150– 200°C under pressure. After the material has cooled the foil is removed. The points printed with binder appear as a pailletine effect. A closed, saturated binder film between textile and foil is necessary for good adhesion of the foil onto the textile. The fastness properties are limited for blotch type foil prints.

An alternative method, even though cannot be considered as a release print but rather a direct print method, involves in using the ink as such and print on to the fabric with fine mesh. A full coverage blotch with plastisol is printed to support with an additional backing. The fabric is directly passed through heated chambers for melting and fixing of the plastisol adhesive. The full print image also can be printed on the fabric directly using suitable screen (which contains enough plastisol quantity and fixing at temperatures of 160-200°C for 15-20 s.

Various kinds of equipments are available in the market for hot pressing and curing purposes for batch wise and continuous application.

Notes:

1. Sometimes when the plastisol ink is used the adhesion doesn't happen well. Please note that the plastisols are hydrophobic. Any moisture present will prevent the adhesion. This can happen even due to the

natural moisture content of the fabric printed. Firstly, in the case of hydrophilic fibre due to the hydrophilic material may not wet the fabric. Secondly when the temperature is raised for transfer steam may be generated (again due to the moisture content) which prevents the adhesion. The solution is probably applying more transfer pressure with some disadvantages of image being pushed through the fabric especially knitted structure loosing the integrity of the design, or giving longer time for transfer (which again loose productivity). A printer may find a balance between the two get the optimum result.

2. After transfer printing the plastisol being still thermoplastic, any ironing has to be done in the reverse side.

5.12   Features of heat transfer printing

Speed and adaptability are probably the main advantages of the heat transfer printing process. Orders can be met at short notice since prints can be turned out in any quantity on demand provided enough transfer paper and undyed fabric are always kept in stock. Similarly, patterns can be modified or replaced to keep pace with changing fashion trends. There is no need to stockpile prints and the danger of being caught at season's end with a warehouse full of largely

Transfer printing machines

unsalable fabrics is very appreciably reduced. In addition, quick and flexible response to customers' needs tend to lead to constantly full order books. Pre-treatment of the fabric to be printed is reduced to a minimum; in many cases the cloth does not even have to be bleached. Heat sensitive fabrics have to be tested for shrinkage to see if a heat setting pre-treatment will be necessary. Since heat transfer printing tends to act very much like the heat setting process essential to the finishing of most synthetic fibre fabrics, printing and setting become a single combined operation.

Heat transfer printing tends to preserve the qualities of a fabric. The handle and "fresh" appearance of light weight fabrics remain unimpaired. The high temperature used in heat transfer printing counteracts the tendency towards barrinessoften encountered when applying dye to polyester knit goods. Continuous heat transfer printing of piece goods results in less waste. The process eliminates all the interruptions as well as all the readjustments of register, repeats and doctor blades which result in flaws when flat screen or rotary printing machines are used. Reduced wastage is a very real advantage when printing small batches.

Advantages	Disadvantages
Operation is simple. No expensive Machine required	Process mainly applicable to synthetic fibres only, (to some extent other fibres)
Low investment capital and personnel costs	Difficulties in procurement of sample material (price) in the case of exclusive designs,
Low space requirement	
No need for water, no wastage of water	
In the case of short metre lengths in terms of setting-up and idle machine times; dye losses no longer occur,	Lack of colour brilliance in the case of polyacrilonitile and mixtures of synthetic fibres with cellulose and wool,
Risk of producing seconds quality is low; no high quality materials are involved,	Articles which require sublimation-fast dyes cannot be used,
Quicker reaction to changes in fashion, little storage of raw fabric, shorter delivery times,	Storage of large quantities of paper over prolonged period of time is a risk in terms of undiscovered faults,
No after treatment, like steaming, washing, drying etc., of fabric required	Colour range is limited
Risk of producing seconds quality is low; no high quality materials are involved	Increased demands on colour technicians; necessity to re-think.
Print is of excellent quality	Cost of Printed Paper very High
Problem-free printing of bonded fabrics, geometrical and horizontal designs are easier to print.	Limited fastnesses in the case of certain fibres, above all with dark shades,
Full use of gravure printing technique (effective shading of full to zero shade),	Less print-through and associated danger of "grinning",
Highly-qualified personnel not necessary in laboratory, print works, steaming facility, washing-off, etc.,	Structured articles become flat, deterioration of handle in the case of certain qualities,
Flexibility during machine changes,	Lower production speeds.
Simple order planning.	
Fewer reproduction problems,	Not economical for small order

A designer's freedom to choose or evolve motifs, patterns and colour combinations for woven or knitted cloth is closely limited by the properties of the fibre and the methods used to convert it into ready-for-use cloth these constraints since the reproducibility of motifs no longer depends on substrate construction. Heat transfer printing liberates the designer from these constraints since the reproducibility of motifs no longer depends on substrate construction. Because the surface of a web of paper is used to apply the dye to the fibre, the creation of motifs and the technique used to reproduce them become two independent operations.

The heat transfer printing process affords unlimited scope in the reproduction of patterns. Because the dyes are first applied to paper by an intaglio printing process, all the halftones and hues of a given pattern can be reproduced in full detail, with sharp contours and bright shades, on a textile fabric or other substrate. The effects obtainable on a substrate can be determined beforehand by using transfer paper samples available. The heat transfer process may be used to produce solid colours in any desired pale to deep shade. The back of the fabric can be left white, printed or dyed in another shade to give double-face effects. Heat transfer printing and dyeing consume no water. This is a real advantage in view of the enormous quantities of water used by the textile industry in general.

Following factors still make transfer printing an attractive proposal: High fashion apparels more depends on synthetic fibres, Responsibility for creativity is devolved to the paper supplier which is easier to make on paper, Non-skilled personnel can be used for transfer printing, risks of producing seconds quality in the case of difficult articles such as georgette, voile, satin drastically reduced.

5.13   Transfer printing on natural fibres

Even though Transfer printing is mainly used for synthetic material, trials have been done for printing on natural fibres. Since in case of most synthetic fibres the heating in the process brings about the fixation of the dyes also, hence it becomes a one-step process. But in case of natural fibres, it is generally possible to transfer the print paste fully, including chemicals for fixation on to the fabric and follow the steaming washing etc., which is not a very attractive process. However there are cases where a resin treatment is initially done on these fabric and the transfer printing done on to this resin layer has been successful but has not become very popular. Panel printing using melt fixing of resins is different than actual transfer printing.

6

Fixation and after treatment of prints

6.1   Drying

Drying is the first operation after printing. In all mechanised and automatic printing machines the drying is an integral part of the machine. Hence in normal printing the drying is also included. But in bock printing etc. normally drying is done as a separate operation or left on heated tables to dry the fabric. Drying has to be done immediately after the print to avoid the spreading of the prints and any marking off during handling. Drying can be done in any drying machines including cylinder dryer provided the printed side does not come into contact with the guide rollers or drying cylinders (cans). The rate of drying is also important especially if the fabric being printed has low moisture regain. If, for example, in block or screen printing nylon or polyester, the colours are seen to be bleeding slightly at the edges then the drying rate is too slow. Drying must be made more efficient by the provision of hot air blowers or by increasing the number or power of existing driers. The idea of replacing the thickening by one of much higher solid content can give rise to another problem. Such pastes gives a much more brittle films on the top of the printed portion and small specks of print paste are apt to jump away during handling. Because nylon and polyester, due to its low moisture regain, can develop static charge when hot the print paste specks become attached to other portions of the fabric. Here they become fixed during subsequent processing and greatly detract from the value and appearance of the print. Drying operation should not get into a baking process which is used for fixation of certain dyes like reactive, pigment etc. Normally, baking is a much slower process and if the full drying and baking process cannot be done in a drier attached to the printing machine and it can affect the printing speed badly. Partial baking is also advisable as the auxiliaries and chemical releasing certain fixation atmosphere may not be available in the actual baking stage. For example, in pigment printing certain acid liberating agents which releasing acid at higher temperature). When printing vat dyestuffs by the alkali carbonate-sodium sulphoxylate formaldehyde ('all in') process, as well as all discharge styles and resist printed styles, it is very important to dry the printed goods quickly and thoroughly if optimum yields are to be obtained. Premature decomposition of the sodium sulphoxylate formaldehyde occurs if improperly dried prints are stored. However if hard dried prints containing sodium sulphoxylate formaldehyde are stored decomposition will also occur due to heating-up. Consequently the procedure should be thorough and efficient drying followed by cooling if necessary and steaming.

Thus the temperature of the goods should not go above 1200C, as the case may be, to about 2.5% moisture regain in the case of cotton. The fabric is further cooled to absorb further moisture from the atmosphere, as it is advantageous to have maximum moisture for fixation. In fact, it is even increase moisture content by different methods like fine water spray to get better colour yield, which will be explained later under the fixation process.

6.2   Fixation

Fixation of prints are done by different the selection of which is based on the class of dye used, available of machinery etc. In printing the dyes along with fixation chemicals, if any is deposited as a film on the surface of the fabric. At this stage if it is washed, the dye will be washed off since the dye has not entered the fibre and got fixed. Thus fixation is the process by which the dyes are transferred to the fibre from the printed film. This process has to be done carefully otherwise one may not get the full colour value, spreading of the colour beyond the limits of the printed design, appearance of specks. Poor fastness properties, staining of the ground etc. We will discuss only the most common methods:

- Ageing
- Normal Steaming
- Curing/Baking,
- High Pressure Steaming
- High Temperature Steaming
- Flash Ageing
- Chemical Pad Development
- Thermosoling

6.2.1   Ageing

Ageing involves in treating the printed materials in steam at atmospheric pressure for long time helping in the diffusion of dye into the fibre and fixation. This can be done in many ways (a) Hanging in large ageing rooms (b) by passing them through an aging chamber continuously where the required conditions for ageing are maintained (c) a short steaming in a continuous steam ager in the presence of moisture or super-heated steam. The ageing rooms were of the past, not much practiced now, was of 10 – 12 m long with a series of wooden bars to hang the fabric in loops. Live steam pipes supply from pipes arranged above the floor made of wood or even concrete and the base of the walls. The steam supply is adjusted and the moisture inside is maintained by wet and dry bulb thermometers. As the printed fabric demand went higher this arrangement of ageing is done for 1-6 days in such rooms has become impractical and rapid agers were introduced.

These agers are iron chambers with top and bottom guide rollers for continuous movement of the fabric. Top is provided with a wooden hood with an exhaust for the spent steam. A steam pipe runs in to the wooden hood. Normally, the fabric enters and leaves at the same side (see fig below). Steam pipes provided at the entry (exit) prevents the water drops forming and falling

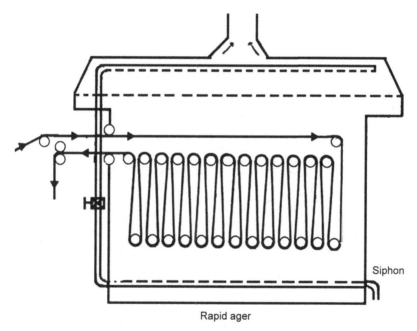

Rapid ager

on the fabric. Live steam at low pressure is supplied by a series of perforated pipes provided at the bottom of the chamber well below the bottom guide rollers. Condensed water formed falls down through the sides of the to the bottom of the chamber and drawn out from the bottom periodically. The temperature is maintained around 65-95°C and aging time is about 3 min at a speed of 50-60 m/m. The conditions of aging and time depend on the class of dye to be fixed. Such agers are not used much now due to many disadvantages like high consumption of steam, long heating up time required, high damages on fabric due to many reasons like condensed water falling on the fabric, uneven steaming etc.

6.2.2   Continuous steaming

This is one of the most common methods of fixation of prints, which is applicable to most of the dyes except Pigment and Disperse Dyes. Normal steaming refers to steaming treatments carried out at in an enclosed space at or only very slightly above atmospheric pressures with temperatures at 100°C (212° F) or a few degrees above this. Steam provides heat, moisture to support the fixing process which will be explained in detail below.

6.2.1   Steaming units

There are many types of steamers available in the market. Since there were many types of steamers available in the market they were divided in to two categories. Machines in which the fabric is treated with steam for a relatively long time (like 45 min to 2 hours) are called steamers, while machines in which fabrics are subjected to the action of moisture and heat for short time (from 1 min to 14 min) are called agers or steam-agers. Today the most common steamer today is the continuous steamer.

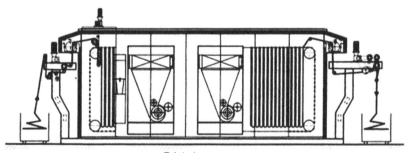

Print steamer

The steamer consists of a (jacketed in most cases) steaming chamber with moving guide rolls or pipe or sticks attached to an endless chain on both ends of the rolls and fabric entry system and delivery system. The rolls or sticks rotate all along the path and change the contact point with the fabric continuously to prevent any fixation defects in contacting points. The chamber can be fully enclosed or bottom open. Closed bottoms are preferable as this gives a better control of process variables (temperature, pressure, moisture, oxygen, steam circulation and exchange) and allows the creation of a slight pressurisation to prevent oxygen from entering.

The fabric (single web or double web as per the width of the fabric and machine) enters at one side of the chamber in full width in to the steam chamber on to a drive system by which the fabric is hung on the first roll to a predetermined length (this can be varied for varying the steaming time) after which the first roll moves to the next position and stop. Now, the fabric moving system hangs the fabric on to this roll to the same length as the first roll and stops, the second and first roll moves to the next position and stops. This process continues till the first roll reaches the delivery end of the chamber. At the end of the path, the fabric gets out of the steaming machine, while the sticks pass in the lower part of the machine and grip another piece of fabric at the entry of the steaming machine. Special inlet and outlet devices, together with a slight pressurisation, prevent the air from entering (max. O2 allowed = 0.3/1000 volume).These type of filling the steaming chamber with fabric is called festooning and hence the machine is also called festoon steamers. By

Loop formation inside a steamer

this time the fabric would have been steamed for a predetermined time adjusted by the speed of the rolls and the length of the fabric inside the chamber. The rods/rolls may be slowly rotated to avoid bar marks due to non-uniform accessibility to steam. With loops (festoons) of up to 5 m in length, long steaming times or high throughput velocities can be achieved A fabric content of 800 m allows an overall throughput speed of 80 m/min with a 10 min steaming time. At this point the plaiting device takes the fabric off from the guide roll and plaits it outside the chamber at the same high speed as at the entry point. The guide roll or pipe moves to the bottom of the chamber and comes up at the feeding position and the cycle continues. By this method the rolls are touching only the unprinted face of the fabric (fabric is fed into the machine in such a way) on the guide rolls and the fabric is freely hanging inside the chamber for the action of steam (steam medium at a temperature of 102-104°C and relative air humidity of 99.7%, the air content being not over 0.4 %) which is filled inside the chamber. The top of the chamber is mostly sopped roof form or sometimes flat. In order to prevent condensation, the ceiling plates are steam heated. To protect the fabric against accidental damage by drops of condensate formed at the entry of and delivery of the chamber they are heated with blind steam is provided and in some cases a heated glycerine chamber.

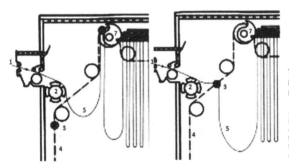

The Process of Loop formation inside the steaming chamber of a Loop (Festoon) Steamer

1. Fabric entry, 2- Peg for pushing the loop, 3- Rod or stick for hnging the loop, 4- Chain carrying the loop, 5- Loop forming, 6- Loop already formed, 7- Unit for disengaging the rod from the carrying chain

The capacity or the output of the machine can be increased by running two layers fabric preferably with a back grey in between the fabric. Since it is difficult to maintain uniform steaming conditions the chamber size is usually restricted than about 800 m fabric content. It can be used for knitted fabric also, but the festoon has to be probably made shorter.

Steam supply

Steam supply to process house is normally from the boiler house which is transported under pressure. This allows the use of smaller-diameter pipes to convey the substantial weights of water vapour from the boiler to the steamer and other steam-using machines. To avoid condensation while transportation the steam is heated further (super-heated steam) so that some heat is lost due to radiation still there is no condensation. This also allows the steam to be at higher temperatures even after reducing the pressure (increasing the volume) to feed for a particular requirement in the dye house.

The requirement of steam in a steamer as mentioned above is about 1000C with 99.7% RH. The above supply of steam may not suffice this requirements. Water vapour at 100 °C and standard atmospheric pressure is known as saturated steam, and as dry saturated steam if it contains no droplets of liquid water. At higher pressures water boils at a temperature above 100 °C. In such cases more heat is required to evaporate a given mass of water and the steam produced has a high temperature. For example, the temperature of saturated steam at 350 kPa above atmospheric pressure is 148 °C. Any cooling would produce condensation, and this is why steam at a certain pressure, and a temperature corresponding to the boiling point of water at that pressure, is known as saturated steam.

The steam supply from the boiler is fed to the steamer at atmospheric pressure the heat produced will evaporate any water content in the steam, and the temperature also may be higher. Usually such superheated steam is cooled by passing through water or by spraying with fine water droplets which will cool and the same time increase the moisture content of the steam, which is a must in steaming. In the lower part of the jackets, the steam, coming from the boiler and passing through the saturator, is caused to expand and boil in water. In this way, the saturated steam at atmospheric pressure raises and heats the walls and the ceiling of the jackets (preventing condensation drops from forming and dripping onto the fabrics, as a result avoiding possible defects). The steam lowers from the top of the steaming machine through the ceiling openings, drives the air away (air is heavier than steam) and fills the steaming machine.

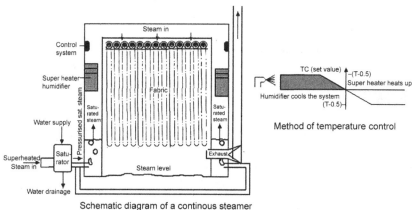

Schematic diagram of a continous steamer

In certain designs of steamers the steam is extracted by fans from the base of the steamer, passed through heatable radiators, sprayed with water as needed, and forced through ducts to the top of the steamer, whence it passes down the folds to the bottom. In this way festoon steamers are increasingly being produced as universal steamers, so that any temperature between 100 and 200 °C may be employed.

The equipment to control moisture and temperature of the steam feeding the steaming machine is positioned, usually in the jackets; the real-time control devices work interactively and start immediately some spray-water humidifiers, each one is cascade-connected with superheaters also assembled in the jackets. Thanks to this system all the variables can be controlled in realtime (the temperature difference allowed is ±0,5°C of preset values, and steam density between 96 and 98%)(see fig above) If necessary, the steam can be heated at temperatures of 170-200° Cat atmospheric pressure passing through the jackets.

Action of Steam in the fixation of dyestuffs

The fixation of dye on to a substrate in printing is also can be same as it is explained in dyeing. Only difference is, before steaming, the bulk of the dyestuff is held in a dried film of thickening agent in the printed areas (localised) and it has to be fixed without any spreading, which probably in acceptable in dyeing. During printing and drying operation of the printed fabric some portion of the dyestuff may have penetrated into the fibre but it may not have got fixed fully to get the final shade. Steam provides the liquid medium (water) and temperature to finish the fixing operation. During the steaming operation the printed areas absorb moisture and form a very concentrated dye bath from which dyeing of the fibre takes place. First, it swells the thickener film, without spreading the print. The concentrated 'dye bath' formed on the fibre exists more in the form of a gel than a solution and any tendency to bleed is restricted The dissolved or dispersed concentrated dye diffuses to the fibre surface, the substrate swells by absorbing the water present and the dye diffuses into the fibre. The temperature provided by the steam accelerates the process of diffusion. The auxiliaries present in the print paste supports all these actions, for example urea helps in quick moisture absorption, dissolution of the dyes. Dispersing agents helps in the dispersion the dye. Swelling agents helps in the swelling of the substrates etc. The function of the thickening agent during this period is to prevent the dyestuff from spreading outside the area originally printed, that is, to prevent 'bleeding'.

If the steam is too moist or the printing paste contains too large a quantity of hygroscopic agent—for example, glycerine, urea, etc.—then the film of thickening agent becomes so diluted that bleeding occurs. Alternatively, if the steam is too dry the film of thickening cannot absorb sufficient moisture and the dyestuff does not fix satisfactorily. In addition to promoting the formation of a localised dye bath the steam may have other functions demanding more exact control on the steamer.

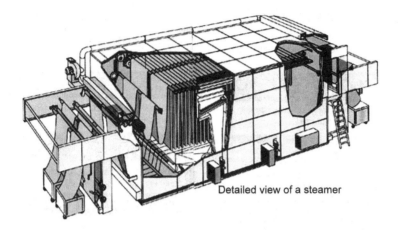

Detailed view of a steamer

Printed dyes are usually fixed by the steaming process, the steamer provides the moisture and rapid heating that enables the transfer of the dye molecules from the thickener film to the fiber within a certain time. The time and conditions for fixation in steam vary with the properties of the dyes and fibers used, which could vary between 10 s to 60 min in steam at 180 to 100 °C.

Once the steam condenses on the cold fabric it gives up its latent heat of 2260 kJ/ kg (539 cal /g), and raising the fabric temperature very quickly to 100 °C. Cotton has a specific heat of 1.4 kJ/ kg/ K (0.32 cal/ g/°C) and, starting at 20 °C and 7% regain, 5.5 g of steam will condense as liquid water on 100 g of fibre. there can be further heat generation in from the exothermic reactions occur along with this process (for e.g., dissolution of alkali is an exothermic reaction; dissolution of anhydrous potassium carbonate in a vat printing recipe is exothermic; the action of reducing agent is exothermic etc. whereas dissolution of urea is endothermic). Even though all the exothermic reactions try to increase the temperature of the fabric since the atmospheric temperature of steam inside the steamer will to allow it to go much above. The resultant temperature of the fabric is a total of all these together. Sometimes the temperature of the fabric goes much higher (110-1200C) than the steam temperature. The print recipe is formulated in such a way as to get the best atmosphere for fixation of the dyestuff to that substrate. Thus, the fabric temperature is raised to the temperature for the fixation process. At this stage the auxiliaries and chemicals present in the in the print paste is released (if it has to be released), pH of the fabric is also brought to the level required, at which the reaction of the dye and fabric takes place (It is not necessary that in cases a reaction takes place) with acid and direct dyestuffs, the steaming

1. Printed, dried and cooled fabric enters the steamer

2. Steam is condensed on the fabric and the fabric temperature goes up

3. Condensed moisture is absorbed by the fabric (Absorption phase)

4. An equilibrium is reached and the tempera- above the fixation temperature

5. The temperature of the steam is increased and moisture in steam reduced

6. Evaporation of moisture is reduced by urea fresh wet steam is circulated

7. Dye diffusion takes place due to the differe- ence dye concentration at different phases

8. An equlibrium has been reached and the dye fixation taked palce

Fixation stages of a print in the steamer with saturated steam

period is usually lengthy, but since only a dyeing operation (or transfer of dyestuff) is taking place. As mentioned earlier the steam should not be too moist due to reasons cited above and air contamination in the steam has to be avoided. Large a volume of cold air being drawn into the steamer may lead to condensation of water on the fabric and thereby cause the print to bleed and in the cases the reducing agents present in the print paste, like vat dyeing, will be lost due to the oxygen present in the air. But slight air contamination is not deleterious in many cases there is no reducing agent in the recipe. Steaming times of up to 1 h are common with these dyestuffs. In the case of reactive dyes there is a reaction with cellulose under alkaline conditions. The reaction takes place in a short time at the temperature in the steamer and hence it needs only a short steaming time where reaction with the fibre as opposed to a simple dyeing operation is involved. Reaction of cold brand dyes will happen in 15s, whereas in the case of hot brand dyes it can happen only in 10 min. In this case, type is unduly sensitive to the quantity of the steam and the presence of air in the steamer does not affect the fixation of dyes of this class. In the case of Vat dyes the process involves in many reactions and conditions of fixation are stringent and hence the steam quality is very important – abundant flow of air-free steam of reasonably high moisture content. Steaming times of 5—12 min are usually adequate, but in the special case of the flash-age process a steaming time of 20 s will suffice. The failure to fix vat dyestuffs or obtain white and illuminated discharges satisfactorily on a small scale can usually be traced to an insufficient supply of steam. When the printed patterns are introduced into a steamer, the air must be very rapidly displaced by an in rush of steam, or the fixation of the dyestuff will be seriously impaired. Air not displaced from the steamer rapidly attacks the sodium sulphoxylate formaldehyde in the printed areas at the high temperature reached during steaming, and decomposes it. Consequently the sodium sulphoxylate formaldehyde is not then available to fulfill its normal function of reducing the vat dye, and/or the dyed ground in discharge patterns, and thus fails to initiate the dyeing reaction which must take place during steaming. It will be appreciated that under such conditions the resulting prints will inevitably be unsatisfactory in appearance owing to inadequate and erratic fixation of the vat dyestuff or inadequate discharge of the dyed ground shade. Indigosol dyestuffs may be developed by a steaming treatment using the sulphocyanide printing process. In this case only a short steaming treatment is required but the temperature of the steamer must be at or near 100°C (212°F) to ensure that the acidic component necessary for development is liberated from one of the reagents added to the printing paste.

Detection of air in the steamer

We have explained the importance of air free steam for fixation process. The presence of small amounts of air in steamers can give rise to a number of problems. When printed or padded vat dyes are fixed by steaming, the oxygen uses up a portion of the reducing agent, either directly or indirectly. It can also happen that an oxygen-containing steam atmosphere can re-oxidise the newly-formed leuco dye, which is then prevented from being taken up by the fibre. It is, therefore, always advisable to ascertain the content of air in the steamer atmosphere under operational conditions when commissioning a new steamer, or when modifying a steamer for the fixation of vat dyes. Since oxygen can lead to the formation of oxycellulose, it may often be advisable to test for the air content of steamers that are primarily intended for other processes. Air can gain access to such steamers in dangerous quantities (a) not usually as entrained air in the goods (with the exception of heavy fabrics), but (b) through leaks in the housing in conjunction with unsuitably arranged steam inlet pipes, or (c) when the steam inlet pipes come too close to the inlet and outlet slits of the fabric, and the jets of steam are directed towards the interior of the steamer, thus drawing in air simultaneously. (d) Practically speaking, no air is introduced into the steam during steam generation. The following methods can be used for the qualitative and quantitative determination of air-content.

The steamer is warmed up and run at the normal speed with dry runners or goods. When the steamer test apparatus is prepared as is shown in the above diagram, the water flows out slowly from bottle 5 to tubing 6. The resulting reduced pressure in bottle 5 draws a portion of the steamer atmosphere into bottle 5 through the tubing assembly 4, 3, and 2.

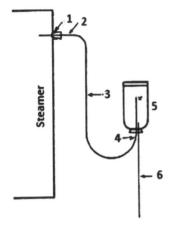

1. Thermometer aperture or drilled hole
2. 4 - 5-mm internal diameter copper tubing with airtight fitting in 1 by means of a rubber bung.
3. Rubber tubing that fits 2
4. 4- 5-mm internal diameter copper tubingwith air tight fit in bottle 5 by means of a rubber bung.
6. A 2-litre bottle filled with water
7. Copper tubing of 2-3-mm (max.) internal diameter and 350-500 mm long. with airtight fit.

The water-vapour component condenses in the cold water of bottle 5, while the entrained air escapes at "x" from tube 4 in the form of bubbles. The air is clearly visible and collects at the top of bottle 5. For the first 2-4 seconds of the test, the air in the tube assembly enters the bottle, and after this the stream of air bubbles usually diminishes. Tube 6 is at least 350 mm long, and its internal diameter should be kept fairly small, so as to prevent air striking back through tube 6 into the bottle.

In a quantitative calculation (given in Laboratory methods for Textile Processors by the same author), values of 0.2 to 0.3% air content or less should be aimed at.

Values lower than 0.01% are attainable We should like to point out the faults and disadvantages that can arise when the air content of a steamer is large:

1. air in the steam atmosphere always means a need for larger amounts of reducing agent;

2. if Hydrosulphite Conc. BASF is used, an excess of reducing agent can compensate for fairly small amounts of air in the steamer, and this is because, in the presence of excess of hydrosulphite,

    (a) the dye can be vatted by the hydrosulphite; and

    (b) it is only some of the excess hydrosulphite that is actually oxidised by atmospheric oxygen in a parallel, independent reaction, and hence the leuco compound is shielded from premature oxidation;

3. The other customary reducing agents usually employed in textile processing do not react as rapidly with atmospheric oxygen under

steaming conditions, and, in such cases, the oxygen in the steamer re-oxidises the newly-reduced dye before it has had a chance to be taken up by the fibre.

4. In vat dye printing, an air content above 0.4-0.5 %0 is already too much. With only 3 %0, the colour strength drops by as much as 30 %.

These per mille data relate to steamers which are usually employed in textile printing works. In these steamers, about 10-40 times as much steam flows through as is theoretically required for heating up the goods. If, however, only 3-5 times of the theoretically required volume of steam were to be passed through the steamer, higher air-content values would be permissible since, in fact, it is not the air content per mille that is responsible for the damaging effect, but the proportion by weight of air, or more precisely, of oxygen, to reducing agent.

Steam temperature	Valuem in l/kg steam
99.09°C	1725
100.00°C	1730
110.00°C	1779
120.00°C	1828
130.00°C	1877
140.00°C	1926
150.00°C	1975
160.00°C	2023
170.00°C	2071
180.00°C	2120
190.00°C	2167
200.00°C	2215
210.00°C	2263
220.00°C	2311

6.3   Machineries

6.3.1   Batch steamer

This steamer is used for batch operation. Batch wise steaming is a technique from the past, and can be found these days only in some places in China or India. But batch steamer shows advantages when colour yields are improved by steaming at above atmospheric pressure or for extended times.

1. Humidifier
2. Rope drum
3. Air evacuation
4. Safety valve
5. Steam chamber
6. Heating coil
7. Support
8. Cover
9. Cover swinging
    support
10. Star frame
11. Displacing carriage
12. Pulling system
    star frame

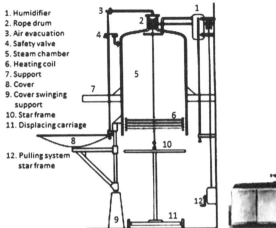

Section through a star steamer showing star frame   Actual star high pressure steamer installation

6.3.2   Curing/Baking

The conventional roller baker or curing oven (see picture) is an arrangement for carrying woven fabrics through recirculated hot air. Shorter times are required at higher temperature, but this could be risky for discolouration. There are some dyes which is fixed by dry heating instead of steaming. Such treatments are often called curing or baking. This treatment involves in heating the fabric This is an enclosed chamber filled with air heated up to 160°-180°C. Rollers at the top and bottom of the baker enable fabric to be run through the

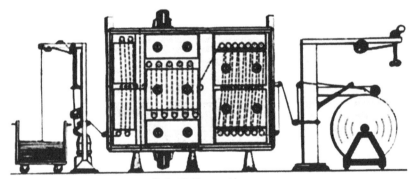

Curing/Baking machine

apparatus continuously over 2–5 min, treatment time. The dyes which can be fixed by curing process are pigment and reactive dyes. Since the chamber is filled with dry air there is least possibility of marking off, hence the fabric is not hung in the chamber as explained under steaming chambers, but it is threaded up and down via guide rolls. This is particularly suitable for some processes like pigment discharges on reactive grounds where you need one steaming process for discharges and then a curing operation for fixing the pigment dyes, both can be done in one machine one after other. But there are steamer cum curing machines the fabric is hung in the same way as in steamer, as usually only one type fabric transport will be available in such machines. There are some disadvantages in this type of machine due to no control over the width, fabric can shrink widthwise. Sometimes, in such cases the curing is done on stenters since the required width can be held inside the chamber with the help of clip or pins. The problem is the difficulty in giving the required time for fixation. Usually the temperature is raised to around 2000C, by which the reaction time can be reduced considerably in most cases. It has to be taken into consideration that it could take up to one min to bring up the fabric to a temperature close to that of the atmosphere around. So do not fix too short.

In case of pigment prints, hot air helps in releasing acid from the acid donors by raising the temperature of the printed fabric and reaches the pH required for the polymerisation of the monomer contained in the binder and ultimately fixes the pigment on the surface of the fabric. Hence these machines are also called polymeriser.

6.3.3   High pressure steaming

High pressure steaming is best employed for disperse dyes on polyester fabrics. Since the pressurisation of steam can be done in closed chambers, it is essentially a batch process.

The steaming vessel can be vertical or horizontal. In a vertical set up (a) fabric to be steamed is sandwiched between a back-grey (Bengaline) and wound on to a perforated metal cylinder which is loaded into the steam chamber. (See fig. below).Direction of steam flow is from outside-to-inside or vice versa, suitable for short runs of fabric, e.g. for small screen printing and hand printing operations; (b) High pressure Steaming Cottage : It is a cylindrical vessel which can withstand high pressure and about 3 m dia and 4 m length to suit the width of the material to be steamed with a provision for exhaust. Each steamer is provided with a carriage (wagon) with wooden rollers carried on the frame work of the carriage. Fabric is interleaved with grey to prevent marking off and arranged in loop form and suspended on the wooden rollers. The fabric and the carriage is run into the preheated ager on rails and closed air tight, and steam is turned on. The exhaust is opened for

Open end of a horizontal steamer

Sectional plan of a horizontal steamer

some time to drive away the air inside the vessel. The Exhaust is closed and the pressure is allowed to develop upto 30 psi. Fabric is steamed for 1 – 1 ½

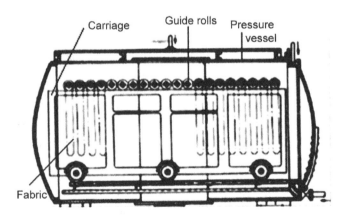

Sectional plan of a horizontal steamer with carriage

hours, steamer is turned off, exhaust is opened and the carriage is removed. (See fig. above) or (c) star steamers: The fabric is firmly hung by its selvages and is fixed to a special backing cloth (Bengaline). The whole is wound on a star-shaped carrier by means of hook-shaped pins. The steam chamber takes the form of a cylindrical pressure vessel, arranged vertically and closed at the top with a door that can be swung into position at its base. The carrier with the fabric is loaded into the steamer; steam is then forced into the heater removing the air from it. The cover is then closed and the pressure is kept constant by introducing or extracting the steam. Once the steaming process has been carried out, the cover is opened, the carrier is unloaded and the fabric is removed. Star steamers are preferred by screen printers and for sensitive fabrics, wool printing, etc.

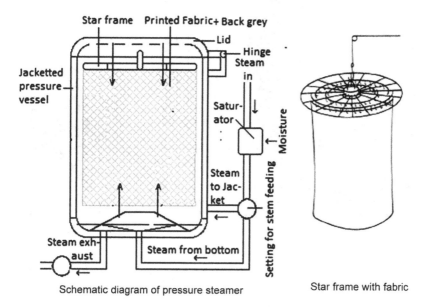

Schematic diagram of pressure steamer   Star frame with fabric

They are also called Star steamers because of the radiating arms used on the cloth carrying frame to wind the cloth in a spiral prior to inserting the frame in the steamer. (See fig below). Due to the poor steam exchange, this system is not suitable for discharge printed fabrics. On the contrary, it gives excellent results with synthetic fibres since it allows working at temperatures ranging between 130 and 135°C and pressures of 1.8 bar (i.e. with dry saturated steam and high temperature). This system gives a good colour rendering but possible head-tail defects (when forced incorrectly, the steam stratifies at different temperatures and with different moisture contents; therefore print dyes are fixed more consistently in the lower end of the fabric). This system is now rarely used since only small lots can be treated (max. 400 m long cloths are treated in 10-60 minutes) and many operators are required, thus entailing high costs.

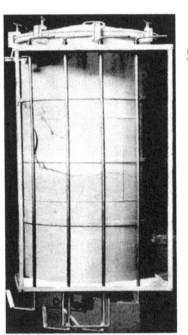

Highpressure batch steamer (Vertical)

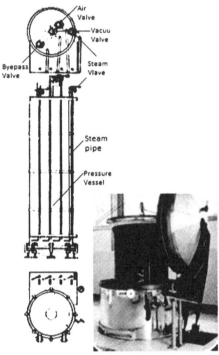

Side elevation   Star frame being taken out after
                                          steaming

In horizontal set up it fabric along with back grey is wound over a frame (see fig.) and put inside the pressure vessel like a beam dyeing vessel and locked to a steam/vacuum inlet. The other end of the frame is covered suitably so that vacuum can be applied safely.

The steamer is connected to vacuum by the central pipe, and by the by pass, and pressure reduced to about 30 mm. of mercury, when steam is introduced. The residual air is immediately swept out through the by-pass when the vacuum connections can be closed and steam pressure increased as desired. The use of steam under a pressure of 0-10—0-25 MN/m2 (1—2-5 atm) accelerates the fixation of disperse dyes on fabrics composed of polyester and cellulose triacetate. When the vapours from the print-colour must be removed, the central vacuum connection is left partly open to ensure the desired flow of steam through the batch. Production in these steamers are very low and it is needless to say, these units can be used for long steaming to the extent of 1-2 hours also.

6.3.4   High temperature steaming

The term high temperature steamers are used for units which are used for the treatment of dry prints in superheated steam at temperatures substantially above 100 °C and at atmospheric pressure. The use of high temperature steam to obtain rapid dye fixation at which is capable of operating, if required, up to 180°-200°C (355°-390°F) is there for a long time. A reactive dye which may take 5 min to get fixed using saturated steam at 1000C or dry heat at 1500C will take only 1 min. when fixed in steam at 1500C provided required amount of urea is used in the print recipe (see explanation below). The use of higher amount of urea makes the print paste costlier and increase of nitrogen content in effluent is a problem with this method. To contain these problems various alternatives are tried. Applying moisture on the fabric without affecting the print is a good alternative without any negative effect. Spraying moisture as fine spray or as foam has been used successfully. This extends the classes of dyes and fibres which may be processed by HT steaming when compared with pressure steaming. More recently, high temperature steamers working on a festoon system have been introduced by many manufacturers (Aioli, Stork). Superheated steam is advantages of faster heating, shorter fixation time and less colour spread; this is the case if the print has not been dried and also in the pad–steam situation, where there is usually more than sufficient water in the fabric. Another advantage of this system is that since the process is done at atmospheric pressure no pressure seals are required at the entry and exit of the material. Hence the steamer can operate continuously. In fact a universal steamer can be used with necessary heating arrangements as HT Steamer.

Action of high temperature steam

Suppose a printed, dried fabric is cooled to room temperature it may take up moisture up to 7-8% depends on the print coverage. When this fabric enters a steamer with superheated steam, the steam may condense on the fabric giving away it superheat. Fabric may get heated more but the condensed moisture will of the same as the saturated steam and its temperature quickly rises to 1000C. There are chances of the this moisture being heated up and evaporated but since the dissolution of urea present in the print being an endothermic reaction (240 kJ/kg , equivalent to 58 cal/g) this will be prevented. This shows the importance of extra amount of urea requirement in the print paste. Further chances of reduction of water content are prevented by urea forming a eutectic mixture with water, and hold the water without getting evaporated. Thus the temperature of the dye–fibre system therefore rises rapidlyto 100 °C, stays at that level (See fig. below) as long as the loss of heat by evaporation is high, and then rises towards the temperature of the steam. The dye dissolves in the eutectic mixture and the dye is transferred faster from this melt into the cloth. This enables the dye fixation to take place efficiently because the fibre

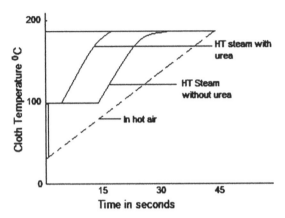

Rate of heating the fabric in hot air and super heated steam

is swollen and the diffusion of dye to ionised sites in the fibre can occur. But this action is not that efficient in the case of viscose and hence the colour yield is not as good as in the case of steaming with saturated steam at 100°C.

HT steam can also be used for the fixation of disperse dyes on synthetic fibres. Fixation can be done by HT steam at 1800C in 1 min compared to the same fixation by pressure steaming in 30 min at 1200C or in 1 min by dry heat at 2000C. This possibility allows fixation in normal festoon steamers with a super heating provision for polyester also. In this case, use of urea may be advantages as it helps in swelling the fibre and forming colour lakes but many disadvantages like fixation of thickener and causes undesirable build-up of deposits in the steaming equipment prevents its usage. 'Fixation accelerators', typically non-ionic surface-active agents of high boiling point and low water solubility, in which disperse dyes are soluble at high temperatures are found more useful.

Even though, the process theoretically happens as described above, the steam atmosphere inside should be accordingly available at all areas. Let us suppose the steamer is full of steam at 1800C. as the fabric enters the steamer there will be movement of the steam from one area to the other. To study this let us divide the steamer in to 4 zones. In zone I (near to the entry) as the fabric enters it is immediately heated to 100°C as the superheated steam condenses on to the fabric and remains at this temperature for 30-45 secs (zone II). and enters the Zone III.

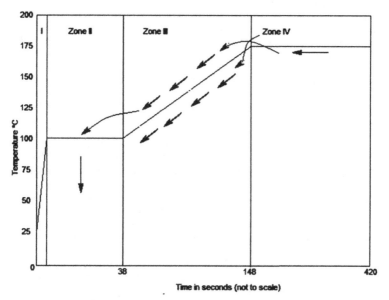

Super heated steaming orofile inside the steamer

Here the fabric temperature rises as the excess condensed water starts evaporating absorbing heat from steam. The fabric temperature reaches 175-1800C, equilibrium with the steam temperature and remains at this temperature for a preset time say, 7 min (zone IV). Maximum energy is required for heating up the fabric and evaporating water which is in zones I,II and III. The heat is absorbed enormously from the steam at zone I and hence the temperature of steam drops down in this area considerably. In zone II and III also temperature reduction occurs and due to this differences in static pressure also happens in these zones (to differences in the specific heat of the steam at different temperatures). Naturally, the steam moves from fixation area to the colder area at the entry zone and back in the order of magnitude of 2.7 m/s and resulting in a large temperature difference in the fixation zone (zone IV) from top to bottom. The machine manufacturers include devices to minimise this difference by automatic recirculation of the steam in efficient steamers and a processor can also adopt following precautionary measures to get better results:

1. One way is to reduce the loop length so that the fresh steam from lower areas immediately occupies the area where the temperature of

steam is reduced. But it reduced the fabric content inside the steamer and hence the output is accordingly reduced.

2. Another way is to increase the fresh steam input and used up steam is released through the chimney. This means an undesirable large energy consumption. If fresh energy of 1 kg is supplied at 1800C and the used up steam of 1 kg at 1050C is discharged

   Consumed energy for heating up of fabric = $(108-105) \times 0.46 = 34.5$ Kcal/kg

   Energy lost through exhaust = 642 Kcal/kg

3. Efficient recirculation and reconditioning of the steam is the best way to save energy and at the same time maintaining the required conditions of steam at the various stages inside the steamer. Thus recirculation is absolutely necessary for ensuring quality steaming and economy of the process. Hence it is very important to understand the steam circulation system and ascertain how efficient it is, before deciding on a particular model of steamer.

The steam can be heated to higher temperature using heat exchangers fed by high pressure steam, gas or oil or electrical heating. Usually temperature up to 1800C can be achieved. The use of Universal steamers is more common these days which are possible to be used as a normal steamer, a curing machine and an HT steamer covering the fixation of all the printed materials using one machine. It is possible to work tension-free so they are especially suited for processing tension-sensitive knitwear. Universal steamers are available with various fabric content like 150-400 m with steam consumption of about 500 kg/hr for 150 m fabric content and about 1000 kg/hr for 400 m content. The loop length varies from 3 m – 5 m.

High temperature steaming with superheated steam has the following advantages:

1. The production rates can be high in case of continuous HT steaming unit. When a loop steamer can produce about 25000m/12 hour a HT steamer can produce about 40000 m /12 hour.

2. Operating cost are much less due to higher productivity and lower consumption of steam.

3. Smaller floor space and less manpower requirement.

4. Better colour yield and higher fastness rating are possible with this machine.

5. Reactive dyes can be fixed by single phase (all-in-one) process.

6. Since the process in atmospheric the steaming units can be designed for continuous production with multilayer feeding or multi end feeding according to the width of the material.

6.3.5   Flash agers (Rapid ager)

This "Flash ager" is mainly used for steaming vat printed fabric by the two phase method. In comparison to the "all in" process, (Print paste contains all chemicals needed like alkali, urea etc. to be fixed by the normal steaming process) in the "two phase" steaming process, the print paste contains only the dyestuff and thickener. Since for this process no urea is needed, it is environmental friendly and it is used in some European printing mills as an alternative to pre moistening the fabric in front of a normal steamer. The steamer has on the bottom a special pad mangle composed by two horizontal cylinders, designed on purpose for two phase treatments. A chrome plated cylinder works on the face of the fabric, while a visco-elastic rubber cylinder works on the back.

Flash-ager can also conveniently used to overcome printing difficulties caused by the very low stability of highly reactive dyes. In this method the fabric is first printed with the dye and thickener. Immediately after passing through padding liquor containing a concentrated solution of caustic soda and sodium silicate, it is steamed in a chamber at for 45-60 seconds. Unfortunately this method suffers from the disadvantage that large quantities of caustic soda and sodium silicate are required, which produces large quantities of alkaline effluent. On the plus side however the process uses much less urea. Sometimes a mini steamer on the basis of rapid ager principle is used (See fig below)

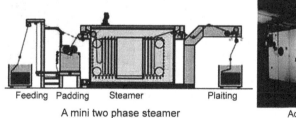 

| Feeding Padding   Steamer   Plaiting |
| A mini two phase steamer | Actual steamer |

This has become a common method for producing high quality prints of vat dyes and is basically a two-stage process. The print paste, consisting of the vat dye (in its insoluble 'parent' form) dispersed in a thickening agent, is applied to the fabric and dried. The fabric is then padded in an alkaline bath of sodium dithionite to reduce the dye in situ, and steamed without drying. The steaming time required is only about 20-40 seconds at 120-1400C, so the steam consumption is economical and the machinery comparatively simple. Fabric is fed in to the steamer in such a way as there is no contact with the face side of the printed fabric during its passage through the steamer in order to prevent marking off. Thickener combinations which coagulate in the alkaline fixation liquor are employed. After treatment like oxidation and soaping etc. are followed. Flash ager is also used in reactive printing where the use of urea can be avoided. Another advantage is that since the dye is applied without the presence of a reducing agent, the dried print is stable and can be stored if necessary before steaming. With a careful selection of vat dyes, prints of excellent definition and possessing a high depth of colour can be obtained.

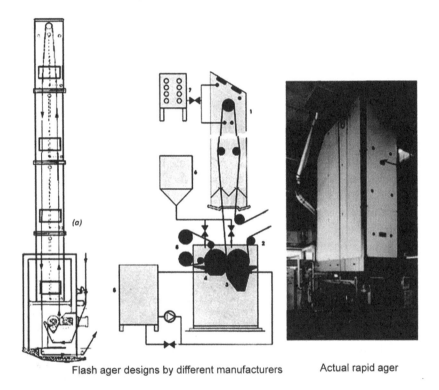

(a)

Flash ager designs by different manufacturers   Actual rapid ager

Flash ager is used mainly for two phase printing where in first stage usually only dyes are printed. In the second phase the fixation chemicals are padded and steamed. For example, in case of reactive dyes printing in the first phase the reactive dyes are printed without fixation chemicals and dried. Phase 2: the printed fabric is padded with an alkali solution containing an addition of electrolyte and steamed. The method has advantages of simple print paste recipe and less hydrolysis of reactive dyes. The other advantage is of low investment of machines and much less consumption of steam due to the small size of the machine. Disadvantage of higher chemical usage of fixation chemicals and hence not environmental pollution as these chemicals padded on the fabric has to be washed offer neutralised during aft treatments.

Steaming time required for fixation various substrates and dyes

Dyestuff Class	Fibre	Material	Fixation medium	Time
Cold Reactive Dyes	Cotton	Woven Or Knitted	Saturated steam	15s - 1 min
Vat Dyes	Cotton	Woven Or Knitted	Saturated steam	30s - 1 min
Sulphur Dyes	Cotton	Woven Or Knitted	Saturated steam	30s - 1 min
Direct Dyes Light Shades	Cotton	Woven Or Knitted	Saturated steam	1 - 2 min
Direct Dyes Dark Shades	Cotton	Woven Or Knitted	Saturated steam	2 - 3 min
Warm Reactive Dyes	Cotton	Woven Or Knitted	Saturated steam	1 - 2 min
Hot Reactive Dyes	Cotton	Woven Or Knitted	Saturated steam	5 - 10 min
Pigments	Cotton	Woven Or Knitted	Hot air (155 C)	4 - 5 min.
Reactive	Viscose	Woven Or Knitted	Saturated steam	10-12 min
Vat Dyes	Viscose	Woven Or Knitted	Saturated steam	10-12 min
Acid/Disperse Dyes	Polyamides	Woven (Liquor application 100%)	Saturated steam	2 - 5 min
Acid dyes	Silk/wool	Woven	Saturated steam	30 min
Discharge prints	Silk/wool	Woven	Saturated steam	15 min
Acid/Disperse Dyes	Polyamides	Carpet (Liquor application 200 -400%)	Saturated steam	5 - 10 min
Basic Dyes	Polyacrilonitile	Woven Pile Fabrics	Saturated steam	5 - 15 min
Disperse Dyes	Polyester	Woven Or Knitted	Super-heated steam 200-220°C	15 s - 2 min
Disperse Dyes	Polyester	Woven Or Knitted	Super-heated steam 170-180°C	3 - 15 min

6.3.6   Importance of moisture in steaming

Three things are necessary to fix a soluble dye:

- dissolved dye
- solvent (water / steam)
- textile fibre

Printing is technically a dyeing process where the dye is exhausted from the dye liquor and fixed on the fibre. Finally, the unfixed dyes are washed off. This difference is in true dyeing is that the material is dyed all over with the same shade but in printing the dyeing takes place in different shades and depth at various places of the material at the same time. The dyeing is happening at ultra low liquor ratio.

The dye has to be dissolved in the solvent, water. The "liquor ratio", i.e. proportion of dissolved dye to water, being very low in many cases where the solubility is limited, one needs even a dissolving asst. to help in the dissolution.

These processes start during drying of the printed goods. In printing the print paste is usually deposited at different areas of the material (unless it is a blotch print or print with 100% coverage) and hence the material is unevenly wetted by the print paste. After printing the material has to be dried to avoid marking off in the next process or handling.

In an effort to fully dry the printed portions to prevent marking off the unprinted areas and the back of the printed fabric are overdried in most cases. When the printed cooled fabric runs into the steam atmosphere, steam condenses instantaneously on its surface. Bone-dry textile material picks up about 4 % moisture from saturated steam. As steaming proceeds the textile material absorbs more moisture. The amount is determined by the equilibrium moisture content and depends on temperature and material. In adjusting to its equilibrium moisture content the textile material draws the moisture it needs from the steam.

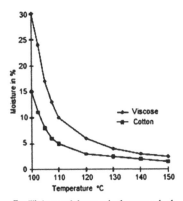

Equilibrium moisture content as a content as a function of the degree of overheating

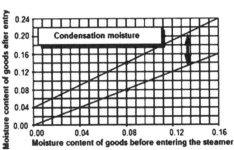

Moisture content of the goods after entering the steamer as a function of the moisture content on entry

The moisture withdrawn leaves the heat of evaporation behind in the steamer. The steam temperature rises and less steam is left to dissolve the fixing of the dye. Another function of the moisture is to swell the fibre so that the dye molecule into the fibre. The urea in the dried print paste starts to dissolve and cools the steam. This endothermic behaviour of the urea dissolving in the condensed steam is responsible for the amount of moisture which further condenses on the substrate. Even with superheated steam, therefore, the desired amount of moisture can be applied to the textile material

If the textile material is wetted before running into the steamer, more moisture of condensation is formed and in this way the material's moisture content after the temperature rise in the steamer due to condensation can be controlled. If the equilibrium moisture content of a given material is achieved via moisture on entry + condensation moisture, an increase of temperature inside the steamer will be prevented. In this way printing can be performed partly or completely without consumption of energy to dissolve urea. With some dyes (e.g. reactive turquoise) the requisite "liquor ratio" can be increased by adding urea and in such cases fixation is improved still further. The amount of moisture of condensation depends on the specific heat of the textile material. And this specific heat in turn is influenced by the moisture content.

Specific heat of cotton (kcal/kg °C)

Moisture content (%)	Specific Heat (Sp. Heat of substrate + Fabric moisture x Sp. Heat of water
0	0.24
10	0.34
15	0.38
19	0.43
28	0.52
36	0.6

Heat balance-sheet studies confirm the practical finding that in the final analysis the moisture content of a material before entry into a steamer determines the colour yield obtained.

Advantages with moistening
Environmental features
• marked reduction in nitrogen content of effluents
• no problems with exhaust air (no "smoke flags")
• no contamination of the dryer

Technical features
• improved levelling properties
• increased brilliance
• better additions in raster prints

- prints are more realistic, show more depth
- outlines sometimes appear more crisp
- reduced paste consumption owing to lower specific weight

Economic features
- yield increase up to 10-20%
- cost savings from urea reduction
- lower cost for energy
- productivity increase due to shorter steaming time
- reduced water consumption in washing off through reduced staining of the wash liquor

6.3.7   Steaming of urea free printing

In textile printing, the (one-phase) fixation of reactive dyes requires the addition of urea into the print paste. 50-100 g/kg for cotton, and 200 g/kg for viscose, in addition viscose fabric is often pre-padded with urea. Urea functioned as moisture supplier, to get a full fixation of the dyes during steaming. After steaming the urea is washed out and discharged with the waste water. Urea contains 47% nitrogen. In order to reduce the compounds of nitrogen in the waste water and therefore in the rivers, some countries restricted the amount of urea, some others banned it completely. During normal steaming (101-102°C) the temperature of the cloth and the steam will increase quite a bit (superheat). This phenomenon has a negative influence on the fixation of the reactive dyestuff. Urea is added to the printing paste in order to eliminate this effect. An environmentally friendly alternative to the use of urea is the moistening of the cloth before it enters the steamer.

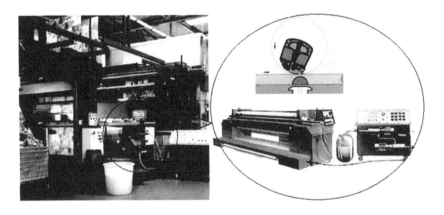

Several technologies to apply moisture on the fabric before steaming are available. One of the common methods is based on foam technology. M/S Stork supply units based on this technology. In this system (FP IIC) foam process or water and air mixed into foam. The foam is then applied with the aid of a screen and a closed squeegee. On the bases of the set process parameters (cloth weight, width and desired moisture content of the cloth) the microprocessor controls the percentage of moisture to be applied, independent of the web speed. Urea is thus replaced by water.

There is one foam generator and one rotary screen for applying the foam. The starting liquor is mixed with air, through which foam is created. This foam is then applied to the goods using rotary screen and a squeegee.

In another system spray application is used where the pickup is very low but without jet nozzles. The system consists of a rotor support, a hydraulic control unit, liquor container and centrifugal pump (See fig below). The liquor is atomised and sprayed with the help of rotors running at 500 rpm or more. It

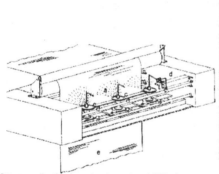

Spray application by rotor dampener (a) Mosture spray, (b) Shutter, (c) Rotors, (d) Sparay chambers, (e) Fabric

Actual rotor dampener attatched in front of a steamer

sprays a part of the liquor as very flat spray in the form of tiny droplets as the fabric moves as vertical sheet. Each spray width is about 100–400 mm. The staggered sprays and overlapping zones result in a spray width of 180–280 mm for approximately 800 ml/m/min. Higher application levels are possible by having several rotor supports, either on one or both sides. The wet absorption depends on the speed, and can be 0–25%, allowing for the product weight. The liquor in the dispensers can be monitored so that they can be topped up if necessary. The residual moisture on the fabric can be measured by means of Microwave moisture or isotopic radiation measurement.

7

Styles of printing

A printer has the task of achieving a print on the substrate same as or nearest to the original print effect given by the customer. He has to choose from different styles of printing to enable him to get the desired. Styles of print are techniques grouped together considering the process involved in a Printing. It can be grouped into mainly three groups. They are called Styles of printing:

- Direct style
- Discharge style and
- Resist styles.

The dyes used may be as per the substrate to be printed, substrates may different but each of these printing styles will have some processes in common. Each style will have the means and methods of applying the print paste on the substrate, and fixation of the print have a common route. The print effect of particular deign may be produced one, two or even all the methods. But under the given circumstances and the availability of the machines the printer will have to select the style which he feels the best suited for the print he has to produce.

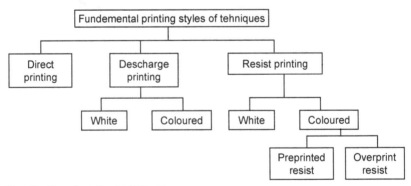

Classification of printing techniques

The above is a broad classification, and each of these styles can again be subdivided into minor styles depending on the dyes used. There can be modifications in the process, or the chemical reactions they bring about may be different but the main operations remain the same. Once the style of print is decided, all the dyes that are capable of being applied and fixed in the same general way may be printed together to get the different colours in a design. The main classifications of printing techniques are given above.

7.1   Direct styles

This is the most the most frequently employed method of textile printing in which print pastes are applied to white or dyed fabrics without destroying the ground colour. It involves in transferring the print paste along with the chemicals involved for fixing by any method of printing and fixed by steaming or any suitable fixing method. This method involves the following steps: printing, drying, steaming and washing. This type of printing is generally used for white or dyed cloths (usually dyed in pastel shades), by applying the sequence of all the colours, until the original pattWern has been reproduced. When printed on white ground the shade produced may be of the same shade as the print shade. But in case of printing on a dyed ground (called over printing) the shade produced may be a resultant shade of mixing the printed shade and the ground shade and such over-printing are called 'fall-on' effects.

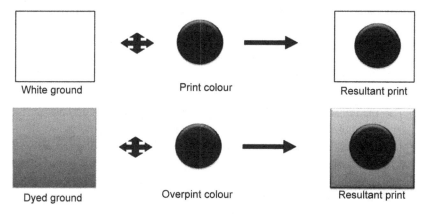

| White ground | Print colour | Resultant print |

| Dyed ground | Overpint colour | Resultant print |

This is possible when the ground colour is not too dark. Tone-on-tone effects, which use colours of similar hue, are often produced in this way, but contrasting colour combinations are also possible. For example, yellow dyes may be printed on to a red/pink dyed fabric to obtain orange areas. Overprinting can have effect on the printed colour and the ground colour also (see above)

Many modern techniques have made the printer enable to produce practically many more designs which was not been able to produce earlier by this method. In this method the print is made in a single operation, other than the fixation and washing processes. It can be used with all the main colour classes of dyes and on fabrics produced with any kind of fibre (some problems may only arise with blends). Pigment printing done by direct printing can be fixed by curing process and without any further after treatment the fabric can be finished. In some cases the ground can be dyed after printing also. Since the process involved is very short and the print paste contains only the necessary for the fixation of the dyes involved this printing technique is the most cost effective one.

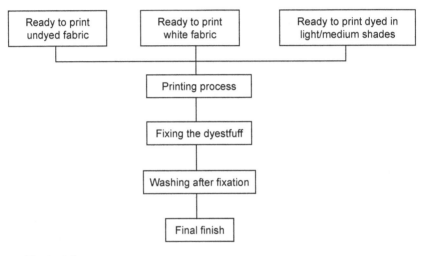

Direct printing sequence

The technical limits of this printing method appear with endless design patterns particularly those obtained with screen printing methods, while no problems occur for roller printing). Some problems may also arise when printing on backgrounds dyed with pastel shades: in fact, this could create problems on several areas of the design to be printed in light shades, thus limiting the number of reproducible pattern variants.

7.1.1   Four-colour printing

In four colour printing process of direct printing can produce all the shades of a print using four colours. In this primary colours (magenta, yellow and cyan, plus black) are used. The different shades are obtained by applying dots of the primary colours in variable densities: this technique also takes advantage of the ability of the eye to combine colours when observing them from a certain distance. Design patterns with different hues and tones can be obtained by using only four printing plates. This method however limits significantly the possibility of pattern varying. This technique is used only for fixed patterns and pure saturated colours cannot be reproduced.

7.2   Discharge styles

Discharge style printing, also called extract printing, is one in which an already present dye is locally destroyed to white or lighter shade (of a dark coloured ground) by the application of print paste during subsequent fixation. (white discharge) If a coloured pattern is to be produced on a coloured base, chemicals which break down the base and discharge-resistant dyestuffs are added to the print paste (coloured discharge). These colours are called illuminating colours. By this method we can print any shade on even the dark dyed grounds, which is a limitation in the direct style printing. Discharge printing has been around for decades. In the early years of discharge printing, the finished discharge print needed to be steamed during the drying process. This discouraged the use of discharge systems in the finished garment arena. The newly developed discharge ink systems are chemically reactive and do not need to be steam-neutralised. This advancement in the process opened the door to discharge printing for the average screen printer. Discharge printing has the ability to make bright, opaque colours on dark fabrics with a soft hand. Years ago, the idea of opaque colours on dark fabrics and soft hand could not coexist.

Many attractive effects may be produced by discharging suitable dyed ground shades, but dyed ground have to be carefully selected to avoid disappointments and the process has to be carefully followed to get the desired effect. Thus, these styles have always required the printer to exercise a great degree of process control and use a good deal of skill and initiative in choice of dyes and printing recipes.

7.2.1   Little history

Indigo dyeing is a long established vat dyestuff and we can find records as old as 1826 about the details of the methods to discharge cotton dyed with Indigo. Process of oxidation was known using products such as sodium bichromate (in conjunction with oxalic acid) or sodium chlorate plus sodium ferricyanide (in conjunction with caustic soda) for oxidation process. This process often attacked the base fabric while oxidation process to destroy the dyes. This process was used for long time but efforts were on for going for reduction discharge process which was successful only after discovery of new compounds, especially hydrosulphites. The process of oxidation discharge of Indigo using chromate is as follows:

Recipe

Quantity	Unit	Addition
810	g	British Gum Paste
160	g	Sodium Bichromate
30	g	Turpentine
1000	g	

Print on an Indigo dyed material, dry and pass through hot bath containing

Quantity	Unit	Additions
60	g	Sulphuric acid 1680Tw
20-30	g	Oxalic acid
920-910	g	Water
1000	g	

With a dwell tine of 30s, and finally washed and dried.

Chromic acid is liberated and the indigo in the printed areas is oxidised to isatin which is soluble in alkali and can be washed off in an alkaline bath. The oxalic acid reduces any excess chromic acid which might be formed and which would otherwise gradually discharge the ground colour. Wherever the dichromate was printed one will get a white design. The problem in this method was the adverse action of the oxidising agent and acid, which will tender the substrate (cotton).

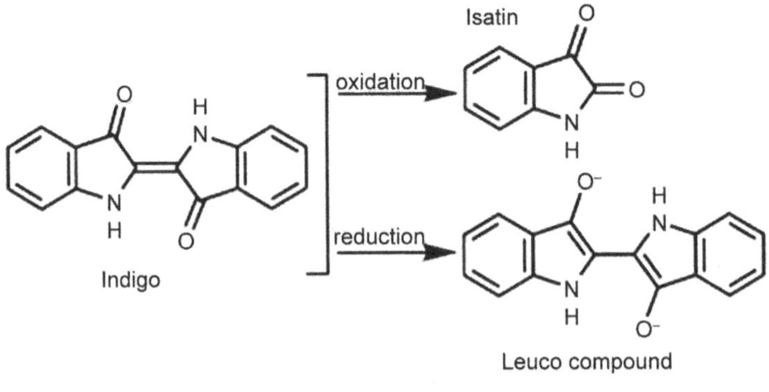

Indigo

Isatin

Leuco compound

For the same dye for a reduction discharge print paste containing stabilised reducing agent, sodium carbonate, anthraquinone and Luecotrope W (see equation below) is printed on a dyed ground, dried and steamed.

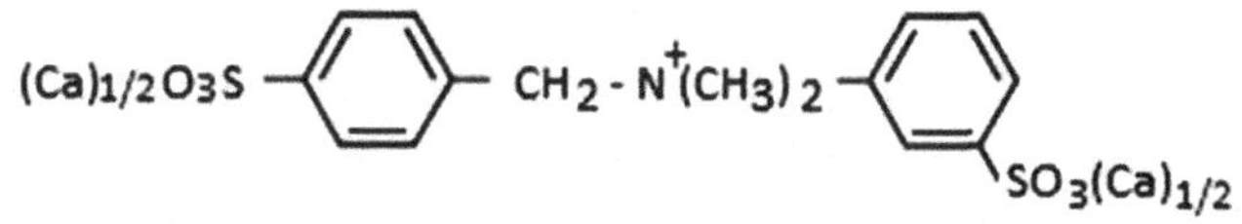

Leucotrope W

During steaming, reduction of indigo takes place to its leuco form which combines with Leucotrope W to form an alkaline soluble orange compound and which can be washed off to give a white discharge. Addition of Zinc Oxide or Titanium dioxide can give a white pigmentation effect of the discharged white. If another compound Leucotrope O is used instead of Leucotrope W, during steaming different orange compound (see below) is formed which gets fixed on the material and cannot be washed off. Thus a coloured discharge is formed without adding an illuminant dye.

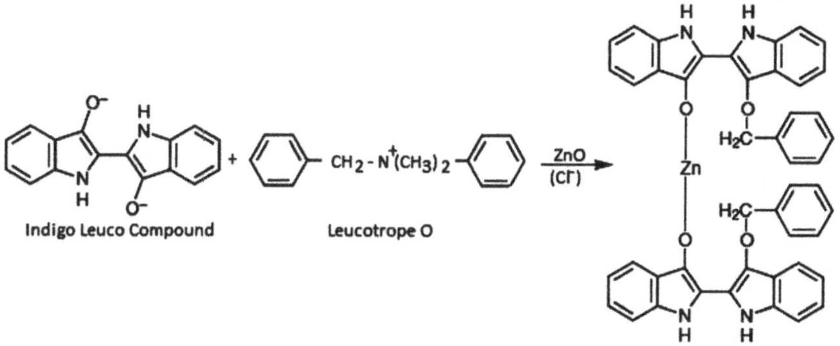

Indigo Leuco Compound   Leucotrope O

Substantive Orange Compound

7.2.2   Chlorate discharge

Oxidation discharge on Indigo using chlorate also was used to be practiced in olden days. Chlorate being very strong oxidising agent, the process is used only for white discharges:

Quantity	Unit	Additions
250	g	British Gum Powder
470	g	Water
200	g	Sodium chlorate
		Boil altogether and cool to 300C and add
50	g	Sodium Ferricyanide
30	g	Ammonium citrate
1000	g	

Print, dry and steamed for 1-3 minutes in the rapid ager with dry steam at 95°-100° C.; then washed in water for 2-4 minutes at 60°-80° C, and, if they contain no yellow, passed in the open width through caustic soda at 4° Tw., or through a bath composed of equal parts of caustic soda and silicate of soda.

7.2.3   Prussiate discharge

This process is founded on the oxidising action of red prussiate of potash [K3FeCN)6] in presence of caustic soda. It is only applicable to light and medium shades of Indigo, as " red prussiate " cannot be kept in solution at a strength sufficient to discharge dark shades. The cloth is printed with thickened solution of "red prussiate," and, after drying, is passed through warm bath of caustic soda, then well washed and, if necessary, soaped.

Quantity	Unit	Additions
200	g	Potassium Ferricyanide
500	g	Water
60	g	Starch
240		4% Tragacanth thickening
1000	g	

7.2.4   Nitrate discharge

Nitrate discharge was another oxidation discharge on Indigo ground was being practiced in olden days. Print with the following paste:

Quantity	Unit	Additions
275	g	Sodium Nitrate
50	g	Sodium Nitrite
675	g	British gum thickening
1000		

Dry, and pass through sulphuric acid 1000 T\v. at 75° C. and soap.

With printers well aware of the wide variety of methods tried as discharges for indigo, it is not surprising that this previous work formed the basis of tests on new dyes that were emerging from dye manufacturers in the early part of the 20th century. Oxidation discharges proved to be of little or no value on dyes other than indigo. On the other hand, reduction discharges were very widely applicable.

7.2.4   Principle behind the discharge prints

The principle of discharge and chemical reactions can be explained discharge printing on a fabric dyed with a synthetic dye containing azo (-N=N-) group as the chromophore (colour conferring). During the reductive discharge, the azo bond is broken into two chemicals which are colourless.

$$A - N = N - B \quad \xrightleftharpoons{4[H]} \quad H_2N - A + B - NH_2$$

Illuminating colours are often vat dyes. This has got twofold advantage, where during steaming, the discharge of the ground and the reducing agent generates the leuco compound that dyes the cotton. Air oxidation then regenerates the parent pigment.

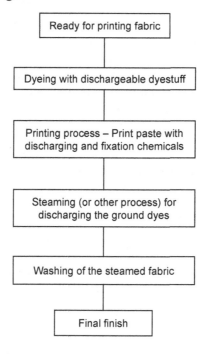

White discharge process

The most common reducing agents are the various salts of formaldehyde sulphoxylate ($HOCH_2SO_2^-$). On heating, these decompose liberating formaldehyde and the sulphoxylate ion (HSO_2^-), a powerful reducing agent. Various researches on the reactions of formaldehyde sulphoxylates as reducing agents and on the effect of different conditions on the resulting prints have proved the most probable are the ones shown below:

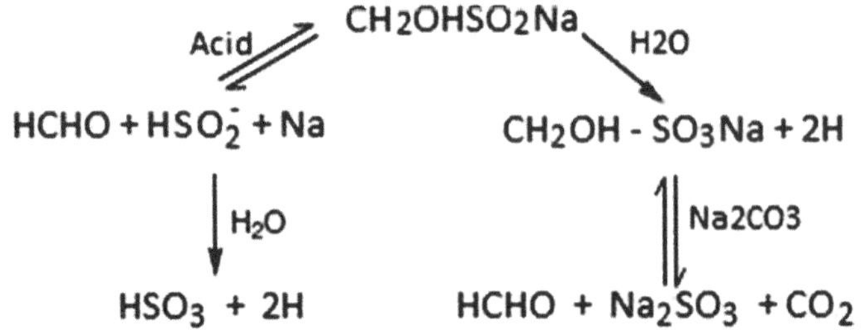

$$HCHO + HSO_2^- + Na \xrightarrow{Acid} CH_2OHSO_2Na \xrightarrow{H_2O} CH_2OH - SO_3Na + 2H$$

$$\downarrow H_2O$$

$$\qquad\qquad\qquad\qquad\qquad\qquad\qquad\qquad \updownarrow Na2CO3$$

$$HSO_3 + 2H \qquad\qquad\qquad HCHO + Na_2SO_3 + CO_2$$

$$A - N = N - A' + 4H \longrightarrow A - NH_2 + A' - NH_2$$

Reactions of Formaldehyde Sulphoxylate in reduction

Thiourea is sold as Manofast, has been used as a reducing agent in certain sectors. The decomposition of thiourea dioxide also generates sulphoxylate. Although chemically inert to many reagents, an irreversible rearrangement takes place when it is heated with alkali and water with the formation of formamidine sulphinic acid. On heating, decomposition takes place to urea and sulphoxylic acid. Sulphoxylic acid is an active reducing agent

$$(NH_2)_2C = SO_2 \rightleftharpoons NH_2C(=NH) - SO_2H$$

$$NH_2C(=NH)SO_2H + H_2O \rightleftharpoons (NH_2)_2C=O + H_2SO_2$$

Another reducing agent, which has been used since the earliest times, is tin(II) chloride. It is a readily soluble compound which reacts with an azo dye as shown below. This chemical has some disadvantages like, tin(II) chloride solutions are used quickly since hydrolysis, which gives a turbid solution, occurs on standing. Also the hydrochloric acid which is formed as a by-product can attack the equipments especially steamers.

$$A - N = N - A' + SnCl_2 + H_2O \longrightarrow A - NH_2 + A'NH_2 + 2SnO_2 + 4HCL$$

Today vinyl sulphone reactive dyes are more used as the ground shades which are easy to be discharged. The reduction can attack at two places of a vinyl sulphone dye. The reduction of vinyl sulphones are as follows:

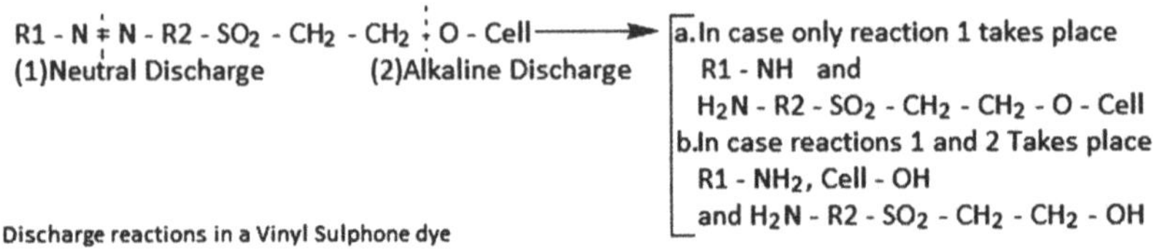

R1 - N ≑ N - R2 - SO₂ - CH₂ - CH₂ ⋮ O - Cell ⟶ a. In case only reaction 1 takes place
(1)Neutral Discharge   (2)Alkaline Discharge   R1 - NH and
                                                          H₂N - R2 - SO₂ - CH₂ - CH₂ - O - Cell
                                                         b. In case reactions 1 and 2 Takes place
                                                          R1 - NH₂, Cell - OH
                                                          and H₂N - R2 - SO₂ - CH₂ - CH₂ - OH

Discharge reactions in a Vinyl Sulphone dye

The insoluble zinc and calcium formaldehyde-sulphoxylates and thiourea dioxide are useful in printing fabrics made of synthetic fibres. Their low

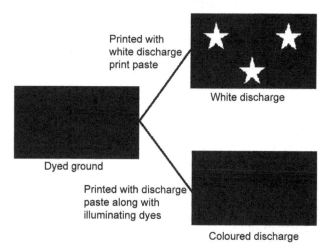

water- solubility minimises capillary flow of reducing agent solution along the non-absorbent fibre surfaces. Such flow reduces print definition and can produce a coloured halo effect. Stannous chloride is also an effective reducing agent for azo dyes. It has the advantage of being a somewhat weaker reducing agent than the sulphoxylates so that it has less effect on illuminating dyes. These are often not completely inert towards strong reducing agents and their colour intensity will be greater when using a less powerful discharging agent.

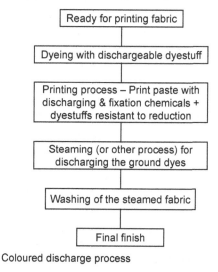

7.2.5   Dischargeable dyes and ground dyeing

To make an effective discharge print the dye used for ground dyeing should be dischargeable. If discharge print has to be done on all fibres, there should be dischargeable dyes of all classes available. Azo group in the dyes are easily destroyed by reduction and hence the dyes which are having azo group (chromophore) has to be selected for ground dyeing. Azo group containing dyes are to be found in various dye classes:

- disperse dyes,
- vat dyes,
- cationic dyes,
- pigment colorants,
- direct dyes,
- acid dyes,
- metal-complex dyes,
- reactive dyes,
- naphthol dyes.

Hence, it is possible to discharge print on all the fibres.

All azo group containing dyes are not to be granted as dischargeable. This may be due the reasons like, cleavage products formed during the discharge process have an affinity for the fibre or small amounts of impurities or side products in the original dye can remain on the fibre, contaminants above a certain concentration of use may give rise to problems. In some cases the dye fragments, if not removed properly (difficult to remove) will give a white discharge which although initially satisfactory, slowly darkens on exposure to light. This appears due to the oxidation of residual traces of A-NH2 or B-NH2, compounds, Dischargeability is also frequently concentration dependent. But dyes has to be selected carefully. Manufacturers give the dischargeability of dyes on the 1-5 rating where the 5 grade is fully dischargeable, and 1 is not dischargeable, which can be used as illuminating dyes. If a particular dye of one manufacturer is dischargeable, the equivalent dye of another manufacturer according to Colour Index no, may not be dischargeable. This may be due the additive and standardisation methods may not be the same for all the manufacturers.

The state of fixation of a dye also has a considerable influence on its dischargeability.

For this reason, dyes which adhere to the surface of the fibre after padding or printing are easier to discharge than those which have already been fixed on the fibre in a dyeing process. Practically, fresh dyeing discharges clearer and better than a dyeing went through redyeing and corrections. The fibre property also plays a part on the dischargeability. Since the reducing agents are mostly hydrophilic and applied through a hydrophilic system, discharges water soluble dyes on an hydrophilic fibre (e.g. wool or cotton) are more easily di naturally due to the easier penetration into the fibre than hydrophobic fibres on hydrophobic fibres. (e.g. disperse dyes on polyester fibres). Furthermore, the type and concentration of reducing agent and auxiliaries, as well as the drying and steaming conditions are of decisive importance here.

In case of coloured discharges the illuminating dyes are added to the print paste containing reducing agent and the dye has to withstand the reducing atmosphere throughout up to steaming conditions. Dyes of anthraquinonoid, phthalocyanine, triphenylmethane, methine or oxazine classes are mainly non-dischargeable. Even though there are many non-dischargeable dyes are available in the market, one has to select dyes considering fastness requirements, reduces the list to a much smaller list restricting the shades achievable by coloured discharge prints. Sometimes two or more dye classes are printed alongside to achieve a better range of shades. Thus, direct dyes, cationic dyes, pigment colorants, vat dyes, acid dyes, direct dyes and mordant dyes can be printed alongside wherever applicable. When using vat colours as illuminant dyes during steaming, the reducing agent generates the leuco compound that dyes the cotton. Air oxidation then regenerates the parent pigment.

As explained above due to the hydrophilicity of the fibre in case of polyester dyeing discharges has to be done after pad drying and before thermosoling. Exhaust dyed grounds are dried at low temperature to remove the creases and printed with discharge paste, dried and passed through HT steaming at 175-1850C for 6-8 min. to achieve good discharge effect. The material can be further washed off.

It is always better to take a trial in order to determine the dischargeability of a given ground because of so many factors controlling the complete reduction to a white ground whether it is a white or coloured discharge. The dischargeable ground dyeing is, in any case, dyed in the same manner as a normal dyeing.

The success of a discharge print lies in control of penetration with humectants, the viscosity of the discharge print paste, the amount of print paste applied (controlled by factors such as mesh size, squeegee setting, engraving depth and so on), steaming conditions, over and above the correct selection of thickeners.

7.2.6   Components of discharge pastes

The main constituents of a discharge printing pastes are

- Discharging agent
- Alkali
- Thickening
- Glycerine, Urea
- Whitening agent (Illuminating dye)
- Zinc/Titanium Dioxide
- Wetting agents
- Reduction catalysts

Discharging agents: Since almost all classes of dyes can be discharged by reduction, today most of the discharge prints are done using reducing agents. The most suitable and common discharging agents used are sodium sulphoxylate formaldehyde. The stability of these compounds is such that only limited losses of sulphoxylate occur during printing and prior to steaming. The use of Zinc formaldehyde sulphoxylate, calcium formaldehyde sulphoxylate, (CI Reducing Agent 2, sold as Formosul or Rongalite C) was established as long ago as 1905, when it was recognised that methods based on this reducing agent offered many Advantages.

Other reducing agents used are Thiourea Dioxide (formamidine sulphinic acid) and tin (II) chloride. Due to some disadvantage of the corrosion of steaming units and even squeegee holders in the printing machines, the importance of tin chloride has been reduced. Over and above this tin salts are not desirable in the effluents. But it has been found useful in the discharge printing of synthetic fibres.

Different reducing agents are used considering different conditions of treatment, fibre, quantity required etc. For example, tin (II) chloride is unsuitable for wool since brown tin sulphate can be formed with the reduced cysteine groups in the fibre resulting in unsightly effects. Calcium formaldehyde sulphoxylate and thiourea dioxide are sparingly soluble in water, hence cannot be used in aqueous pastes where higher quantities are required. Tin (II) chloride is applied under strongly acidic, and zinc formaldehyde sulphoxylate under mildly acidic conditions. Thiourea is used under both mildly acidic as well as alkaline conditions. All other products are used in the weakly alkaline range. Strongly alkaline sulphoxylate discharges are not particularly suitable for the production of discharge effects on alkali-sensitive fibres, e.g. wool, silk, acetate. Since thiourea dioxide is effective under neutral or mildly acidic conditions, it does not cause any damage to alkali-sensitive fibres. Choice of the optimum discharging agent for a particular application is based on the product which is capable of giving the best effects with appropriate additives under the given conditions of fibre type, dye class and processing technique. The selected product must not cause damage to the fibre nor influence the colour fastness of the dyes used. A prerequisite for the reaction is a sufficiently high reactivity of the discharging agent. This is characterised by the redox potential, i.e. the more negative the redox potential, the more reactive the discharging agent (dependent on temperature, pH and concentration). Sulphoxylates are stronger reducing agents than tin (II) chloride, and can be used to discharge a greater range of dyes. On the other hand, since very few dyes are absolutely resistant to reducing agents, tin (II) chloride is preferred with illuminating dyes. It would, therefore, seem logical to use both types of reducing agent for a pattern with white and coloured discharges, but most printers prefer to use only one reducing agent for simplicity of operation. The soluble sulphoxylates can give haloing problems on the synthetic fibres, because of capillary movement in between the fibres along the yarns even though the fibre is hydrophobic. In such cases insoluble reducing agents like zinc and calcium formaldehyde sulphoxylates or thiourea dioxide. Thiourea dioxide has been found considerable success in the discharge printing of acetate and triacetate, due to its low tendency to haloing and also because it is effective under acid conditions, which do not saponify the fibres as an alkaline reducing system can do. The quantity of reducing agents used in a print pastes depends on various factors like, depth of the ground shades, quantity dyes to be discharged, the fabric being printed etc.

Hydrotropic auxiliaries accelerate or make possible the discharge process during dye fixation and also dissolve cleavage products thereby facilitating their removal in subsequent after washing. These include products capable of making otherwise difficult to discharge dyeings (e.g. vat dyes) dischargeable, and which assist the discharging action and fixation of the coloured discharge illuminating dyes. In the main, these are products based on quaternary ammonium compounds. Hydrotropic auxiliaries also ensure the presence of a certain quantity of water on the fabric during steaming thereby increasing the efficiency of the reducing agent. Urea, which is the most frequently used hydrotropic auxiliary, performs the following functions:

- increases dye solubility in the discharge print paste or pad liquor,
- accelerates dye fixation during steaming,
- improves dye colour yield,
- prevents the fibre drying out during steaming.

Thiodiglycol is also used as a dye solvent and hydrotropic agent,

Penetrating agent is another auxiliary often added in the print paste which during steaming, especially with white discharges to ensure that the discharge paste thoroughly penetrates the fabric and to prevent any 'grinning' or show-through effects, especially on knitted fabrics. Glycerol is used as penetrating agent and is used as a hygroscopic agent and solvent(e.g. for wool and silk). They are also hydrotropic agents and examples are glycerol, ethylene glycols and thiodiglycol. The actual amounts used must be carefully determined under local conditions as too little would give a poor discharge effect, but too much could result in flushing and haloing. A penetrating agent is not always necessary with coloured discharges as the illuminating dye tends to mask any incomplete discharge. In fact, any auxiliary that improves penetration of the fibre can improve the discharge effect. Therefore, carriers and fixation accelerators are often added when printing illuminated discharges on synthetic fibres; in some cases, they improve a white discharge on such substrates.

Wetting agents are also employed in discharge print pastes to reduce the surface tension between water and air so that the aqueous medium can penetrate air-filled capillary spaces in the fibres more easily.

White effects produced by white discharges can also be improved by additions of zinc oxide and titanium dioxide (often as mixtures of both), as well as fluorescent brightening agents, to the discharge print paste. It is necessary when printing on fabric of low absorbency which may be coated with dried film of thickener from the preliminary dyeing operation, as in the 'discharge-resist' process Illuminating dyes are added to the print paste for coloured discharges which are capable of fixing at the places where the print is done during the discharging process. Since this dye is added to the liquor or print paste containing the reducing agent and must therefore be resistant to reducing agents. Discharge-resistant or coloured discharge dyes are mainly anthraquinonoid, phthalocyanine, triphenylmethane, methine or oxazine types. Whilst numerous discharge resistant dyes were recommended by various dye makers in the past, the number of suitable dyes has been reduced to a minimum today because of their generally poor colour fastness properties. It is therefore almost impossible to cover all the desired shades for coloured discharges with one dye class due to the relatively limited selection now available. For this reason, up to 4 different dye classes can be combined alongside each other to produce one coloured discharge design. Acid and direct dyes are the most widely used among the following types:

- cationic dyes,
- pigment colorants,
- vat dyes,

- acid dyes,
- direct dyes,
- mordant dyes.

Cationic dyes withstand reducing agents such as 'Formosul' but do not have sufficient fastness to be of interest today on cellulosic fibres. They are used for certain specialised styles on wool and natural silk.

Thickening agent selection for discharge prints are more important as it has to have all the characters necessary for a direct print at the same time it has to withstand the additions of all the chemicals necessary for the discharging agent, especially reducing agents. Thus, non-ionic thickeners are necessary and anionic thickeners, such as the carboxymethylated types, should be avoided. Thickening agents like Gum Tragacanth, British gum, Gum Senegal, emulsion thickening etc. can be used for discharge printing. The thickener must be stable to the reducing agent and non-ionic types are better, particularly if using stannous chloride. The latter coagulates many types of gums with anionic groups or by hydrolysing (at a pH of 2-3 of tin chloride) the thickener thereby reducing the viscosity. It is important that the paste consistency allows penetration of the reducing agent into the fabric to avoid any residual coloured surface fibres that might be visible. Water-retaining chemicals called humectants, such as glycerol, are therefore often present in the paste. The degree of penetration of the reducing agent into the yarns depends on the print paste viscosity, the amounts of solids and humectants present, and theprinting and steaming conditions. If the print paste is made thin for helping in the penetration there are chances of the paste can spread over to the unprinted areas which is called flushing which in turn cause loss of patter definition. This often produces a visible 'halo' around the printed design, which may be coloured if an illuminating dye is present. Haloing problem also can happen Haloing is an uncoloured intermediate zone which develops between the printed colour and the ground in coloured discharge. In case of white discharge this is often called flushing. The effect is caused either by incomplete drying of the printed fabric, steam which is too wet during steam fixation of the printed colours, other than excessive dosing of the discharging chemicals. This printing defect is prevented by pre-padding the fabric with sodium m-nitrobenzene sulphonate (a mild oxidising agent, e.g. Ludigol) before discharge printing, adjusting the quantities of discharging chemicals to suit the particular ground dyeing, the use of appropriate thickening agents in print paste preparation, and the use of less soluble or less hygroscopic substances in the discharge pastes. The oxidising agent reacts with the 'scum' of reducing agent and renders it ineffective. The amount of oxidant is insufficient to influence the high concentration of reducing agent in the printed areas.In some special cases, 'haloing' may be deliberately made as a part of the design. It is usual the discharge design contain intricate patterns, which has chances of flushing and bleeding. It is better to use low-viscosity thickeners and high solids content. In such cases non-ionic locust bean gum ethers, sodium carrageenates, starch ethers and crystal gums may be used.

Anthraquinone and some derivatives are 'reduction catalysts'. The reducing agent first generates the anthrahydroquinone. It is more used for azo dyes which are difficult to discharge (not very common today due to the restrictions in the usage of azo dyes). This then reduces the dye, the anthraquinone being regenerated. This cycle of reactions continues until reduction of the dye is complete. Why this should be more effective than the reducing agent alone is not clear. A number of quaternary ammonium salts with a sulphonated benzyl group improve the discharge effect and renders them more stable in air by retarding any oxidation of fission products.An example is ρ-sulphobenzyl-dimethylanilinium-m-sulphonate (see below).It is also observed that it promotes reproducibility in fluctuating steaming conditions but, to prevent subsequent discoloration. This is an efficient benzylating agent and presumably reacts with the amines

ρ-Sulphobenzyl-dimethylanilium-m-sulphonate

generated by the reduction of azo dyes. Washing readily removes the water-soluble benzylated amines produced.

$$R2N - CH_2C_6H_4 -SO_3^- + Ar - NH_2 \longrightarrow R2 - NH^+ + Ar - NH - CH_2 C_6 H_4 SO_3^-$$

The print paste includes a variety of other chemicals besides the reducing agent

7.2.7   Difficulties in discharge printing

One problem in discharge printing is that the pattern printed onto the coloured ground is not usually visible especially dark shades of the ground. Hence it is difficult to identify printing mistakes in printing stage and take corrective action. The actual design can be seen only after steaming, when it is too late to do any correction. Sometimes a small amount white pigment or marker dye is added in the white discharge paste to identify the design. Deliberate inclusion of titanium dioxide, or an optical whitener, makes the printed design visible and improves the white discharge. Printing pastes containing a high proportion of insoluble matter can give rise to the difficulties of blocking of screens (which cannot be noticed while printing) unless finely ground powders with soft particles are used. In this respect, certain forms of calcium formaldehyde sulphoxylate are considered to be better than the insoluble zinc formaldehyde sulphoxylate, as they have a softer, 'talc-like' consistency as well as being more stable. A low pH like a print paste where tin chloride (II) is used can also rise to specky prints, due to aggregation or precipitation of certain dyes, notably some disperse dyes. Controlled rapid drying of discharge prints is imperative if loss of reducing agent is to be minimised.

7.2.8   Imitation discharge prints

Instead of discharging the ground shade using pigment dye containing zinc oxide or titanium dioxide and of course a binder, can make a print completely masking the ground shade giving almost a discharge effect. This a much cheaper and shorter method than a discharge printing. The main disadvantage is the harsh prints. A, printing method applying a relatively thick film is required, e.g. screen printing or rotary printing has to be employed. No reducing agent is employed and hence no destruction of the ground shade is involved since the white 'discharge' effect is due entirely to the pigmented white paste masking the ground shade. Where a ground shade is relatively pale and it is required to print illuminating colours on it, there are times when discharging may be avoided. The printing recipe is adjusted taking into consideration of the available ground. But for some dark shades on light or pales ground shades, one may be able to print directly over the ground (especially when both colours are not heavily contrasting). For example, a pale yellow ground shade to be illuminated with an orange, brown, red and black would merely be overprinted with these colours since the ground shade would usefully contribute to all of them. This technique would be unworkable if a green dye ground was required to show a bright scarlet as well. A straight overprint with a soluble scarlet dye would be unacceptable since it would be dulled by the green and a true discharge would be necessary. The only alternative would be to try and mask the green using a pigment scarlet, but only trials would confirm that this was possible.

The new plastisol dyes being opaque can be printed directly on dyed ground, without the addition of ZnO or TiO_2. Successful light-on-dark printing with plastisol relies on increased pigment loads, fillers and other additives to block out the colour of the ground.

7.3 Resist printing

Historically resist printing were practised in printing methods like batik printing. These were physical resist printing where (hydrophobic) products or printing pastes were applied to the fabric to avoid contact and penetration when the fabric was subsequently immersed in the dyeing liquor. Now, main resist printing systems are chemical resist printing where pastes containing chemicals, which avoid fixation of background dyes (particularly for "reactives on reactives" applied on fabrics made of cellulose fibres).

Examples of resisting chemicals include acids, alkalis, oxidising or reducing agents. These chemicals react with the dye, the fibre, or with the dyeing auxiliaries and prevent dye fixation. Resist printing style can give a discharge effect in a different way. It involves two steps.

- The fabric is first printed with a resisting agent which function either mechanically or
- chemically or, sometimes, in both ways and resist the dye penetration to the substrate or dye fixation.
- Piece dye the fabric with the ground shadeby an appropriate 'dyeing' technique, such as dyeing, padding or overprinting where by the substrate gets dyed at the areas which is not covered with the prints.

Advantages of this style of print are that since there is no discharge is involved almost all dyes can be used, hence dyes of great chemical stability, which cannot be discharged can be resisted and prints of high fastness standards can be produced. Thus shade ranges become much broader and the end results of these prints are often indistinguishable.

Some of the printing methods are detailed in the following:

(a) Resist printing on covered background: this refers to the removal of a dye previously mechanically applied to the fibre by the subsequent application of a printing paste. These may be viewed as chemically-enhanced "displacement prints"; it is not only a mechanical displacement which takes place, which at best gives rise to tone-in-tone brightening. Special chemicals or auxiliary agents, which possess dye or fibre affinity, given the relevant pH, may in addition prevent the fixation of the dye of the previously applied background dye. The background can hereby be applied using a nip pad or, using a blotch printing template if required. The process involves in a pad dye is applied and dried; the printing is carried out with printing pastes containing products avoiding the fixing of background colour (but they do not avoid the fixing of any brightener used). The fabric is then dried, steamed and washed (this is the most diffused resist printing method).

(b) Resist printing by overdyeing: the operations of the resist printing method previously detailed are carried out in reverse sequence; therefore the fabric is first printed and then covered.

(c) Resist printing by overprinting: this method is similar to the previous one, but the covering operation is replaced with the roller printing of the background.

(d) Printing on polyester: polyester printing must be carried out applying the resist-discharge printing method. Printing pastes containing both the discharge and resist products applied on covered background must be used.

(e) Overprint/applied print resists:

The methods can be subdivided into single-stage wet-on-wet processes with or without intermediate dye fixing and two-stage processes, by which printing is carried out first, and the goods are subsequently continuously or batch process overdyed.

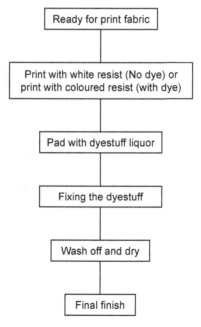

Process route of resist style printing

Since the print paste is printed of the fabric surface, and dried before the dyeing operation, the resist paste on the fabric surface must obviously be stable under the dyeing conditions.It is therefore preferable to use materials that are not too readily soluble in water and to modify the dyeing process.

Nip pad or over printing application of dye reduces the time of contact and therefore the bleeding of soluble resist agents. The dye liquor is padded on the fabric by relatively mild method. Nip-padding is one method of applying the ground colour without large amounts of water being present. The fabric does not pass through the pad bath but directly between the pad rollers. The lower roller dips into the bath, picks up a somewhat thickened solution (some thickening of the pad liquor may be necessary to increase the volume of liquor carried on to the fabric) of the dyes and transfers it uniformly to the fabric. This avoids the fabric passing through long pad bath there by the print getting affected and distorted affecting the print pattern. Other precautions are also taken to avoid softening or bleeding of the print paste like temperature of the dye solution must also be kept low, and it is often necessary to dry the fabric immediately after application of the dye. Use of an engraved cover roller, known as overprinting roller or blotch printing using a rotary screen, provides an alternative to nip padding and one stage in the process may be eliminated if drying of the resist is not essential

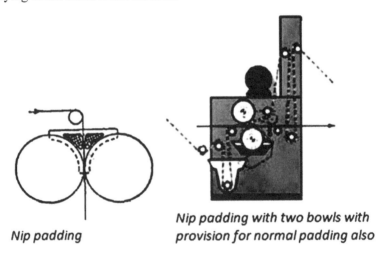

Nip padding   *Nip padding with two bowls with provision for normal padding also*

There are two different types, pre-printed and over-printed resists, depending on whether the resist print paste was applied before or after preparation of the fabric with dye solution. In both cases, the dyeing of printed areas is prevented. The resists used in resist or reserve printing can be broadly grouped in to three categories:

1. Mechanical resists (paste resists): wax reserves (resin, wax, etc.) block dye access to the printed areas mechanically.

2. Chemical resists: break down the chemicals required for dye-fixing on printed areas.

3. Chemical-mechanical resists: combination of I and II. As well as breaking down the dye-fixing chemicals, insoluble salts are usually formed as well as an extra layer of protection.

Mechanical resists use resisting agents like waxes, fats, resins, thickeners and pigments, such as China clay, the oxides of zinc and titanium, and sulphates of lead and barium. Such mechanical resisting agents simply form a physical barrier between the fabric and the colorant. They are mainly used for the older, coarser and, perhaps, more decorative styles in which breadth of effect and variety of tone in the resisted areas are of more importance than sharp definition of the pattern. A classical, and nowadays almost unique, example of a purely mechanical resist is to be found in the batik style, using wax applied in the molten state. In a true batik the wax is applied by hand, but the process has been developed and mechanised for the production of those styles which now come under the general heading of 'Africa prints'. Since the wax cannot be mixed with dye coloured resists by wax method is mostly impossible. But in a laborious process, the white resist is done with a wax resist and after removal of the wax, another colour can be printed within the resisted area or dyed in a different colour where by the white resisted area will have the colour dyes and other portion will have mixed with. A mechanical resist is usually used in conjunction with a chemical resist, so improving the overall effect.

Wax resists consist mainly of natural resins (pine resin, colophony), beeswax, ceresin, stearin, spermaceti, tallow, etc. melted with turpentine oil or rectified petroleum. These waxes can be applied on the substrate by block printing or in some cases with heated rollers with deep engravings on a roller printing machine. The fabric can be sprinkled with kaolin, diatomaceous earth or fuller's earth to avoidsticking and smearing. Where the crackling effect is required fabric is printed with a wax resist, then cooled down quickly in cold water and subjected to a mechanical breaking action in order to achieve the typical crackle effects (fine veining).It is subsequently dyed, usually with cold dyeing dyes like selected acid (silk) , reactive, basic, direct or mordant dyes as well as indigo. If the wax is removed by solvent (e.g., benzene), the dyes used should not soluble in it. Many times the wax is removed by dipping in hot water. After squeezing-off or centrifuging, the fabric is dried (steamed if necessary) and hung. Cationic dyes are dyed under neutral or weakly acidic (acetic acid) conditions in 2–4 h (and treated in cold tannin and tartar emetic baths if necessary). Acid dyes are dyed with 20% formic acid 90% based on the weight of dye for 30 min followed by another acid addition and dyeing for a further 30 min. For discharge batiks, dyed and batiked silk is allowed to lie in a cold bath containing 20–30 g sodium dithionite, removed after 30–45 min, 4–5 ml/l sulphuric acid 78.5% added, and the fabric treated for 15 min, after which it is rinsed and the cracked wax resists are dyed with another colour in a fresh bath (using acid or substantive dyes).

Paste resists are also comes under mechanical resists where pre-printed resist made of many high solids thickeners, thickener containing a weighting agent (e.g. kaolin, zinc oxide, lead carbonate, etc.) used in dyeing with indigo and vat dyeing.

Chemical resists: Chemical-resisting agents include a wide variety of chemical compounds, such as acids, alkalis, various salts, and oxidising and reducing agents. They prevent fixation or development of the ground colour by chemically reacting with the dye or with the reagents necessary for its fixation or formation.

Chemical-Mechanical Resists: a resist usually used for indigo resist can explained to illustrate a chemical - mechanical resists using indigo for the ground. Resist paste will contain high solid thickener like British Gum or gum Senegal, China clay and tallow (if necessary) together which acts as a mechanical resist together with a copper salt like copper sulphate which acts as an oxidising agent, which acts as a chemical resist. The resist paste is printed on an RFP fabric pre-treated with starch and calendered fabric to

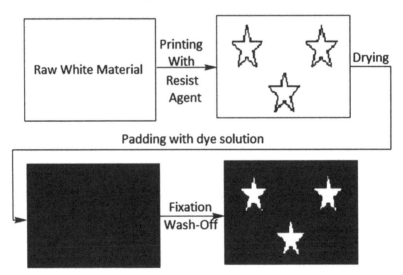

give a flat surface upon which the resist print would form a more perfect protective cover, and dried. On thorough drying the printed portion forms a resist barrier. The printed, dried fabric is dyes with indigo dye vat several times oxidising the dye in between to get the required depth. As indicated earlier the resist print acts in two ways: the mechanical part of the resist prevent the indigo leuco vat to reach the fabric at the same time the oxidising agent would convert the leuco form into an insoluble form which again deposit inside the resist print making the resist more effective than earlier.   The fabric is oxidised for the indigo to fix on unprinted areas and then the resist is washed off from the fabric after soaking thickener well.

Resist printing pastes usually have a high solids content. Because of this, they have a tendency to stick in fine engravings or block screens. It is common practice to strain the paste through a screen finer than the printing screen to avoid the latter problem. Stork (rotary screen manufacturer) suggests to use Penta 155 mesh screen in their range for blotch printing (Over print resist) and for discharge and resist printing.

Coloured resists equivalent to coloured discharges are possible by incorporating dyes or pigment and binder along with the resist paste. A coloured resist with pigment dye on a reactive ground possible by adding pigment and binder in the print paste containing acid. After printing, the material is nip padded or overprinted with reactive dyes and bicarbonate solution or paste fabric steamed without drying. The acid in the printed portion prevents the fixation of reactive dyes at the printed patterns and at the same time the acid present in the print paste help the pigment and binder to form a resistant film while steaming the print. The alkali present in the unprinted portion help the reactive dye to react with the cotton and fixation while the printed areas have the colour of the pigment held on the surface by the binder film.

The reaction of sodium bisulphite on vinyl sulphone dyes open the ways to resist print a reactive dye over a reactive dyes which otherwise appear to be impossible. The same reaction is not applicable to chlorotriazine dyes makes the above printing possible

$$Dye - SO_2 - CH = CH_2 + HSO_3^- \longrightarrow Dye - SO_2 - CH_2 - CH_2 - SO_3^-$$

In practice, the print paste containing the chloro-triazine reactive dyes, sodium bicarbonate and sodium bisulphite is printed and dried. The material is over printed with a paste containing the vinyl sulphone dyes and alkali. The fabric is dried and steamed.

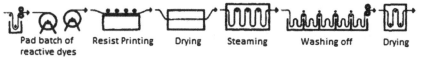

Pad batch of   Resist Printing   Drying   Steaming   Washing off   Drying
reactive dyes

Wet on wet Resist printing (Reactive on Reactive)

During the steaming the chlorotriazine dye reacts with the cotton aided by the sodium carbonate that is formed in the resist printed areas but the sodium bisulphite also present inhibits the reaction of the vinyl sulphone dye. The latter only colours the ground where bisulphite is absent.

7.3.1   Discharge resists

Discharge printing involves removing the dye from a previously dyed layer in the places to be printed. The term discharge resist printing is applied to the technique whereby dyes are locally removed prior to fixing, so that they are not present on the fibre at the fixing stage. In other words, in contrast to normal resist printing, fixation of the dye to be resisted is not only prevented but also simultaneously destroyed by the discharging agent. In practise it is a highly skilled technique calling for strict control on the process than a normal resist process achieving sharper and brighter colour effects. In the original process of yesteryears it was aniline resist, but the aniline liquor used to be padded before the resist print, hence the name discharge resist. This process had to be controlled strictly in many aspects like, the aniline black liquors must be freshly prepared and drying of the padded material must be carried out with great care. The dried material should have only light yellow colour and by any mistake in process, if the goods are with some trace of green (like overdrying), white or light coloured resist won't be successful on such grounds because the green is undischargeable. The fabric is cooled to avoid any oxidation due the available heat on the fabric and printed immediately with the resist print. During the print drying also utmost care has to be taken to avoid overdrying. Soon after this the steaming is done to develop the ground aniline black. This method may not be practised much now, probably during the oxidation the strength of the substrate is also affected, but the beautiful colours which can be obtained in a fast, unattainable opaque, cheap black by other methods makes it attractive.

A typical process is as follows:
It was found that alkaline salts and other neutralising agents can, therefore, be employed either alone or in combination with reducing agents, to prevent the formation of the aniline black in the printed areas.

Aniline black solution for padding is prepared by mixing the following A, B and C solution and making up to 1 litre.

Solution	Quantity	Unit	Additions
A	80	g	Aniline Hydrochloride
	5	g	Aniline oil
	50	g	Gum tragacanth (6%)
	250	g	Water
B	50	g	Potassium Ferrocyanide
	250	g	Water
C	30	g	Sodium Chlorate
	200	g	Water
	1000	g	

Pad the fabric, dry carefully without over drying, cool and print with following resist paste containing reducing agent and alkaline salt.

The resist print paste (Stock paste for white resist)

Quantity	Unit	Additions
125	g	Sodium acetate
125	g	Zinc Oxide Paste (1:1)
10	g	Sodium thiosulphate
740	g	Low Viscosity Alginate
1000	g	

After drying the print is steamed to develop the black and oxidation is completed by an after treatment with hot acid dichromate solution. Such an after treatment also serves to reoxidise vat dyes used as illuminating colour in case of a coloured discharge like vat colours.

7.4   Special printing methods

There are many special printing methods which may not be used in large scale. Some are used as cottage industries, some are not practiced in modern days due to its laborious process and labour intensive. Batik printing, which is practiced only by selected few areas around the world has been explained under the chapter on historical background. There may be many other methods of printing which may not have been explained here.

7.4.1   Half tone printing process

Half tone printing can be done by two methods - half-tone resists and half-tones produced with film. Half-tone resists are obtained either by using several screens with progressively weaker print pastes or by pre-printing or overprinting halftone resists on full depth print pastes. Resist effects are produced by pre-printing a colourless print paste, i.e. containing no dye (with or without the inclusion of resist agents) followed by overprinting with a normal coloured print paste. A lighter colour is produced in those areas of the printed design where the second print paste falls on the first. By different fixation levels of pre-printed or overprinted colour, the resists cause weaker printed shades in these areas. In the second method, half-tones produced with film after exposure will allow all the print paste to pass through for the full tone but only allow part of the print paste to pass through the half-tones thereby breaking up the printed surface into smaller areas tonal gradations. By varying the proportion of white area lying between the printed half-tones, a reduction in the depth of colour is obtained. The individual elements within the printed areas can be uniform in shape, in which case they are known as autotype gradations, or irregular in shape with different grained effects (e.g. circular grain, worm grain or Idento half-tones). No special screen-making technique is involved in the half-tone process since the half-tone cross-screen grid is already present on the negative or positive and transferred in this form on to the screen gauze by the methods described. Using laser engraving precise half tone screen can be made much easily.

7.4.2   Flock printing

Flock printing or flocking is a printing process in which short fibres of rayon, cotton, wool or another natural or synthetic material are applied to an adhesive-coated surface. This adds a velvet or suede-like texture to the surface. Since the fibres can be dyed, flocking can also add a colour to a printed area. The fibres used in the process are known as flock. Flock printing is an old technique which was already in use in China around 1000 BC.

The principle of flock printing is: An adhesive is printed in a design on the fabric. Next, the fabric is covered with cut fibre known as flock. The fibre is then embedded in the adhesive by one of various techniques such as compressed air, the shaking process, or the electrostatic process. Once the fibre is embedded in the resin, the resin is cured to firmly fix the fibre. This technique produces a three-dimensional pile surface effect in a specific design on the fabric.

7.4.2.1   Flock

Flock is prepared from both synthetic and natural fibres as required. From synthetic fibres it is prepared by cutting them accurately into pre-determined lengths using a suitable machine. And from cotton and wool it is usually prepared by grinding and lengths may be as short as 0.3-0.5 mm and 1.7 – 22 dtex diameter. Flock can be white of coloured as per the printing colour requirement.

7.4.2.2   Base Fabric

Base fabric can be any fibre or even blend and flock can be same fibre or different. For flock printing the fabric to be printed should smooth and as compact as possible. For fixing the flock the base fabric can be fully coated with the resin adhesive or in the form of pattern. Prerequisites for the base fabric are: well desized material if, e.g. woven cotton; unlevelness of the surface results in unsatisfactory flock effects).

7.4.2.3   Adhesive

A wide variety of flock adhesives are available, both single part and two-part catalysed systems. Adhesives used to capture the fibres must have the same flexibility and resistance to wear as the substrate. There are many types of adhesives used for the flocking – some of them are silicone elastomers, isocyanate crosslinking polyurethane adhesives, halogenated polymers, a self-crosslinking acrylic emulsion etc., which can be solvent free, water based or 100% products. Silicones are known to have excellent properties regarding heat stability, low temperature flexibility, media resistance and durability. All these positive properties can be used for textile compounds as well by adjusting the adhesion properties to the substrates in use. Adhesives are generally water or solvent based. Some are air drying, others temperature or catalyst curing. Adhesives are usually applied by brush, roller, spray or screen printing.

7.4.2.4   Application of adhesive

A very heavy deposit of adhesive is applied on flocking using a screen or film. While applying this step care should be taken to apply adhesive as it should not be too thin otherwise it will not hold the printed flock. Coating can be done by squeegees, rollers, screens or spray methods. The adhesive (bonding binder) must make good contact with the textile surface without penetrating too far into the material (to maintain flexibility). Consumption of bonding agent is approx. 130g/m2. When coating or printing (mainly using rotary, in small scale probably flat screens) a thick film of resin should be transferred to the base fabric for better results. The resin used should be flexible and soft to handle at the same time stable to washing and capable of anchoring the flock. Full coating is mainly used for manufacturing of high quality suede, velvet, velour or imitation fur effects as well as floorcoverings (see Fig.). Once the adhesive has been applied to the entire surface of the substrate, i.e. including the motif, and this has been produced in the flocking zone using templates, sharpness and resolution power are established by flocking.

7.4.2.5   Flocking process

There are many ways to apply flock. It is also known as flock pile process and flock printing. The process use special equipment that electrically charge the flock particles causing them to stand up. The fibres are then propelled and anchored in to the adhesive at right angles to the substrate. Flock prints are both durable and permanent. Multicolour flocking equipment has one print station for applying the adhesive and multiple stations for applying the flock. It uses a flat metal screen that is coated with an emulsion and exposed with each of the design elements, the same as it would be for screen printing. The flock is placed on the metal screen, which acts as the high voltage electrode, and a rotating brush precisely dispenses the flocking material. When the screen is lowered to the proximity of the adhesive coated substrate, the flocking fibres are propelled into the adhesive, as determined by the stencil on the metal screen. Since the electrostatic field strength is controlled, and because the metal screen and the adhesive-coated substrate are brought close together, the flocking material is prevented from attaching to the adhesive except where the stencil is located, regardless of the size of the adhesive coated substrate

7.4.2.6   Mechanical process

Mechanical process is based on spraying the flock on to the adhesive applied on the fabric. The flocks are sprayed using an air compressor, reservoir and spray gun similar to the one used in spray painting. The resulting finish using this method is similar to a thin felt coating, as most of the fibres will be lying down in the adhesive. It is primarily used when large areas needs flocking. It

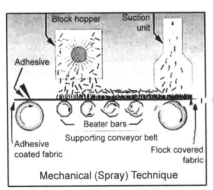

Mechanical (Spray) Technique

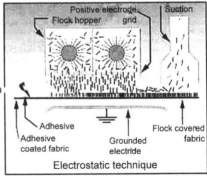

Electrostatic technique

is an untidy process because some of the flocks become airborne. Flocking is also applied while rapidly vibrating the substrate mechanically, while the flock fibres are dispensed over the surface printed with the adhesive.

7.4.2.7 Electrostatic method

Electrostatic flocking is entirely different from mechanical or pneumatic methods. This method needs an electrostatic generator which is used as a source of high tension which, when connected to an alternating current supply (100–250 V), can deliver a high tension direct current (approx. 20–100 kV) and can be regulated steplessly. It is a high-voltage, direct current grid connected to a power generator and a grounded substrate. The electrostatic charge is generated that stimulates flock fabric designs to get deposited on the adhesive. This results in forming high-density uniform printed flock coating on fabric materials.

7.4.2.8 Principle of the process

When a high voltage is generated and connected to two plates and if charged particles are introduced in between it has a tendency to align with the surface of the opposite charges based on the opposite charges attract and like charges repel principle. Here the flocks between these pole-plates will endeavour to close the circuit and spring from one plate to the other in an intense to and fro motion which results in them becoming aligned vertically. The tiny particles of flock are charged with a high voltage by contact or ionisation is introduced between two plates causes the fibres to move in the direction of the oppositely-charged electrode.Condition of humidity has a great influence in flocking by this process and support the charging of the fibre. Experience has shown that a 15% R. H. gives best results. To facilitate the generation of high voltage on the short fibres, it is advisable to treat them with a hygroscopic salt (e.g., ammonium chloride). Electrostatic forces are thus produced on long-shaped particles, e.g. flocks, which are capable of carrying electrical charges.

7.2.2.9 Flocking process

The fabric to be flocked is applied with resin adhesive is introduced between the plates and earthed. The difference in potential between the flocks and the base material is made to propel the flocks into the freshly applied adhesive surface. During this process, the electric force aligns the individual flocks which enter the adhesive layer at right angles and are in that way attracted to the base material at one end of their short length (0.5-1.00 mm)where they are anchored in the aligned state. As the fibres are closely arranged and held in the adhesive firmly at one end, the resistance to rubbing is greatly improved.

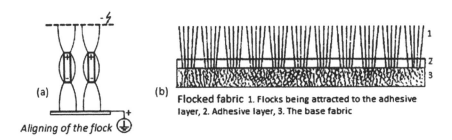

(a)

Aligning of the flock ⊕

(b) **Flocked fabric** 1. Flocks being attracted to the adhesive layer, 2. Adhesive layer, 3. The base fabric

In principle, a high voltage is generated and connected to two pole-plates, one of which is earthed the other being, effectively insulated. (See fig (a) above). During this movement, the fibres are oriented in the direction of the force line and are separated from each other due to the repulsion of their like charges. The electrical charges are discharged over the adhesive film. Those fibres which have not been able to penetrate into the bonding agent fly back to the other electrode after charge reversal has taken place and participate in the flocking process once more. The purely electrostatic process is mainly used for the flocking of made-up garments or moulded articles. Because of the vertical adhesion of the flocks, the desired velvet-like appearance is achieved. A collection device removes excess flock from the treatment area and returns it to the machine for re-use. There is no loose flock floating in the atmosphere as in the case of either spraying or sprinkling methods of application, and the room in which flocking operation is being carried out is completely clean.

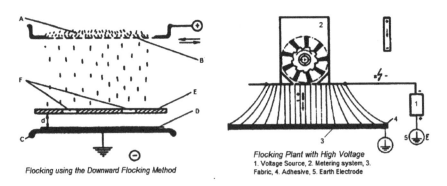

Flocking using the Downward Flocking Method

Flocking Plant with High Voltage
1. Voltage Source, 2. Metering system, 3. Fabric, 4. Adhesive, 5. Earth Electrode

Flocking may be performed in a downward or upward direction with the parameters of the flocking process determining the sharpness of the contours of the flocked motif. When the flock motif is produced by downward flocking. Flock A is on the upper electrode B, which is in the form of a screen. During flocking the screen makes oscillating movements on a horizontal plane. The substrate with the adhesive D is on the lower electrode C. The template E with the openings F for the flock is between the substrate and the high voltage electrode (see fig above).

The base fabric is applied with the adhesive paste which may contain resin, catalysts, white spirit, China clay, etc. By coating or printing by rotary screen and the flock is immediately applied before even the thinnest skins has had the time to form on the coating. It is natural, that the adhesive will penetrate into the base fabric to a certain extent, but if the material is of a relatively open structure and exceptionally hydrophilic, penetration will be excessive. Thus the coating of adhesive will be thinner and the flocks less firmly anchored. This drawback can be effectively counteracted by increasing the viscosity of the adhesive paste or by applying a water repelling agent on the fabric previous to the application of the adhesive. A heavy calendaring prior, to coating help to prevent undue penetration.

In combination processes, i.e. electrostatic/mechanical resp. electrostatic/pneumatic, the characteristic features of both techniques are combined.In the first case, the electrostatic process is assisted by dosing the flocks with a brush system (See Fig. above) and vibration of the substrate (using beater rolls or vibrators). In the 2nd process, the transport of flocks in the electrical field is supported by a current of air. The electrostatic/mechanical combination is the standard flocking process for continuous webs of fabric. Because of their geometry, the electrostatic/pneumatic process is often used for flocking hollow articles where the use of purely electrostatic flocking would result in unsatisfactory flock densities

7.4.2.10 Drying

The drying conditions are dependent on the requirements of the particular bonding agent used. Drying is generally carried out at approx. 700C (solvent-based adhesives) for 3–8 min. Suitable drying units include steam-heated warm or hot air tunnel driers, air jet driers, chamber driers or loop driers, as well as cylinder or plate-contact driers and infra-red radiation in combination with hot air.

7.4.2.11 Curing

After drying the material is cured for five minutes at 130° to 150°C for the resin to polymerise and permanently fix the flock. Drying of the flock printed good by radio frequency driers was found to give good results probably because the drying starts from the centre outward to the surface.

7.4.2.12   Final treatment

Once the adhesive has been fully cured and set, the flocked fabric is given a soft brushing or beating to remove the unfixed flock together with a mechanical suction unit which is sucked out and sent to the flocking machine.

Energy consumption is generally low and less than 200 W/h. Various models are available commercially, either as single units or integrated units in complete plants. Electrostatic generators are now supplied with a safety relay which provides protection against short stoppages, accidental contact, and overloads by means of an automatic circuit breaker.

7.4.2.13   Flock Transfer Printing

Usually done on materials which are directly used without any cleaning or after treatment (e.g., T Shirts) In this process the flocking done by the individual application patterns. . No after-cleaning is required although the preparation is laborious as far as the printing technique is concerned: a flocked release paper is screen printed with a single or multicolour design first. (See figs. below)

Next, a cover adhesive is then also applied by screen printing. It is most important here to ensure that the flock is only surface printed with adhesive in this operation, i.e. the adhesive layer must not penetrate through the flock. Immediately afterwards, a hot-melt granulate is strewn on to the adhesive layer whilst still wet before intermediate drying. Excess granulate is removed by brushing after the material has been dried (See below fig). The print paste/s and dye/s must then be fixed by condensation since the fastness achieved after subsequent transfer to the garment by hot ironing is dependent on this stage. The final transfer is obtained by placing the flock print (paper side upwards) on the substrate and subjecting it to heat in an ironing press (1200C for at least 10 s). The spent paper is then peeled off leaving the flock transfer behind.

These days flocking of printed designs technique these days are used in a number of other areas like automotive industry, where they are used inside cars to reduce rattle noise or any other unpleasant noise in the vehicle. Similarly, in the textile industry, it is preferred by fashion geeks and they make experiments by using printed flock fabrics designs in stoles, scarves and

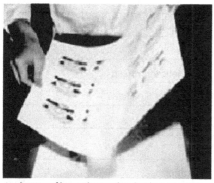 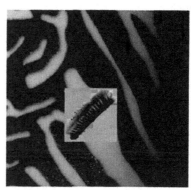

Application of hot-melt granulate (and intermediate drying of the adhesive print)

Intermediate cleaning (dry) to remove the excess granulate

socks. Electrostatic flocking is used extensively in the automotive industry for coating window rubbers, glove boxes, coin boxes, door cards, consoles, and dashboards. Rally cars usually have their dashes flocked to reduce reflections and to provide an as new finished to a modified dash. Flocking is proving successful in a number of artistic ventures including the decoration of jewellery, ceramics and pottery. Using suitable adhesives flock can be applied to an endless range of materials including plastic, metal, wood, rubber and fibre glass.

7.4.3   Special Tone in tone printing with glucose resist

Batik like print can be produced using glucose resist having glucose at various levels.. The process is as follows:

7.4.3.1   Wet on wet process for batik effects

Print the glucose resist screens first which may also have an engraved raster to enhance gradation of the tonal effect. Insert a crush screen to press the resist pastes into to fabric. Over print with either an open blotch screen or other engraved screen using MCT dyes containing urea, alkali etc. Partly printing the print pastes on the resist paste and partly on the fabric without resist paste to create tonal resist effect as required. For lighter resists tones increase the glucose (100g/Kg) for darker resist tone reduce the amount of glucose powder (30g/kg)

Print –Dry -Steam – Wash Off - Dry

Recipes

Stock Paste

Quantity	Unit	Additions
500	g	Water
50	g	Synthetic thickening
450	g	Water, Mix 5 minutes
1000		

Resist Paste

Quantity	Unit	Additions
800	g	Stock Thickening
10-100	g	Glucose powder, Mix 10 min
1000	g	

Make two (or more) paste with 10% and 4% Glucose/kg and printing in following order of screen.

Screen with 10% Glucose – Screen with 4% Glucose – Crush Screen for penetration of the paste – Colour Screen.

Print – Dry – Steam – Wash off – Dry.

7.4.3.2   Wet on dry process

Recipes

Stock thickening same as above

Resist paste

Quantity	Unit	Additions
800	g	Stock Thickening
30-100	g	Glucose powder, Mix 10 min
1000	g	

Procedure for Wet on Dry method (Sharper print than wet on wet)

Print the glucose resist screens and dry - over print either an open blotch screen or other design screen using MCT dyes containing urea , alkali etc.

For lighter resists increase the glucose (100g/Kg) for darker tone resist reduce the amount of glucose powder (30g/kg)

Print- Dry- Over Print –Dry - Steam - Wash off –Dry

Steam for 8 – 10 minutes at 102°C for both methods and wash off

Wash off

    1 × Cold rinse (<30°C) soft water

    2 × Hot wash (90-100°C) - plus 2g/l Suitable Detergent

    1 × Warm Rinse (60°C)

    2 × Cold Rinse

    1 × Neutralise, if necessary

    1 × Cold rinse

8
Digital printing

Digital Fabric Printing is by far one of the most exciting developments in the textile industry. Not only does it open up endless opportunities for customisation, small run printing, prototyping and experimentation but it also puts textile printing within the budget of your average illustrator. From small runs to big productions, all can be printed with ease while maintaining the demanding parameters of textiles.

Digital printing refers to methods of printing from a digital based image directly to a variety of media. A dye-sublimation printer is a computer printer which employs a printing process that uses heat to transfer dye onto medium materials such as a plastic card, paper or fabric. Technology has been developed to print preproduction sample and now even production prints on textile material using high-definition ink-jet printers in conjunction with computer-aided design (CAD) software. Manufacturers are developed a number of wide-format, ink-jet printing machines, together with suitable inks, are now being marketed and these can yield prints of acceptable quality and fastness properties on most textile materials.

Digital Textile Printing is the technology that consists of printing the desired pattern with its individual colours is built up by projecting tiny drops of 'ink' (special dye liquors) of different colours in predetermined micro-arrays (pixels), onto the substrate surface, directly from your computer, with no other additional step. This means that after you are finished creating your designs, and once you have them in repeat, you can print on fabric just like you print on paper. It is considered to be the next generation printing and is different from Traditional Textile Printing. Usually a set of inks is used consisting of at least three or four primary colours, namely cyan (turquoise), magenta, yellow and optionally black, the so-called CMYK inks. The three primaries are either printed as dots of varying diameters (amplitude modulation) or as uniformly sized dots of various randomised density arrangements (frequency modulation). Less commonly so-called spot colours may be pre-mixed to match the specific shades in the pattern, as is done in conventional textile printing. The only requirement of the Digital Textile Printing is that the fabrics which have to be used must be pre-treated to confirm about their ink holding capacity. Apart from this, a wide range of colours can also be obtained through the pre-treatment process.

Textile printing has seen a number of innovations in printing methods since hand block printing were first superseded by machine methods. Successively, the textile printing sector has absorbed engraved roller printing (intaglio or gravure) printing, the invention of silk screen printing, used in manual screen printing (table printing) and now in automated flat screen (or flat-bed) printing, rotary screen printing, and dry-heat transfer printing. The global market for dyed fabric is very large, but for textile printing the world market is smaller and more dependent upon fashion. However, it still amounts to around 30 billion m2/ year. It is estimated that around 11 - 13% of textile products are printed each year with a forecast annual growth rate of 2%. The dominant textile printing method is rotary screen printing (about 50%), followed by flatbed printing (28-30%). The remainder is printed by transfer printing (5%), intaglio printing (3%), and hand screen printing (about 1%). Digital ink jet printing (6-7%) is as yet a very small global market, estimated to be only 200- 300 million m2/year, printed by limited number of firms, each with limited jet printing machines. So why there is so much interest in new digital ink jet printing and what are the advantages and limitations of this innovative technique? To understand and to set these developments in context, it is necessary first to consider the competing conventional textile printing technologies, which are based upon analogue technology.

Nowadays the markets/customer requirements change rapidly as per fashion trend and commercial pressure. Manufacturers have to submit to customers' requirements in order to survive. It is usual that the customers expect a wide variety of patterns and colours especially in the printed field. Conventional printing methods are not able to satisfy customer's requirements especially in niche markets without increasing the cost and waste. But commercial pressure limits this. Thus ink-jet technology, which was being used in the paper printing, is now being used in the fabric printing market, more and more as it meets the demand in the market, especially in the area of wide variety of patterns and colours. It has been understood that the key market drivers for shifting towards digital textile printing are the need for economic short print runs, fast and frequent design changes, increased demand for personalisation and increased number of niche products. The new technology can be handled by artists and designers without special knowledge or skill in textiles.

Today the digital printing is primarily used in the following areas
Technical Textiles
- Automotive
- Specially Designed

Home/Furnishings/Interior Design
- Drapery
- Upholstery
- Linens
- Carpets/Rugs

Fashion Apparel
- Dresses, skirts, blouses, shirts, scarves, neckties, fabric

Graphic
- Sign/banner, trade show/exhibit, retail

Today's printing volumes by different methods

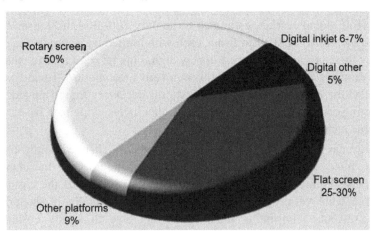

8.1   Conventional textile printing methods

When machine printing methods were developed, from engraved roller (intaglio) printing onwards, the printed textile market was aimed at the economic production of long runs of fabrics printed to the same design in a number of different colour ways. An important factor that is dominating the market is the rapidly changing nature of fashion worldwide. Now there may be a change of fashion seasons on a regular basis, sometimes as many as five or six changes a year. As a consequence of this, and because retailers now wish to retain only minimum stocks and re-order more frequently to replenish their depleted stocks of fast moving items, printing companies have had to develop competitive strategies for industrial survival. Engraved roller printing suffers from a fabric-width limitation, because above a width of 120 cm the rollers bend under their own weight, leading to printing distortions. Many fabrics are now produced at 150-160 cm width or greater and rotary screen printing is the most favoured high-production route, with printing speeds up to 80 m/ min, but with some restrictions with regard to the design repeat and the print design that can be printed. Flatbed printing generally operates at lower speeds, typically 30-50 m/min, and the screen size can be up to nine metres width for some more specialist fabric end-uses, e.g. flags. Dry heat transfer printing is a technique that is dependent upon transfer of the print design from a printed paper substrate onto a textile material, under the action of heat. This method is dependent upon disperse dyes that will sublime (that is spontaneously change from the solid state into the vapour state, without going through the intermediate stage of melting). It also requires fibres, such as polyester, that are substantive to disperse dyes. Dry heat transfer printing became popular in the 1970s with a global market of about 500 million m2/year, but the fashion change towards cellulosic fibre based fabrics has favoured other traditional printing methods. The major problem with all of the traditional printing methods, block printing, rotary screen, flat bed, engraved roller, etc., is that these methods rely upon the production of a design on a block, screen, or roller. The production of high quality screens or rollers is a complex, costly, and time-consuming process that is not amenable to just-in-time manufacturing methods. Screens or rollers are costly to make, costly to' store, can become blocked or damaged and need replacement, and must be cleaned and dried after every production run. For all these screen and roller printing methods, over half the total production time is expended on engraving and sampling. Sampling, in particular, is the bugbear of the textile printer. For every five designs sampled perhaps only one will be selected for bulk production runs. The production of short sample lengths for approval by customers is thus a very costly exercise, and textile printers worldwide are looking for rapid response machines that could also function for the printing of short- to medium-production runs. Within Europe, the average production run is now only 500 metres or less, accentuating the concerns of the textile printers.

Conventional textile printing of short runs by rotary or flatbed printing is uneconomic because of the high downtime, high wastage of fabric and print paste, high engraving costs, and high labour costs. The set-up time for production must include colour matching, print paste preparation, sampling, design, and registration. Design sampling and proofing are part of a very lengthy and costly process (see flowchart). Legislative demands on the reduction of environmental pollution add further costs, with screen engraving, print washing, screen washing, and disposal of waste print paste becoming increasingly expensive.

Thus the benefits of ink-jet printing are:

1. Avoiding most of the steps involved in conventional sample making (strike off) and bulk production a very quick customer response for both strike-off and bulk prints and wastage on pre-production sampling minimised.

2. Since screens are not involved considerable savings in screen inventories with major savings in storage space (patterns can stored on CD-ROMs and retrieved immediately and print can be done instantaneously.

3. The actual dyes involved is 4-8 but the no. of shades which can be produced are un limited as against the no. of colours depends on the number of screens and as the no of colours are increased the capital investments are high.

4. The same way as above the size of the pattern is limited to the screen circumference and it is also limited due to practical difficulty. But in digital printing size of the pattern are virtually unlimited enabling the production of very long repeats (e.g. fully bordered bed sheets) and full tonal (photorealistic) prints.

5. Instant fitting of patterns at start-up, thus minimising fabric and paste wastage.

6. Minimal downtime, because pattern changes and also colour changes, when using CMYK inks, are virtually instantaneous.

7. Only the ink required for the design is laid down, thus eliminating any waste of print paste.

8. The amount of ink applied to the substrate is far less than that used in a screen printing process.

Digital printing technology is considered to be a new technology. Originally it was developed and used for paper printing. The first patent for an inkjet printing system was received by Lord Kelvin 9n 1867 (See milestones of inkjet printing technology below). But in 1970s or after more than 100 years only this technology was found to be used for textiles. It was only used in carpet printing industry because the technology could print only with relatively low pattern definition. Still in 1980s only it was possible to produce laser versions of electrophotography transfers of T-shirt and other garment accessories.

Milestones in the development of digital printing

1878   The principal mechanism of inkjet technology (Lord Rayleigh)

1960s   First inkjet system (Continuous Flow Inkjet System)

1972   Piezoelectric D.O.D. heads by Clevite Corp in Ohio

1975/76   Millitron Printing System by Milliken - Carpet and upholstery fabrics.

1979   Thermal D.O.D. inkjet heads. (HP and Canon -bubble jet)

1980's   Desktop Publishing

1984   HP thermal D.O.D. desktop printer

1988/89   Advancement of CCD (charge-coupled device) for flatbed scanners. Iris Continuous Flow Inkjet Printer by Iris Graphics – paper proofing.

1990's   Screen printer, Photo LAB, Sign Printer – Moving to Digital

1994/96   Epson piezoelectric D.O.D. desktop printer Seiren Viscotex System (Production inkjet printing on cloth) Encad TX 1500 series (Thermal D.O.D. heads)

1998/99   Wide Format Printer (Epson, Roland, Mimaki) – graphic, photography and textile proofing Development of archival paper ink

2000's   Industrial Digital Printing- Archival Colorants (UV, Solvent, Textile, Material depositions)

2003   Production Inkjet Textile Printers (Reggiani, Konica/Minolta, Robustelli, Mimaki, Honghua, Zimmer) Flat-Bed Garment Printers (Kornit, Brother, Mimaki)

2005   Archival ink for consumer photography market (Epson UltraChrome K3 ink)

2010's   ITMA 2011 High Speed Production Textile Printers (EFI/Reggiani, Dover/MS, Stork SPG, Konica/Minolta, Durst, Zimmer, Epson/ Robustelli, Mimaki, Kornit, dGen, Arioli, Honghua, DGI, Ichinose / Toshin, Roland)

2015   Single pass inkjet textile production printing system (ITMA 2015)

2016   Onwards, quest for production machines of higher output

Digital printing defines a set of technologies that could be used to transfer an image in the digital form to a target surface (or substrate). Since different target surface are of different characteristics, not every printing technology can be applied for any given substance. The printing head plays an important part in getting the desired pattern on a particular surface. Thus developing print head suitable for textile requirement was the most important for achieving digital orienting on textile substrates.

8.2   The technology

The inkjet technologies available today can be classified as follows:

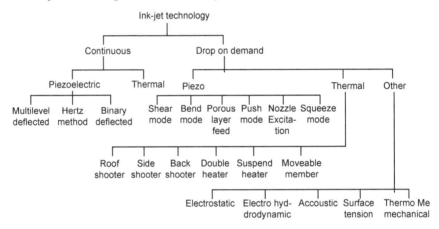

Different classes of ink-jet technologies under continuous and Dod groups

8.3   Continuous inkjet technology

The oldest technologies are the continuous ink-jet or CIJ. Its operation is based on a physical phenomenon called Rayleigh instability (Plateau-Rayleigh instability) that explains the behaviour of a stream of liquid with respect to the droplet size under the influence of surface tension. Basically, it means that it is possible to break a stream of liquid into a stream of fine droplets. If they are charged, the direction of the ink-jet can be controlled when droplets are passing through the electromagnetic field. Thus in this technology ink is squirted through nozzles at a constant speed by applying a constant pressure. The jet forms into droplets shortly after leaving the nozzle. The drops are Either allowed to go the surface of the medium or deflected to a gutter as per the action of the imaging signal by electrically charging the drops and applying an electric field to control the trajectory. The deflected droplets fall into a gutter (preventing it to fall on the medium surface and allowed to recirculate depending on the image to be printed. Since the ejection of the droplets (ink) is continuous at all times it is called continuous jet technology. This printing technology has several advantages. First the system can run at extremely high speed. Secondly, it is a non-contact process, the printer can print on many substrates of different shapes.

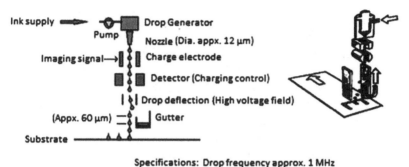

Specifications: Drop frequency approx. 1 MHz
Drop diameter approx. 20µm
Drop speed approx   40 m/s

Continuous Inkjet Technology

In the CIJ technology the size of the bubble, which is formed immediately coming out of the nozzle, is accurately controlled by a combination of the jet velocity and frequency of the excitation. Next, the droplets are electrically charged to the extent that required by the image to be printed and then through a deflection field. There are two ways of deflecting the droplets in piezoelectric-driven CIJ.

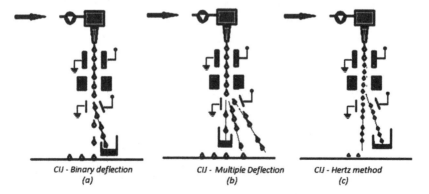

CIJ - Binary deflection
(a)

CIJ - Multiple Deflection
(b)

CIJ - Hertz method
(c)

There are many type of CIJ units. It is basically done based on the deflection of the ink drops:

1. Binary: In the binary deflection method the droplets are directed either to a single pixel location in the medium or to the recirculating gutter. [See above (a)]

20 In the multiple-deflection method the deflection is variable so the drops can address several pixels. Charged drops are variously deflected onto the print media, allowing one nozzle to print several dots across a larger area. [See above (b)].

3. There is a third method of deflecting the droplets called the Hertz method [see above (c) Several dots of small size can print on a single area. In this method the amount of ink deposited per pixel is variable. This enables variation in print intensity, appearing as "grey scale" for black colour on white media, or "gradient" in colour printing. This type was used in the early ink jet printing for medical monitoring, where such variation was particularly useful. This is achieved by generating very small drops (of the order of 3 pL) at speeds of about 40 m/s with excitation frequencies of over 1MHz . The printing drops are slightly charged to prevent them from merging in its path. The droplets which has not to be fallen on the medium is deflected away to the gutter.

4. Microdot, or drop "modulation." This can vary the size of drops, allowing for variations in print intensity as well as variations in charges carried by the drops.

In this technology (by Kodak), each nozzle has an annular electrical heater that is pulsed at a certain frequency. The heat generated raises the temperature of the ink jet in the vicinity of the nozzle and locally lowers the viscosity of the ink. Because the heating pulse is periodic in time and the jet velocity is constant, the resulting jet breaks up into equally sized drops in a reproducible way. (See fig below). The thermal CIJ uses different methods to deflect the droplets like air deflection, or by applying different energy to two heaters placed on diametrically opposite sides of the nozzle, and steering the jet as required etc.

CIJ technologies are complicated and hence costly, but it has advantages like the operating frequencies of these devices are typically at least an order of magnitude higher than in DOD systems and hence faster. Thus, CIJ systems are more used in industrial applications. The Stork Trucolor TCP 400 is a CIJ employing Hertz method and the first commercially available ink jet printer for textiles. Continuous ink jet was also the basis of the Osiris "Isis" machine, and has adopted the CIJ principle because of its ability to turn out higher productivity. This machine has been credited as the first ink jet printer for textiles that was capable of "industrial speed"

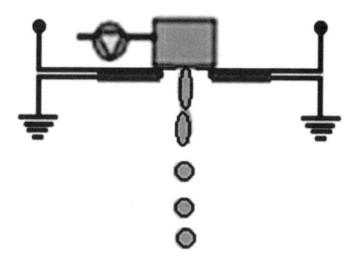

Thermal excitation and conttrolling the size of the droplet

8.4   DoD inkjet technology

In this technology, instead of the deflecting the drops when not required, the drops are ejected only when the image formed requires. The two main drop ejector mechanisms used to generate drops are piezoelectric ink jet (PIJ) and thermal ink jet (TIJ). In thermal ink jet technology, an electrical heater inside the nozzle heats up the ink to form a steam bubble forces the formation of the bubble whenever required, where as in a piezoelectric technology the volume of the ink chamber inside the nozzle is reduced by means of a piezo electric actuator which in turn forces the bubble formation whenever required. Even though there are other bubble formation technologies are available piezo electric and thermal jet technologies are more common.

Most inkjet printers are using the drop on demand approach or impulse printing, when a small drop is being separated from the printer head by means of heat or piezoelectric component. The pressure is applied to the reservoir only when a droplet is needed, so no deflection is needed, which simplifies the construction of the printer. The difference between the two approaches DoD is in the heat that is being transferred to the ink. In order for the ink to get transferred, it is heated to produce a bubble. For that reason water based inks are usually used for the heat based DoD Printers.

Unlike the thermal ink jet systems, piezo ink jet printing systems allow a broader variety of ink formulations. Hot melt inks are commonly used for multicolour printers. A very high resolution at very high speed is achieved by piezo technology makes it more popular with the Ink-jet printer manufacturers.

Inkjet technology has proved to be a sustainable technology for printing textile substrates. As the technology of ink-jet is ever evolving area, many innovations are being made continuously. The main directions of development for printing textiles are print heads and colourants. Increasing the number of operating frequency of the print head nozzles will allow achieving greater speed of an inkjet printer, which is another area of development being ensued. Better image quality will be accomplished by smaller drop size, grey scale capability and expanded colour gamut.

There are three methods of drop generation under DoD technology
Thermal, Piezoelectric and others (see Different classes of Inkjet Technologies above).

8.5 Thermal inkjet (HP technology, Bubble-jet) technology

A small quantity of dye in aqueous solution is heated at 300-400°C by an electric heater is typically built inside the nozzle, usually by microelectronic device fabrication, inside a small container. The steam bubble created causes the drop of ink to be forced through the nozzle and adhere to the fabric. Once the bubble starts forming and starts expanding, there is no point in continuing to provide power to the heater because the bubble is a poor thermal conductor. Thus, the pulse is usually tailored to stop shortly after bubble nucleation. As the bubble expands it cools and its pressure (which starts at over 70 atmospheres in water based inks) drops quickly. The bubble reaches its maximum size and then, just as violently, it collapses, retracting the meniscus to a region inside the channel. After the bubble collapses, capillary action drives the refill process, which continues until the channel is full again, ready to fire. Because of its explosive nature, there is little control over the process beyond the pulse length and power applied. Techniques of providing a short pre-pulse (or train of pre-pulses) to pre-warm the ink in the vicinity of the heater are sometimes used. Sometimes, quick cooling of the ink creates a concentration of dry ink particles and an immediate stop of the drop. With these techniques, one can control or modify in a limited way the total ejected ink volume. A very high jet frequency can be obtained with this system. This technique makes it possible to use plates with a very high number of nozzles at quite low cost but one of the problems lies in the application of special inks that can be used at high temperatures. (See fig. below)

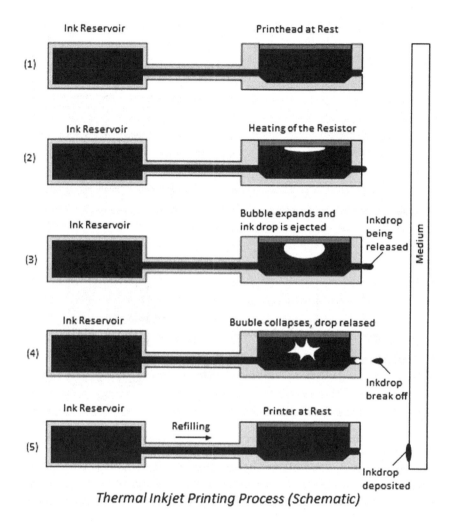

Thermal Inkjet Printing Process (Schematic)

There are many methods used for TIJ drop ejectors for forming the droplets for printing. (1) Roof shooters: A heater is placed parallel to the nozzle plane and the bubble formed pushes the droplet out of the ink container. (2) Same way, in a `side-shooter' the heater is placed perpendicular to the nozzle pale and hence the droplet is ejected sideways to the bubble formed. (See the fig below) (3) There are also `back-shooter' drop generator designs where the heater is located on the back side of the nozzle plate, as shown below (a). (4) There is another design called multi-heater type (by Canon -1997) where three heaters are placed in series below the Nozzle which can also function as drop modulator also by one or two or three heater working as per the size of the

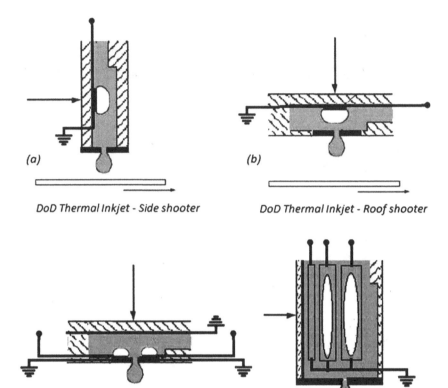

(a)

DoD Thermal Inkjet - Side shooter

(b)

DoD Thermal Inkjet - Roof shooter

(a)

DoD Thermal Inkjet - Back Shooter

(b)

DoD Thermal Inkjet - Multiheater

droplets required. (See figure below (b). (5) Yet another type droplet generator uses two independently operating side by side heaters. This is also a roof shooter type droplet generator by manipulating the heater operation the direction of the ejected drop also can be controlled, which is an added advantage. [See Fig. below (a)] (6) Suspended heater is also in operation which is supposed to be more energy efficient [See fig.below (b)]. Due to the fact that the heater is embedded in the ink, a larger portion of the total heat generated during the fire is transferred to the ink as against the other models where the heater is in the body of the nozzle or tank which has got the chances of heat loss (Sony). (7) In yet another model by Cannon a drop ejector design with a moveable member that, pushed during the vapor bubble expansion, prevents the ink from flowing into the ink reservoir through the rear channel region. This feature would be expected to enhance the energy efficiency of the drop ejector. (See fig. below)

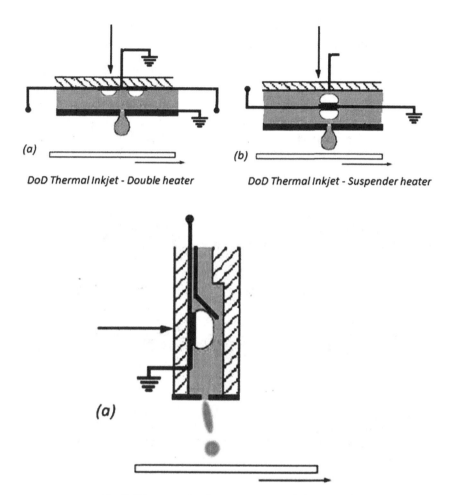

DoD Thermal inkjet - Moveable member

The nozzles can be subjected to continuous and quite rapid deterioration due to deposits (originating from the decomposition of dyes and/or precipitation of salts) produced by high temperatures in the steaming unit. We can briefly conclude by saying that this technique is scarcely reliable due to the colour variations connected with deterioration, and also because when a single nozzle does not efficiently work in a printing head featuring a very high number of nozzles (even though a single nozzle is quite inexpensive) the whole printing head must be replaced thus entailing process interruptions and higher costs.

8.6   Piezoelectric Inkjet (Epson, Canon, Roland, Mimaki, etc. technology)

When certain kinds of crystals are subjected to an electric field, they undergo mechanical stress, i.e., they expand or contract. This is called the "piezoelectric effect," and it's the character of these crystals are conveniently used in digital printing print heads of certain printers. When the crystalline material deflects inside the confined chamber of the print head, the pressure increases, and a tiny ink droplet shoots out toward the paper (See fig. below). The returning deflection refills the chamber with more ink.

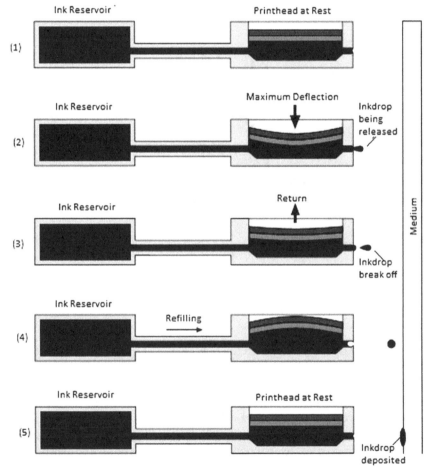

Piezoelectric Inkjet Printing Process (Schematic)

Example of such a piezoelectric element is typically made of lead zirconate titanate (PZT). The element can be attached to an ink chamber forming the ink tank or may be the chamber itself. The technique is essentially used for ejecting drops of ink contained in a small unit by effect of the deformation of a crystal subjected to the action of an electric field. This technique is more precise and reliable that the previous one because crystals are much more strong and hard-wearing than resistances and also because the system reliability remarkably improves thanks to the elimination of the deposit problem; furthermore less sophisticated (and therefore less expensive) dyes/inks can be applied with this system. This technique makes it possible to vary the size of the dye/ink drop by varying the intensity of the electric field. It is still impossible to use plates

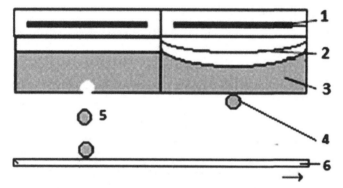

**Example specifications: Drop frequency approx 10-20 kHz
Drop diameter approx. 30 μm**

**1. Adjustable piezoelectric crystal, 2. Pressure forma-
tion inside the ink storage, 3. Ink stock, 4. Forming drop,
5. Ejected drop, 6. Substrate, 7. Imaging signal.**

Piezo Ink Jet

(printing heads) featuring a very high number of nozzles, and each head is much more expensive than the previous ones. (See fig. above)

As in the case of thermal inkjets there are different designs of inkjets mainly according to the geometry of the drop ejector and/or how the piezoelectric element operates. The classes, shown in Fig. below are shear mode which of two types as per the electric field is horizontal or perpendicular to the poling direction. In the former, the application of this field produces a

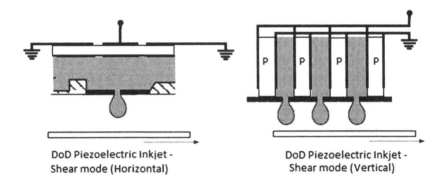

DoD Piezoelectric Inkjet -
Shear mode (Horizontal)

DoD Piezoelectric Inkjet -
Shear mode (Vertical)

shear motion in the piezoelectric material which is placed outside the chambers that makes the membrane move like oil can. Where as in the latter case the firing chambers are grooves diced into the piezoelectric material and the electrodes are placed inside the chambers.

Another design of the inkjet print head is called bend mode in which the electric field and poling directions are parallel. The piezoelectric material is placed on the membrane and the membrane moves like an oilcan. See fig below (a) and in yet another type called push mode in which the electric field and polarization vectors are also parallel but the membrane is placed in the expanding direction of the piezoelectric material (See fig. below (b).

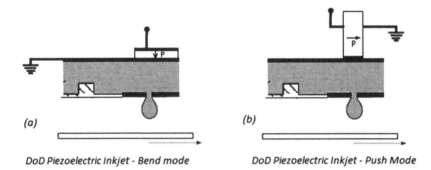

(a)

(b)

DoD Piezoelectric Inkjet - Bend mode

DoD Piezoelectric Inkjet - Push Mode

There are other versions like squeeze mode, where the drop ejector is a hollow tube of piezoelectric material and upon the application of an electric field, the inside volume of the tube (firing chamber) decreases its radius and ejects the ink in the direction of its axis (see below fig. (A)), Nozzle extraction where the piezoelectric elements are mounted on.

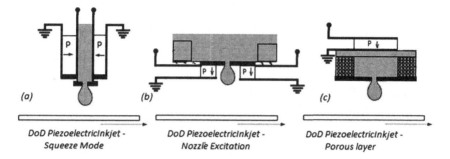

(a)	(b)	(c)
DoD PiezoelectricInkjet - Squeeze Mode	DoD PiezoelectricInkjet - Nozzle Excitation	DoD PiezoelectricInkjet - Porous layer

The nozzle plate (see fig above (b)), Porous layer type where the actuator chamber is Made out of a porous metal layer (e.g., sintered stainless steel) and the ink is fed to the chamber through this porous material [see fig above (c)] etc.

8.7   Other inkjet technologies

There are other inkjet technologies like DoD electrostatic inkjets, acoustic inkjets, thermo mechanical inkjets, electro-hydrodynamic extraction inkjet, surface tension driven ink jet etc. which may not be that important in textile digital printing and hence we are not going into details of these inkjet technologies.

8.8   Comparison of the results of piezo and thermal inkjet technology

The Piezo technology gives a better clarity of print than the bubble jet technology. Example of a print done by bubble jet and piezoelectric inkjet is shown below. Last column (the word Sie) shows the difference in the clarity of prints. As you can see below DOD Bubble jet does not give a clear dot on print, hence DOD piezo technology is more suited for Textile printing. Print head technology is different for different requirements. As described above, out of the different print head technologies, Drop on Demand Piezo is best used for Textile printing. The Drop-on-Demand technique, i.e. the direct application of a drop of dye on the fabric is the most diffused digital printing technique applied to the textile field. It is based on the approach, that a stream of ink droplet is projected on a substrate in order to form a pattern. The precision of the droplet is achieved by using the electromagnetic field. In particular, the nozzle technique used to spray the drop of dye determines the size of the drop itself, the spraying frequency, the accuracy, the evenness and partly also costs and reliability related to the machine. The type of nozzle used also influences the choice of dyes/inks used.

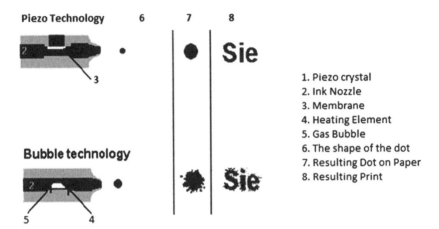

1. Piezo crystal
2. Ink Nozzle
3. Membrane
4. Heating Element
5. Gas Bubble
6. The shape of the dot
7. Resulting Dot on Paper
8. Resulting Print

There are many advantages of thermal inkjet technology over piezoelectric inkjets, like cost, possibility to integrate the electronics necessary to drive the heaters into the print head etc., still piezo electric ones score over them due to some disadvantages of thermal inkjet and of following advantages.

The bubble jet printers has inherent problem of air dissolved in the ink can nucleate at rough surfaces and sharp edges and fine particles suspended in the ink can also lead to air bubble nucleation. Another problem is the presence of these trapped bubbles forms part of the ink delivery system that can be difficult to fill in the priming process. PIJ waveforms typically tend to create areas of low pressure in the ink in portions of the firing cycle which tend to exsolve air through a process called rectified diffusion. Rectified diffusion occurs because the rate of diffusion of a gas toward the liquid during the compression portion of the cycle is smaller than the rate at which the gas leaves the liquid in the low pressure portion, causing the bubble to grow. Finally, the heating of the ink during the firing pulse in TIJ devices also causes air ex-solution.

In Bubble jet or TIJ drop ejectors, there are some other problems which makes it life much less than the PIJ. The chances of depositions of ink related substances and other additives in the ink solution on the heater plates which makes it less conductive and non-uniform heating, and ultimately cause failure of the unit. Coating of the heater surface to avoid depositions and additives in the ink recipe which reduces the depositions are solutions to these problems but PIJ gives a longer performance as there is no heating involved in the process.

Several important differences in textile printing and paper printing will impact the applicability of ink-jet engines.

- The volume of ink required to print a square metre of fabric is significantly higher than that required for paper. Estimates suggest that typical fabric printing will require at least 30 ml of ink per square metre at current colorant loadings. This difference has importantimplications in speed, drop size, drop frequency and nozzle life for ink-jet printing of fabrics.– Issues of image quality are also vastly different in the two types of printing. Image quality is a function of both the resolution (dots per inch or dpi) and the number of levels of colour (grey levels) that can be achieved at a given point in the image. In ink-jetpaper printing, improvement in image quality has been generally in the direction of increased resolution with some ink-jet printers now delivering 1440 dpi. Colour has been produced by mixing four primaries: cyan, magenta, yellow and black (processcolour). Textiles have traditionally been printed at much lower resolution (rotary screen mesh of 125 to 150) but with individually formulated colours (spot colour) which in essence gives an extremely large number of grey levels. As a result, the colour gamutrequired for textile printing is greater than the gamut traditionally obtained in printing on paper.

- The number of picks and ends in fabrics will impose a structure on the image that will limit the resolution that can be achieved without regard to the printer's capability. It has been estimated that with spot colour a resolution of 200 dpi will give very good image quality in printed fabrics, but process colour may require in excess of 360 dpi.

- In addition to these considerations, demands for fastness properties are quite different in both systems. In textile printing high fastness levels, particularly the abrasion, rubbing, light, wash and dry cleaning fastness are mostly required.

These differences in paper and fabric as substrates for ink-jet printing and different approaches taken to achieve image quality have important implications in the design of print machines for textile printing.

There are also a couple other alternatives for the inkjet technology Thermal DOD and continuous flow printing. However, it is piezo electric DOD technology is one that is primarily used for digital printing on textiles at the moment. But printing on textile substrates needs some adaptation of the traditional printing technologies in terms of fixation of the ink (dye) on the textile substrate. The requirement comes from the fact that the fabric has to pass washing and other fastness requirements. Thus, there are two methods for digital printing on textiles: indirect inkjet heat transfer printing and direct ink jet printing. Since the PIJ can fire aqueous and solvent, within a given range of operatingviscosity and surface tension whereas with TIJ normally can be used in an aqueous system only. This makes PIJ more versatile. Another advantage of PIJ is its capacity to control the volume of the drop through the shape of the waveform. In case of TIJ, dissipation of the heat generated continuously in the printer poses a problem. On continuous operation of the printer the temperature of the head increases and the ink viscosity decreases which in turn badly affects the print quality.

In indirect ink-jet transfer printing uses a special transfer paper with a layer of plastic to print to print an image. After that the image can be transferred to the fabric by applying heat to the transfer paper. As a result of heat of heat application, the plastic layer will be transferred to the fabric surface.

For the direct inkjet printing, colour fixation is also can be applied at high temperature to the target surface, or print on a fabric already treated with fixation chemicals (say, reactive dyes). It can be done by means of steam in a steaming device or by means of heat in the oven (curing). The particular heating approach is chosen based on the chemical components that were used for printing.

8.9   Development of print head technology

The print head is the probably the most important part of an inkjet printer. The print head has to take care of many problems happening during printing like, ink clogging, ink drying, excess ink, printer maintenance etc. In general, there are two types of print heads used – fixed and disposable.

A fixed head is an integral part of the machine throughout its life hence requires a well- defined cleaning and maintenance mechanisms especially when clogging and drying of the dyes. In this type of machine only ink has to be changed and not print head and so cost effective. But in the case of disposable head, both ink and print head has to be changed time to time. Many consumer oriented printers are with disposable print head and ink cartridges. In these type printers the head is changed when it is damaged or clogged beyond cleaning etc.

DoD ink jet printing heads for printing textiles usually operate on piezoelectric or, alternatively, by thermal pulse principals. These print heads are controlled by software and only deliver a drop of ink when it is required for printing. However, other types are based upon continuous ink jet printers which produce a continuous stream of ink drops, with the drop selection controlled by electrostatic charging.

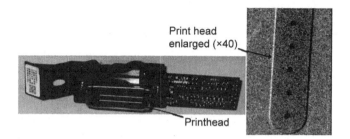

Print head (Epson 10000) – 180 Nozzles per channel, 360 Nozzles per colour

A print head has following components:
   Ink chamber
   Ink channel
   Ink outlet, orifice or nozzle
   "Actuator"–thermal heating element or piezoelectric material
   Filters and sensors to monitor the quality of the ink drop

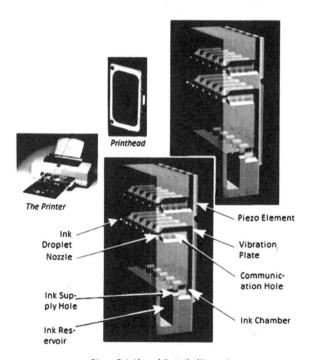

Piezo Printhead Details (Epson)

8.9.1   Ink chamber

Ink from the reservoir comes to the ink chamber and from this the ink is ejected through the nozzle of the print head to the textile substrate. Since it has to fill the same quantity of the ink which has to been disperse through the nozzle it should have a sensor that checks and controls the ink level in the chamber, temperature of the ink, and a filter to clear any particles and air bubble coming from the reservoir which can affect the drop formation. In case of TIJs the ink chamber carries the heating element for bubble formation. There are different designs of this chamber according to PIJ or TIJ and way in which the drop is delivered. Degassing has to be done for removing undesirable air bubbles which can hamper the pressure formation in the chamber. In case of inks which needs agitation (e.g., Pigments) there can be agitators also incorporated in the ink chamber.

8.9.2   Ink channels

Ink channels are the capillary which connects the ink chamber and the delivery point of the nozzle from where the droplet is ejected. These channels are often the actuator and are made of piezo electric materials in most cases. These channels should avoid any depositions and should always carry maximum ink without any air bubble debris due to drying out which can affect the droplet formation.

8.9.3   Nozzles

Nozzles are the part through which the droplet comes out from and only visible part of the print head. Since the shade, size uniformity of the droplet is to some extent controlled by these microscopic orifices, they are made with high precision. Because of the size of the orifice there are chances of them blocked by debris in the ink, ink drying which as to be monitored. There can be an array of nozzles and one or more arrays in one print head as per the designs of the print head. The position of the nozzle orifices, relative to one another in the print head, and relative to nozzles in other print heads, is precisely controlled to achieve uniform printing of ink colour and coverage and to avoid "banding" as uneven ink density leaves visible white or dark "lines" in the print. Nozzles also can be arranged at an angle, diagonally across a smooth fabric or on the "bias," to can improve uniform coverage by printing at alternating angles, covering the print of the previous pass as the array moves back and forth across the substrate.

8.9.4   Other aspects of print head – Print head array and passes

The explanation above print head was on the basis of one nozzle per head. But actual print head will have many nozzles arranged in multiple lines. This is called an array. In most of the printers there will be print head array. A print head array is an arrangement of multiple print heads within the print "carriage" mounted on an arm or "beam". Both nozzles and print heads are generally arranged in brick to avoid any gap in the coverage. like repeat. Print heads, like nozzles, are often arranged in a brick-like repeat so as to eliminate any gaps in coverage, particularly likely at the extreme ends of the print heads where there are no nozzles. Usually, the arrays are arranged on a carriage or beam which moves to and fro width way (of the fabric) but it can also be arranged along a beam which extents full width of the fabric which may be called a full width array and can be printed without moving the print heads to and fro. Even though it appears to be a better option but not practiced often due to the high cost of many print heads involved.

Costlier printers may have two or more scanning arrays, working at the same time but traveling in opposite directions to increase productivity. In such cases inbuilt software takes care of the synchronisation of print heads and the movement of the substrate, quality of print, detecting missing or misfiring of the nozzles and compensates them. In case of movable carriages and multiple print heads there are possibility of printing during forward and reverse pass across the width of the fabric. since printing ink is given of one strength the print head print in multiple 'passes' over the fabric to build up desired intensity of the colour wherever necessary. In some printers with multiple heads can have one set heads working on one type of ink say reactive, and another set on another say, disperse. This will help to change one substrate to another without much down time. Another advantage is that if a print head has to be removed due to any reason, the printer can be programmed to run the print with the remaining but with a lower productivity. In full width arrays without moving of the print heads it is obvious that such adjustments are not possible and there will be difficulty in accessing the print heads for any maintenance.

8.10   Selection of a print head

When the print has to selected the following criteria may be taken into consideration

8.10.1   Clarity of the image

In digital printing the clarity of the image is decided by resolution, drop volume and grey scale. In general, finer prints (finer details) can be done with lower the drop volume. It is needless to say that the drop volume decides the size of the printed dot. Even though the drop volume of the earlier printers used to be of around 90-100 pL, to day printers of drop volume of $1.5 - 2.0$ pL is very common. ($1pL = 10^{-12}$ liters)

Another important factor which decides the clarity of print is the resolution which is often measured as dot per inch. In digital world the resolution of a digital, bitmapped image is determined by how many pixels there are. If you have a scanned image and can count 100 pixels across (or down) one inch of the image, then the resolution is 100 pixels per inch or 100 PPI, which is normally used in case image files, monitors, and cameras. But in relation to printers it is DPI because in case of an inkjet, the printer's software translates the pixels into tiny little marks or dots on the paper. When an image indicated as 1600×1200 image in means it is of 1600 horizontal and 1200 vertical pixels. And the total pixels of the image is $1600 \times 1200 = 1920000$ pixels or 2 Mega pixels [$1920000/ (1024 \times 1024) =$ approx. 2 Megapixel]. When the resolution of a printer is mentioned as 2880 x 1024 dpi it is always the maximum resolution it can print. But the printer will have multiple modes that allow for more than one resolution setting; naturally, only the maximum is advertised. The smaller the resolution numbers, the faster the printing, but the lower the image quality. The first number is maximum number of dots the printer can cram into one inch across the paper, or in the direction of the print head's travel and the second number is the is the maximum number of dots the printer can place in one inch down the material (in the direction of the fabric feed). Please note that these are not separate little dots standing all alone; they are frequently overlapping or overprinting on top of each other. Horizontal numbers are always higher because it's a lot easier to position the print head precisely than it is to position the material precisely. The print head typically doesn't actually lay down a dot every 1/2880th of an inch in one horizontal pass. What happens is that different nozzles on the print head pass over the same line or row to fill it in. It might require up to eight passes to print all of the intermediate dot positions and complete the row. Each of the manufacturer has its own way to arrive at the maximum resolution numbers. However, it is not necessary a higher number of resolutions give a better print, even though higher printer resolutions produce finer details and smoother tonal gradations.

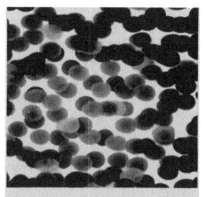

Printed dots, some overlappig to give secondary colors

 EPSON   Eight channels, 180 Nozzles, Variable Drops

 Konica/Minolta Two Channels, 256 Nozzles, Grey scale

 Seiko-Epson   One Channel 501 Nozzles, Grey Scale

 Xaar   One Channel 256 Nozzles

 APRION/HP   Seven Channels 512 Nozzles, Constant Drops

 Spectra   Six Channels, 304 Nozzles Grey Scale

When the resolution goes higher than a limit results in slower printing speeds and increased ink usage, but we may not be able to really find the difference in the print results. This due to the fact that the dot diameter is much larger than the resolution, the print quality improvement is insignificant (and other problems related to drying time and speed could be generated).

The third character of the print head to be taken in to consideration regarding print quality is grey scale, which is the ability to generate drops of variable sizes from the same print head). In grey scale lower number can achieve better image quality (smaller dots) while higher number can achieve higher productivity. The decision has to be a number considering the best print quality required by the printer to achieve without much loosing on the productivity.

8.10.2   Productivity

Output is always a consideration of deciding on any machines used in textiles. Productivity is the multiplication of the number of nozzles in the head by the operating frequency and the drop volume. The productivity will be reduced if the required printing resolution is higher than the native resolution of the print head since more passes will be needed (or the print head would have to be placed so that the array direction is not perpendicular to the printing direction) but the productivity definition stated above still limits the maximum throughput. Considering productivity continuous inkjet (CIJ) scores better than DoD Inkjets, since the CIJ operating frequency in much higher than the operating frequency of DOD inkjet (10 times approx.).

There are different designs of print heads. For e.g. Aprion/Israel and HP/ USA - 7 Channel 512 Nozzles Const. Drops, Konica/Japan - 2 Channels 256 Nozzles, Grey Scale, Seiko/Epson Japan 1 Channel 508 Nozzles Grey Scale, Spectra/USA - 6 Channel 304 Nozzles, Grey Scale, Xaar/UK – 1 Channel 256 nozzles, Grey Scale)

8.10.3   Life of the print head

The DOD piezoelectric system is based upon the software imposing an electrical potential across the piezoelectric materials. This causes a contraction, in the electric field direction, and an expansion, in the perpendicular direction, which leads to the ejection of an ink droplet. Capillary action is used to refill the ink chamber from the ink reservoir, when removal of the electrical potential causes the piezoelectric material to return to its normal dimensions. The drop volume ejected from the tip of the nozzle is usually smaller from piezo based print heads than that produced by thermal-pulse based systems. However, the cycle time, which is limited by the ink replenishment rate to the nozzle, can be higher, e.g. 1400 drops per second. Thermal-pulse systems (bubble jet printers) utilise a computer signal to heat a resistor to a high temperature, typically > 350°C. This rapid heating action creates a vapour bubble in a volatile ink component, which causes an ink drop to be ejected from a nozzle. With this system decomposition of ink components can give rise to problems. This can lead to inferior heat transfer and I or nozzle clogging. Rapid thermal cycling can also lead to resistor failure.

Care may be taken in the following area to increase the life of the print heads

1. A head strikes is the most common cause of premature head failure (another cause is constant flushing; the flushing seemingly wears out the nozzle system). A single head strike may wipe out only a few nozzles, or may kill the entire print head. Head strikes may be occasioned by a diverse variety of situations.

2. Improper loading of the media, which make cause buckling, because the media is caught, or not going through the printer properly.

3. Thin media can curl, thereby causing a head strike on the curled part
   • Edge guards, which work on thin materials may be raised too high

4. If media is absorbent, too much ink can make the material bubble up

5. If media is curled or bubbled by heat; the head can hit the raised part

6. If media is defective to begin with, or uneven, the head can hit the raised part

7. If adhesive pulls off the material the adhesive may get stuck on the nozzle plate of the head

8. For a textile printer, an additional cause of print head failure is the fuzz of the threads which may stick up and rub the nozzle plate

9. Some material is like sandpaper to the nozzle plate, some papers, and metal (and the metal edge is another danger to the print head nozzle plate)

8.10.4   Problems and maintenance of print heads

Even though the software and machine architecture takes care of achieving best quality print, maintenance of the unit, especially print head, is very important. The most important is blocking of nozzles or misfiring, ink drying. Though the machines are having inbuilt precautionary methods, like cleaning, to avoid the nozzle blockage, hard to clean blockages has to be taken care of. These types of blockages will result in misfiring ink out of sequence or inaccurately. Nozzle blockage can be caused by debris or air bubbles, unsuitable ink types or viscosity, incorrect firing information from the printer software (even faulty connecting hardware), or problems with the related actuator mechanism, whether piezoelectric or thermal. TIJ has the disadvantage of dyes or chemicals getting deposited on the thermal electrodes which can later block the nozzles if it is released, which has been mentioned earlier. In an array of nozzles neighbouring actuators can also trigger misfires, due to their proximity in the small space of the print head. In many cases one may have to do special cleaning to clear the blocks. The inbuilt cleaning processes are called Flushing and Purging. Normally, it is done at the parking station which is normally at the RHS of the machine the purged ink is absorbed by a sponge which is connected to the waste bottle. See figure below.

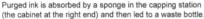

Purged ink is absorbed by a sponge in the capping station (the cabinet at the right end) and then led to a waste bottle.   Printhead height can be slightly varied say 2-3 mm

In normal course, with textile inks the print heads are rated at six months or more with production usage. How long your print heads last depend on what kind of ink you use; how often you flush and wipe, and in general how you take care of your heads. Flushing and wiping wear out your heads. Some manufacturers give vacuum pumps, for suction cleaning the print heads.

The distance between the substrates and print head has to be adjusted as per the substrates. Many manufacturers provide print head alignment. The print heads are in a row (in line), with the nozzle rows perpendicular to the printing path of the carriage. The distance from the print head to the fabric surface is adjusted by means of mechanical or electrical adjustments for small distance like 2-3.5 mm, which is the distance the ink droplets must fly to reach the substrate (see fig. above).

8.10.5　Cleaning of print heads

Cleaning is an automatic operation in most of the inkjet printers using, ink and some special cleaning liquids which may be different by different manufacturers. In many cases the print heads nozzles will be kept covered to avoid any drying when not in use. In case of sponges used for cleaning the head has to be periodically checked for soiling and any hardness formations which can damage the print heads. The cleaning liquid is a mixture of a solvent (e.g., alcohol), surface active agents, distilled water, which all will help that help this breakdown the dried ink and dissolving in the distilled water. There can be different mixtures for different dyes. The cleaning is more important when one printer is using two type of dyes (e.g., Reactive, Pigment) because one type may be predominantly printed than the other. The heads which are not being used frequently will have chances of dyes getting dries up. In such cases, care has to be taken to clean, the heads periodically, even though not in use. If colour type is changed or even one brans is changed for another brand in the same class dye, thorough full cleaning is a must.

8.11　Digital ink jet printing

It can readily be seen that the analogue approaches in conventional textile printing methods now have some severe limitations for the modern printed fabric market. Printers now have to 'think digital' if they are to survive and prosper in an extremely cost-conscious and cost competitive global market.

Digital ink jet printing has some considerable advantages over conventional textile printing methods because no screens or rollers are required. In digital ink jet printing, print heads, containing banks of fine nozzles, fire fine droplets of individual coloured inks onto a pre-treated fabric. The print design is created digitally and the ink droplets are mixed together on the fabric surface to create the final colour (so-called 'spot'

colour). By contrast, in conventional printing the final colours are premixcd with print thickener / auxiliaries and applied directly to the fabric surface. Because the print design concept is manipulated digitally the set-up time to produce short runs, sampling, or proofing is dramatically decreased. All the costs that are generated due to screen engraving, paste making, strike offs, downtime and wastage are also completely eliminated. Further, there

are no screen registration problems and the design repeat distance along the fabric length is no longer restricted by the physical dimensions of the screen but is unlimited, giving the print designer much greater freedom. The application of dyes by ink jet printing no longer needs the use of conventional print thickeners, because the dyes are supplied in colour cartridges by the dye maker and, once connected in the printer, are ready for instant use. Ink jet printing is a low water consumption and low energy consumption process, compared with conventional screen printing methods. A typical comparison of digital ink jet printing with rotary screen printing is illustrated in below Table.

8.11.2   Classification of digital printing

Digital printing machines can be classified on the basis of Technology, printer head, Application, Ink used etc.

1. Application

   Garment

   Home Furnishing/Dress material

T-Shirt

Flag & Banner

Niche

2. Print method

Direct / Transfer

3. Printer Design

Roll-to-Roll (Print on Fabric) / Flat Bed (Print on Garment)

4. IJ Head

Piezo / Thermal

5. Ink

Acid Dye / Reactive Dye / Disperse Dye / Pigment, etc.

8.11.2   Advantage of digital textile printing

The concept of digital printing on fabrics is very advantageous; it has opened new opportunities for designers, merchandisers and salespersons. The process is not only time effective but also cost effective. It permits customers to control the design and printing process from remote locations. Eliminates colour registration of plates or screens. It makes Just In Time (JIT) delivery and Quick Response possible. Greatly reduces the need for inventory. Reduces risk. Digital printing can virtually eliminate the threat of design theft before market release. It facilitates the increase in the number of fashion seasons. It can print directly from easily stored, transmitted, and transported computer files. Lastly it reduces the space necessary for archiving art, films, plates, and screens and reduces proofing time from weeks to hours, thus accelerating design and product development.

Digital printing permits customisation and personalisation. Changing each print on the fly does not increase costs substantially. Allows for design correction and modification at any time without significant schedule delays or cost increases Reduces the over-run waste which analog volume-print pricing promotes. Eliminates the design and process distortions associated with on-contact analog printing. It is cleaner, safer, and generally less wasteful, and less environmentally hazardous than analog textile printingmethods. This method can free up production printers from the sampling process thus increasing their productivity on products that can make the printer money.

In digital printing, small droplets of dyes, in aqueous solution, are sprayed on the fabric to reproduce the pattern. The dynamics of droplet formation impaction of the drop, viscosity of the ink, capillary of the medium (textile

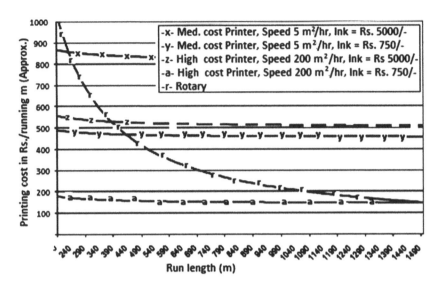

Printing cost as a function of print run and printing paste cost

material) all play an important part on the printed profile. Taking the case of drop formation and impaction of the drop, the size of a printed dot on the substrate in inkjet printing which greatly affects print quality, is determined by spreading of an ink drop when it impacts the substrate. The drop size is less than 100 microns and high impact speed of around 5±20 m/s may not be visible with naked eye, but has definitely an impact on the final print. When the droplet from a printer meets the substrate, the outcome greatly depends on many factors like drop speed, drop volume, liquid physical properties (viscosity, surface tension, and density), solid surface energy, drop/surface interaction, and surface characteristics. However, print quality is related to spreading of an ink drop when it impacts the substrate. There is lot of difference between printing a smooth surface like paper and a textile material. Considering the drop size mentioned above and considering the fibre size of about 25 micron (dia) The ratio of the diameter of typical inkjet drops to the width of the rayon fibers ranges from 0.8 to 3.2. Hence in case of a textile substrate much of the liquid flows in the filament axial direction rather than strictly radially, because of 'roughness' (striations of most of the fibres) elements blocking the direction perpendicular to the filament axis. Out of the parameters affecting the dot formation on given substrate and given printer, processor can manipulate only liquid physical properties and solid surface energy. The area of liquid physical properties is also limited as the ink has to be within certain limits to be accepted for a printer. Thus, in inkjet printing also, the main area of attention has to be in the preparation of the fabric as in the case of any other processing.

The viscosity of the ink has to be moderate to low and therefore, to strictly keep the pattern profiles (this problem originates from the capillarity of textile material) and allow a good definition of the pattern, the fabric must be prepared with a special pad-wetting process with thickeners and auxiliaries (typical of traditional printing) and then dried. The printing operation will be followed by a steaming process (to fix the dye) and washing, like traditional printing.

This technology uses large format digital inkjet printers. Practically it is the same technique used by desk top inkjet printers to print paper. Due to which digital Textile printing has various advantages over traditional printing. With the digital printing technology photographic & tonal graphics with multiple shades as well as colours can be printed on textiles. Unlike Traditional printing techniques (like Rotary & Screen printing) there is no limitation on number of colours as with this printing technique any number of colours can be printed in fabrics. There is no limitation on repeat size as in case of traditional printing methods. It offers faster processing speed where everything that is required in the print can be prepared on computer digitally. It allows user to print as little as possible. Therefore there is no minimum order quantity as such. High Precision printing is possible, which is usually a drawback with other forms of printing. No costs to produce screens, rollers or other printing equipment, and the possibility of printing patterns in a few hours (not days =or weeks) in all the desired colour variants. Furthermore, limited yardages do not excessively increase costs and patterns can always be slightly modified or customised. For environmental protection and cost reasons, it is also worth considering that this system applies controlled quantities of dyes, thickeners and auxiliaries, avoiding printing paste residues (usually excessive when prepared with traditional printing methods) and their consequent disposal. An inconvenience due to ink-jet printing is the low output efficiency and difficult reproducibility (by changing program or printer), the reduction of the colour spaceand above all the poor penetration of the dye onto the fabric, with evident discrepancies between the two sides of the fabric. All these problems make this printing system suitable for specific uses.

8.11.3   Comparison of digital printing and conventional printing (Rotary)

If one is using 100-300 Mesh screens for printing a design, the printed design will be comparable digital print of 254-770 dpi (100 μm down to 33 μm drop length) resolution. For textile printing of a normal rotary printing designs, 720 dpi digital printer is sufficient. These printers can give 80 - 100 μm print dot length and a drop size of 5-25 pL

Description	Digital textile printing single pass multipass	Traditional textile printing
Design Transfer	Directly from computer to machine	Computer to screen (in laser engraving) or to exposing and screen then to machine
No. of colours used in printing	4-5 liquid colours (dyes), also use of 7-12 colours depending on print head	Range of dyes
Dyes used	Special liquid dyes prepared for digital printing	Any commercial dyes can be used
Flexibility	Greater flexibility	Less flexible
Easiness of application	Easy Applications	Less Convenient in application
Versatility	Can be used for Versatile purposes	Can be used for Limited Purposes
Printing process	Fixation chemical is first applied on RFP fabric before printing	Fixation chemicals is printed along with the paste (in most cases)
Number of Colours per Design	Theoretically an unlimited number and range of colours for printing available.	Limited to no. of screens used
Working atmosphere	Special inks in containers supplied by manufacturers and can be directly used (tidy)	Aqueous Paste made in Large quantities on shop floor
Ready for printing fabric	Fabric to be treated with fixing chemicals before printing	Fabric printed with fixing chemicals along with dye paste
Maximum dimensions of Design	Unlimited	Limited to Screen size
Resolution of design	1440 dpi or more	Screens theoretical limit , (normally, 100 dpi in special cases up to 225 dpi
Repeat size	No restrictions on repeat	Repeat restrictions

Half tone printing	No limitations	Limitations due to screen hole size and not straight forward
Fabric contact	No Contact with fabric while printing	Contact with fabric
Provision necessary for repeat production	No Screens, designs stored in memory for reprint, no screen storage etc.	Screen cost, Screen making cost, (engraving cost), washing, storage, larger inventory
Ecological effects	Very Less	High amount of energy required for the treatment of waste water , excess dyes
Wastage of dyes	Only wastage in cleaning, negligible	Paste quantity cannot be exact to the requirement, large quantity is wasted
Minimum Quantities	Any minimum Quantities depending on Machines	May be 2-3000 m per colourway
Software involvement	Fully Integrated with CAD Design directly Printed on to fabric avoiding all in between processes	Digitised Design to Laser Engraver and design transferred to Rotary screen and Print
Down time to colour change	A colour change on the fabric does not require a physical colour change on the machine	30-60 minutes
Design Possibilities	Present a variety of exclusive textile designs	Having less no. of varieties regarding designs.
Strike off/sample scheduling	1-3 days	1-3 weeks
Strike off to bulk correlation	Strike off may differ from bulk print as they are done on different machines	Strike off and bulk if done on the same machine will have no difference
Consistency in printing quality	Very consistent	To be checked periodically
Printing speed	75-100 m/min / 1-10 m/min & increasing	50 m/min
Investment	Higher at present	Lower

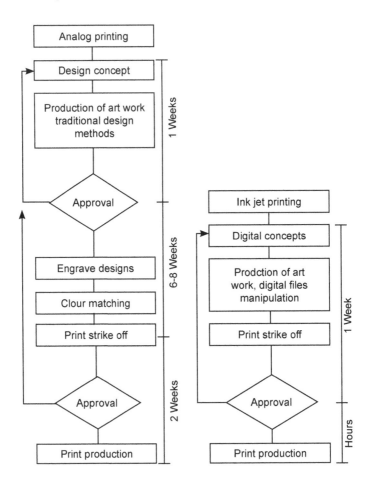

Comparison of screen printingand digital
ink-jet printing production routes

At present, ink jet printing still has a number of limitations which are restricting the commercial expansion of this innovative technique. Fabric for ink jet printing must be specially prepared with a pre-treatment prior to printing. Pre-treated fabric can then be printed at open width and the droplets of liquid dyes will then be rapidly absorbed where they strike the fabric, giving clarity of line. For ink jet printing the number of dyes available with a suitable low viscosity and satisfactory colour fastness standards is limited, mainly to reactive, disperse, and acid dyes and pigment dispersions. More colours are still not available and special effects, such as pearl and metallic colorants, are still required.

Digital printing efficiently produces designs at run lengths as low as one yard of fabric without need of screen changes. Digital textile printing eliminates the substantial amount of water and electrical energy one requires for rotary screen preparation, printing and clean-up. Even greater water and power savings can be achieved with disperse sublimations and pigment digital textile inks, which only require a heat fixation step for post treatments. Digital textile printing results in significantly less ink usage and waste production less relative to screen production. Thus printing digitally offers a greener advantage for printing. Digital textile printers can print large designs (e.g. cartoon characters on sheets and blankets) on roll of fabric without the usual rotary screen printing limitation in pattern repeat size. Digital textile printing permits the option to print a design at will. This means that the manufacturers with an integrated digital printing system in their production chain can keep a stock of unprinted textiles on hand to print as required. This reduces the need for pre-printed inventory of fabric that may or may not be used. Sampling and production on same printer. By able to print samples (strike offs on the same printer one uses for production , digital print shops can present their customers with proof samples of design that will exactly match the final printed material. By not having to prepare and store customer screens for future use, the production foot print for digital printing is a fraction of the size one requires for rotary screen print facility.

Printing houses utilising both digital and screen technologies can choose to print a small quantity of designs with different colour combinations (colourways) first with their digital textile printing solutions for the test market. They can later opt to print higher volumes of the most desired colour designs using rotary screen technology. Digital textile printing provides the option to print photographic/continuous tone images, spot colour pattern designs on a combination of both. This explains the creative printing alternatives for fashion and interior designers. The relatively low capital investment to set up a digital textile print shop especially compared to rotary screen printing production makes it possible to start small and expand as business grows.

8.11.4   Problems / Limitations associated with digital printing

Digital printing is new technology as far as textile is concerned and it has its own problems and limitations. Till recently digital printing was used only for strike offs and sampling, hence the problems was limited to that area only. But now, it is being used for even bulk production also, the problems are more serious and has to be dealt with.

Metallic colours cannot be printed by these machines. In case of flat colour printing, there can be a gamut of colours which the machine cannot produce. No sure shot formula to achieve desired results in case of photographic files. Attaining good results in digital textile printing for photographic files is an art where good desired results can be achieved only after correcting the file after various rounds of strike off. It takes few attempts / trials, before the optimum results are achieved. There is always a minor difference between the colours of the screen or artwork in comparison with the printed fabric.

The major factor limiting digital ink jet printing of textiles is, however, the general low printing speed. In 1991 the printing speed was around 6 m2/hour. Most ink jet printers being used commercially are now operating at 20-30 m2/hour or more, still considerably less than rotary screen printing at speeds from 30-80 m/ min. Now jet printing speeds of 150 m2/hour or higher productions were put in the market by different Manufacturers, which opens up greater possibilities for short- to medium- production run lengths.

Another factor is the necessity to use conventional fixation methods such as dry heat fixation, or steaming, followed by washing-off and drying. These treatments consume large amounts of water, electricity, and steam. Ideally, what is required for the future is a method of digital ink jet printing (preferably with no need to pre-treat the fabric) coupled with a fixation method that gives complete dye fixation and requires no wash-off and subsequent drying stage. It is possible that techniques from other industries that use ink jet printing technology (such as graphics, paper and packaging) may prove useful in this respect. Printing with UV-curable inks could open up even more opportunities for ink jet printing, particularly for garment printing, e.g. for T-shirts.

One of the major problems the digital printers facing is the so called 'Banding'.

Banding can happen due to following reasons

• A blocked nozzle or a nozzle that is otherwise dead

• A deflected nozzle that fires in the wrong angle

• When the ink lines and printhead become contaminated by air

• If the heater(s) are set too high.

• When the media feeds incorrectly

If the media feeds too fast, it leaves a white line.

In digital printing, print heads goes back and forth (or left to right) on a straight line above the textile substrate. The substrate is moved orthogonally to the print head path and the images are transferred line by line. If the media feeds not far enough, then the new printing path overlaps and prints on top of the previous printing path. This leaves a dark line across the print. These banding lines form a regular pattern, so they are painfully evident. In order to print a continuous image the alignment of mechanical parts and the control over the substrate movement should be perfect. If this movement is defective narrow stripes of unprinted bands may appear. Sometimes, some textile material like three dimensional structure compared to other substrates, have characteristics which may reduce the banding problem, 'naturally'. Clogged nozzles can be solved by cleaning them (purging, sucking, wiping, etc.) or by replacing dead or defective nozzles.

Banding in the printed cloth

There another problem arising due to misfiring or clogging of the nozzles. These type of problem result in missing pixels, partially coloured patterns. Newer version of printers has inbuilt systems to avoid/minimise such mistakes.

8.12   Inks for ink jet printing

Fabric, unlike paper, is a three dimensional structure and the ink and colorant requirements vary over a large range. Practical limitations exist on the range of fabrics and colours that can be produced with a single ink set. On some fibers that are absorbent, like wool and cotton, the ink is absorbed quickly and easily, so bleeding of the water-like ink jet ink is minimised even without a pre-treatment. Unlike the thick, paste-like ink used in conventional screen-printing, these water-like inks will bleed badly on non-porous fibers like polyester and nylon. A mechanism to control bleeding must be incorporated to avoid the ink wicking along the non-porous fibers of the textile. This also is important in applications that require print through on the design to give nearly equal colour on both sides of the fabric. In traditional printing this is controlled by the high viscosity of the inks used. With ink jet printing pre-heating the textile or addition of a fabric pre-treatment may help control these effects. The binding mechanism of the pigment to the textile and the reaction of the dyes with the fibers usually require a complimentary pre-treatment chemistry and/or post treatment to achieve the optimum result. The bottom line is that the ink, textile and the printing system must be designed to control bleeding while achieving the hand, correct colour and fastness required by the intended application.

Both the class of colorants- dyes and pigments - are utilised as inkjet inks. The dyes are soluble in water, water-organic solvent mixtures, only solvents or oils. The pigments are organic mainly but, depending on the application, possibly also inorganic; the bulk of black ink is based on carbon black. Improvements in inkjet printing are mainly due to recently emerging technologies in pigment surface treatment, pigment milling and new especially synthesized dispersing agents for ink formulations.

The dyes have high colour brilliance and can be used for a large gamut of colour. They have a good penetration power and penetrate into the layers. They produce stable inks, and abrasion resistance is good. Good transparency can be achieved with inks based on dyes. Formulations are easier when dyes are used. With dyes, one can have a large choice. The disadvantages of dyes are lower performance and diffusion. In contrast to dyes, pigments have food light stability and no diffusion. They have good water fastness. Generally they do not interact with layers and are also suitable for printing on uncoated surfaces. Pigments-based inks are difficult to formulate, have poor ink stability and are prone to abrasion.

Pigment-based inks have low colour brilliance. Nozzle-clogging is very common in these inks. Their cost is also high. Many disadvantages encountered in the early years when using pigments in ink-jet inks have now partly been overcome. These improvements in inkjet printing are mainly due to recently emerging technologies in pigment surface treatment, pigment milling and new especially synthesised dispersing agent for ink formulations. Solubility of dyes is generally taken as the basis for classification. Majority of the soluble materials ionise in water and thus ionic dyes are soluble, and the non-ionic dyes are water-insoluble. When anion bears the coloured entity, we

Classification of dyes

Solubility	Ionicity	Types of Dyes
Soluble	Anionic	Direct Dyes
		Acid Dyes
		Acid Mordant
		Solubilised Vat Dyes
		Solubilised Sulphur Dyes
		Reactive Dyes

	Cationic	Classical Type
		Pendant Type
Insoluble Dyes	Non-ionic	Disperse Dyes
		Oil Soluble Dyes
		Solvent Dyes
		Self Dispersing Dyes
		Polymeric Dyes
		Metal Complex Dyes
		Vat Dyes
		Azoic Dyes
		Pigments

call the dye to be anionic. If the cationic unit is coloured, then the dye is called cationic. Anionic dyes are traditionally called acid dyes while cationic dyes are called basic dyes. The more general classification scheme is presented in Table1. The classification based on ionicity does not reflect necessarily the potential use of the dyes in different inkjet applications. The applications of dyes other than in textiles are photo imaging, graphics, large format display, office printing, different industrial applications, textile printing, and other non-image printing. Many dyes, given a correct formulation, can be used in more than one type application. Compatible dyes can be suitably mixed to get different colours. Another equally important way of classification of dyes is based on the chemical structure. The chromophoric unit is used in naming them. A major part of dyes belong to the azo class and thus it is justified classifying them into azo and non-azo. Further subdivisions are presented in the following Table:

Classification of dyes according to chromophore

Functionality	Dyes Type
Azo Dyes	Mono-azo Dyes
	Bisazo dyes
	Trisazo Dyes
	Tetraazo Dyes
	H-acid Dyes
	Gamma Acid Dyes
	Heterocyclic Dyes

	Metal Complex Dyes
Non Azo Dyes	Formazans
	Anthraquinones
	Condensed Heterocyclic Dyes
	Phthalocyanines
	QuinacridonesIndigoids
	Metal Complex Dyes
	Sulphur Dyes

The earlier dyes used in ink-jets, the first generation ink-jet inks, are the readily available off-the-shelf commercial dyes. The four dyes used were:

CI Acid Yellow 23 (Tartazine yellow)

CI Acid Red 52 (Rhodamine B)

CI Acid Blue 9

CI Acid Black 2

These dyes are known to produce fairly bright images; they had, however, disadvantages in practically all other desired properties, especially in light and water fastness. Several reactive, direct and acid dyes were often used for the yellow, magenta and black channels. Metals containing dyes such as chromium, cobalt-containing acid and direct dyes were often used for the black channel. One of the important requirements is that the dyes need to be specially purified and made salt free for ink-jet applications. Important applications for the ink-jet dyes are:

- high water solubility
- diffusion fastness
- thermal and chemical stability
- hue and chroma
- physical properties
- light stability and performance
- dark solubility
- purity
- toxicology

In view of the above requirements, new dyes are needed to be developed with certain properties of ink-jet printing to achieve the picture quality and printer reliability.

Considerable research activity has been undertaken in the field of inkjet inks.

Different types of Inks used in Digital Printers and their applications

Ink Chemistry	Fibres	Post Processing	Areas of Applications
Solvent Based Inks			
Dyes	Polyester	None	Soft Banners
Pigments	Vinyl, Polyester, Nylon	None	Outdoor Banners
Water Based Inks			
Acid Dyes	Silk, Wool, Nylon	Steaming, washing	Fashion textiles, indoor soft banners
Disperse Dyes (Sublimation)	Polyester (flame retardant or non-flame retardant)	Heat fixation	Fashion textile, indoor & outdoor soft banners, home textile (wash fast and durable.)
Reactive dyes	Natural Fibres	Steam/Wash, can be dry cleaned	Fashion textiles, indoor soft banners
Direct Dyes	Cotton, Viscose, Linen, Silk	Steaming, washing	Fashion textiles
Pigment without Binder	All Fibres	Dry Heat	Indoor and outdoor soft banners, home textile
Pigments with binder	All Fibres	Dry Heat	Indoor and outdoor soft banners, home textile, affects handle of substrate

The inks used in ink jet printing must be very carefully formulated to ensure that they attain the correct balance of physical and chemical properties for application at high speed through the fine nozzles of the print head. Typical characteristics of suitable reactive dye inks are summarised below:

Physical characteristics of suitable inks

Parameters	Range for Digital Inks
Surface Tension (mN/m)	21-60
Viscosity (mPa)	2.3 - 5
Conductivity	6 - 12 (MCT)

	20-32 (VS)
Particle Size (μm)	<1

Typical compositions of water-based inks

Components	Function	Concentration (%)
Demineralised water	Aqueous carrier Medium	60 - 90
Water Soluble Solvent	Hygroscopic viscosity control	5 - 30
Dyestuff or Pigment	Colouring agent	1 - 10
Surfactants	Wetting and Penetration	0.1 - 10
Bioside	Prevent Fungi and Bacteria growth	0.05 - 1.0
Buffer solution	pH Maintenance	0.1 - 0.5
Other Additives	Dissolving, Complexing, defoaming agents	< 1

The formulators of inks for ink jet printing of textiles face formidable technological challenges. The ink must be compatible and possess satisfactory solubility or dispersibility and purity for the print engine and also have satisfactory physical properties and provide reliability of performance. The ink must provide a satisfactory colour and end-use performance and print and image quality. Factors important for reliability are the drop volume, jet velocity and straightness, and shelf life. Problems with inks could arise because of evaporation, foaming, air entrapment, or contamination.

The dyestuff concentration in commercial printing inks is about 10%. However, the maximum amount of ink applied, with current piezo-technology, to a fabric is only 20 g/ m² leaving 2 g/ m² of dyestuff on the fabric, compared with up to 100 g/ m² colour uptake applied in screen printing. Thus screen printing is capable of applying some five times more dyestuff on to the fabric than digital ink jet printing. Hence, in ink jet printing of reactive dyes, the maximum colour depth corresponds to a colour of a 2% dyestuff concentration in screen printing. To increase the colour depth, therefore, either the nozzle diameter has to be increased or, most commonly, the printed design must be printed several times to build up the depth of colour. The development of insoluble pigment based inks and of sparingly soluble disperse dye-based inks presents different challenges to the ink formulator. Different properties are required for specific print-based technologies. A particular difficulty is the production of sub-micron milled dispersion inks of adequate stability is to be maintained.

Dyes used for digital printing inks and related fibres

Name of Digital Printing Inks	Types of Fibres Used
Acid Inks	Silks and Nylons
Disperse Inks	Polyesters
Dye Based Inks	Not used for Textiles , Used for Photography
Reactive Inks	All Cellulosics- Cotton, Linen, Rayon

Main Characteristics (Requirements) of Digital Printing Inks

1. Purity – Should be pure.
2. Particle size – Fine Particle size (for disperse and pigment inks)
3. Viscosity (flow must be high enough to avoid starvation of nozzles and not so high that it flows out of the nozzle plate).
4. Surface tension (must be such that ink wets the capillary channels, flows through the nozzles, and forms droplets correctly).
5. Droplet formation (ink drops come with a tail or ligament which has to coalesce with the head)
6. Stability
7. pH value as per requirements.
8. Low Foaming properties

Digital printers deposit far less dye onto the textile than screen-based processes, regardless of type. As explained earlier, the amount of ink to be deposited, mixing of different colour dots, no. of drops at appoint etc., all are decided by digital computer programs, which control the ink jet printing process. They translate the design files into files containing the technical information needed to print the design onto specific substrates with specific inks. Colour matching, either from a sample or a colour as seen on a screen, is the most important in preparing a design for printing. The fundamental differences of colour achieved by a "subtractive" mix of cyan, magenta, yellow, and black inks (CMYK) as opposed to that represented on screen by an "additive" mix of red, green, and blue light (RGB) are not always easily overcome. Programs such as Adobe Photoshop or special textile based programs can be used for this purpose. Software is also an important part in achieving robust results.

As mentioned above there are two systems of colour mixing to achieve a designated shade – Subtractive colour mixing or CMYK (or a greenish-blue "cyan"; a bluish-red or pink "magenta"; a yellow; and a black, represented by K), and additive colour mixing or RGB (or red, green, blue). The first system mixes these colours as pigments or dyes for colour printed onto a surface. Printing usually begins with a white substrate. An area printed with a particular colour ink reflects only that colour's wavelengths of light as colour, absorbing or "subtracting" everything else. Ironically, the colour produced is the colour it rejects, or reflects. This is known as "subtractive" colour mixing. It starts with all of the colours and removes or subtracts them until only the colour needed is left. The second system mixes these colours as light for the emission of colour to a viewing screen. The computer monitor starts from no light or "black," and adds colours in as light, mixing to make colours, then emitted as light on a screen as individual spots of red, green, and blue, so closely packed that the observer only sees their combined effect. This is known as "additive" color mixing. It starts with nothing and adds in colour as needed.

Manufacturers give different shades in place of CMYK or RGB, hence there is always a limit of the gamut they can achieve by the colours. Gamut is the range of colours capable of being reproduced by any system of colour mixing. They can mimic each other rather than exactly reproduce each other's range of colours. Many printers use "light" versions of the CMYK inks or an additional set of primaries, such as red, orange, blue, and a deep or dark black to extend the range of possible colours.

Many design software programs use three variables in an interface, often in two steps to allow users to select colours. These often correspond to three principle activities involved in producing a colour: Hue. Saturation, Value.

Hue is a common word even in day to day life, which the software uses to adjust the proportions of these primaries, even when adding "secondary" colours such as purple (red+blue), orange (red+yellow), or green (blue+yellow), or tertiaries: aqua = primary blue+ secondary green (blue+yellow) to achieve the final shade required. It is basically by colour mixing. Saturation is to make a colour brighter by making it more intense; saturation is also sometimes used to refer to the brightness or dullness of a colour. Value may be the amount of light a colour absorbs or reflects, or how light or dark.

8.12.1   Reactive dye inks

In reactive printing on Rotary, normally an all in Process is followed – this is dyes and fixation chemicals are printed together and the steamed. But there is difficulty in following this route in digital printing, because of the following reasons:

1. Readymade dyes are used on digital printing machines and any additions or dilutions etc. are not allowed.

2. Even if 'all-in' inks have to be prepared by the manufacturers, it is not stable, especially for prolonged storage.
3. Some chemical can destabilise the ink.
4. Some of these chemicals may damage the print heads, if not immediately but on prolonged usage.
5. Low viscosity ink is required for printing.
6. Additions of 'all-in' chemicals may disturb the standard parameters requirements of the Ink.

Due to the above issues a two phase process is adopted. The ready for printing fabric is pre-padded with the chemicals other than the ink (dye) is padded and in the second stages the ink is printed on it. This method also helps in the improvement in the quality of printed image.

Major requirements of dispersion-based inks include the necessity for operability (runnability) testing with a specific print head and its resistance to crystal growth that could subsequently lead to jet blockages in ink jet printing. Jet blockage is common and on the latest Mimaki ink jet printer, jet blockage is checked automatically by software. Other factors for dispersion-based inks include the need to avoid any drying-out of the ink and the attainment of low particle size and low viscosity, dependent on the print head technology. In addition, the miscibility of different colours in spot colour mixing on the fabric surface must be satisfactory. The sub-micron particle size is required for satisfactory printing from the extremely small diameter jet nozzles used in print head technology. These can be less than 100 nanometres in diameter and deliver very small ink droplets in the range 10-80 pico litres (i.e. 10-80 × 10-12 litres). With pigment-based inks there are other considerations, because the pigment ink composition is dependent upon the print head technology used. In particular, the pigment must be bound to the textile fabric surface using a binder. In screen printing the binder is put into the print paste but in ink jet printing the binder could be supplied:

Pigment ink requirements for different print head technologies

Pigment Ink	Low Viscosity Piezo Head	High Viscosity Piezo Head	Thermal Print Head Bubble jet
Without Binder	Yes	Yes	Yes
With Binder	No	Yes	No
With Specific Binder only	Possible	Yes	No

In response to these competing conditions, manufacturers, has developed two ranges of pigment ink that contain no textile binder. Thus the Helizarin P inks are designed for low viscosity piezo print heads and the Helizarin H inks for 600 dpi (dots per inch) thermal (bubble-jet) Drop On Demand (DOD) print heads.

Ink chemistries and compatible fibres

Fibre	Acid	Disperse	Pigment	Reactive
Nylon	R	P	P	P
Nylon/Lycra	R	P	P	P
Polyester		P		
Silk			P	P
Cotton			R	R
Polyester /Cotton Blends			P	
Viscose/Rayon			P	R
Linen			R	R
Wool	R		P	
R - Recommended, P - Possible				

Inks colours may vary from manufacturers, but basic colours available are 4 colours and black, blue for better versatility additional colours are also available. General range of colours include – *Basic* - Cyan, Magenta, Yellow, Black, *Additional* - Deep Black, Blue, Orange, Red, Grey, Absolute Black, Pepper Red .

Some manufacturers provide special penetration fluid for Reactive, Acid and Disperse inks, for through print where the image quality on the backside of the fabric matches the front side (for printing materials like Flags). Solvent for printhead cleaning is also provided.

8.13   The printing process technique

Textile Digital printing has been adopted from paper printing. Paper is able to capture prints of high quality primarily because of the surface characteristics. But in the case of textile, it is much different than paper and moreover, each type of the fabric will have its own surface characteristics, due to its composition of fibres, textures, structure of fibres and water resistance. Also textiles will have much more absorbency than papers. Textiles are also flexible, stretchable, often highly porous and textured, which may vary in each case. Hence the printer has to be designed adaptable enough to be able to handle these fabrics during the printing process. The printing also should not affect, the handle of the fabric.

Colour production or matching is a complex subject in digital printing. Here we explain the procedure to manage and adjust the colour to ultimately reach the desired colour and design as a print.

8.13.1   Design development and display

The use of computer aided design (CAD) is almost standard procedure in textile designing field. If the design is made manually by a designer of artist it can be captured in to soft copies using a flat or rotary scanner or camera. Camera produces images in "raw" (RAW) format, but otherwise including its own colour determinants whereas scanner converts to RGB, based on manufacturer's set of default colour specifications. Image captured will be in the form of pixels and the clarity of the image will depend on the number of pixels in a unit area. The data will represent the location and colour of each pixel. When we observe the image on a monitor the design may then be tidied up and edited. Such software does the manipulation of the scanned image of the original artwork so as to define the number and precise shades of the design colours (usually a maximum of eight to 12, with several colourway combinations) and information about individual colour separations and the required design repeats, fall-ons, and optimum fitting. It is necessary to ensure the image observed in the display units , whatever it is, is the same for practical reasons. A software known as a colour management system (CMS) is used to achieve this. Visual display unit can display an infinite number of shades which may not be able to be achieved by any class of dye applied to a textile and also very much wider than is possible when printing with cyan, magenta and yellow jet inks. Each monitor may show the shades differently due to the characteristics of different makes of the VDU. A standard monitors can display very much brighter shades than CMYK printers in the red and blue regions, yellows can be of similar brightness whilst in the cyan/bluish green region CMYK prints may be marginally brighter. Thus the colour gamuts of Display systems and digital printers are different. Both computer and monitor usually run according to their manufacturer's set of default colour specifications.

Display systems, scanners etc. are based on the additive colour system, is different in that if you add colours, you ultimately end up with white, not black. The additive primaries are Red, Green, and Blue. The secondaries are Cyan, Magenta, and Yellow. With subtractive colour, light is reflected off objects that absorb some of its wavelengths and let others continue on. Our eyes interpret those remaining wavelengths as colour. In other words, the colour of an image on a piece of paper is what's left over after the ink and the paper have absorbed or subtracted certain wavelengths. If green and red are absorbed, what we end up seeing is called blue, which is in the 400–500 nm range. If cyan and magenta are absorbed, we see yellow. If we keep subtracting wavelengths by piling on more dyes or pigments, we end up with black. Printers work on substractive colour systems the Additive colour system (RGB) has to be transformed into substractive colour system (CMYK) (or the software has to tell the printhead where and how to place the dots and remapping the RGB colours to CMYK or whichever subtractive colours are used.)

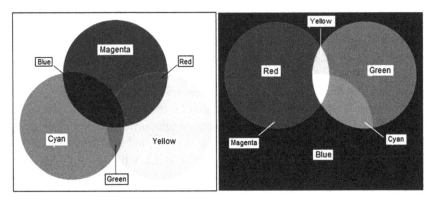

8.13.2   Colour management system (CMS)

A CMS is a software solution to the problems facing all digital imagers. It's a way to smooth out the differences among devices and processes to ensure consistent colour all along the art production chain. In 1993 International Colour Consortium was formed to create and encourage the use of an open, cross-platform colour-management system to make consistent colour reproduction a reality. The ICC Colour Management System comprises three components:

1. A device-independent colour space, also known as the Reference Colour Space. CIE's XYZ and LAB are the two related colour spaces chosen by the ICC; XYZ for monitors, LAB for print devices. To get consistent colour across different devices, a transform (a fancy word for a mathematical process) is needed to convert the colours from one device to the other. Monitor to printer, for example. But what actually happens is that the transformation takes place through an intermediary colour space or PCS (Profile Connection Space).

2. Device profiles that characterise each device. An ICC device profile is a digital data file that describes a device's capabilities and limitations.

If you characterise (or profile) any input, display, or output device by relating its specific colour space values to a known reference space, then any image file moving from one profiled device to another can be rendered so that the image looks the same (has the same values). This can apply to scanners scanning images, monitors displaying images, and printers printing images, and there are ICC profiles for each situation. The profile is actually a fingerprint of the device or process, and it helps each new device in the chain understand what that image is supposed to look like—objectively.

3. A Colour Management Module (CMM) that interprets the device profiles and maps one colour gamut to another. CMMs are also called colour engines, and they use device profiles and "mapping" any out-of-gamut colours into a reproducible range of colours by the next device.

Thus, design and colour is manipulated using a specific softwares and a colour management system is a must to ensure colour consistency between input devices, monitor displays and the final printed output. There are many colour management systems available provided by the ink-jet printer suppliers suitable for textile purposes for example, CAD based designing software like TreePaint (BTree/Duagraphics), Inspiration (Sophis), Image 4000 and ImageBox (Stork) and Vision Studios (Nedgraphics). software for device calibration softwares like Harmony (Sophis), PrintMaster (Duagraphics) and Vision Easy Colouring Pro (Nedgraphics) and other softwares like RIP for which translates pattern (RGB) data into printer (CMYK) data and then drives the jet print head causing each of its jets to fire at the appropriate times. As mentioned above since different monitors display shades differently colour communication software like GretagMacbeth's ColourTalk or Datacolor International's (DCI) ImageMaster to specify shades purely from on-screen appearance. A fully integrated colour management and recipe prediction system can even be downloaded online from www.ewarna.com. Such a system has a particular advantage over the provision of individual software packages in that new product data and software updates can be instantly transmitted to customers. For the standardisation of scanners and printers the provision of a standard test card, on which are printed graduated colour patches, is the first requirement and in practice the 240 shade, IT8.7/2 reflective colour targets (produced by Kodak, Fuji and Agfa to the ANSI standards) are commonly used. In colour management, an ICC profile is a set of data that characterises a colour input or output device, or a colour space, according to standards promulgated by the International Colour Consortium (ICC). Profiles describe the colour attributes of a particular device or viewing requirement by defining a mapping between the device source or target colour space and a profile connection space (PCS). This PCS is either CIELAB (L*a*b*) or CIEXYZ. Mappings may be specified using tables, to which interpolation is applied, or through a series of parameters for transformations.Every device that captures or displays colour can be profiled. Some manufacturers provide profiles for their products, and there are several products that allow an end-user to generate his or her own colour profiles, typically through the use of a tristimulus colorimeter or preferably a spectrophotometer. Once different units in the system are calibrated colour communication or manipulation of colour data becomes easier. All mathematical conversion of colour data between any two devices is carried out via an intermediate, device-independent, colour space. This can be understood from the figure below.

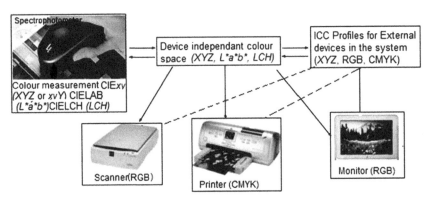

The manipulated soft design can be saved in one of the various formats (PDF, TIF)and sent to the printer. Printer software converts the file to CMYK or other system required for that printer. Ink is used, even though the ink may differ manufacturer to manufacturers, duplicate the design which has been made. For this a basic calibration process is necessary by checking against a definitive scale used to synchronize all devices. One important thing to be noted is printer prints based on "raster" — that is, the picture is built with individual colour dots arranged in a grid format and is not a not a simple substitute for screen-based printing. If the bulk production has to be done on screen based printers, the colours has to be separated for screen-based printing, such as a smaller range of solid colours, restricting the number of screens, still keeping soul of the design intact.

Some of the software suppliers

1. ERGOSOFT Texprint colour management software
2. SOPHIS CAD-CAM system with 5 levels

Level 1: Viewing station

Level 2: Design station - Colouration

Level 3: Design station - Coditex

Level 4: Design station - Colour expert

Level 5: Design station – Liberty - enables you, starting from a scanned image, to make your own separation, reconstruct and colouration

3. STORK - CMS 2000 Colouring Expert, incl. MTP CPS (Colour Physics System)- connect from design over Ink-jet to traditional printing.

4. NedGraphis Software: Design and Colour management and calibration

    Chemistry: Ink and coating development and production

    Colour physics

    Digital printing

    Application development

5. DGS (It)

6. AVA CAD/CAM (UK)

7. Dr. Wirth (D)

8. Honghua Computer Techn. (China)

9. Wastach

8.13.3   Using the digital informationin digital printer

Whatever may be the type of the printer the generation of the individual droplets for printing follows the same procedure and the round droplets falls on the surface of the material as per the design requirement. Droplet may be of different size according to the printer and the drive energy used, but it does not affect the overall appearance of the design. The pattern and shade manipulated as per the customer requirement as explained above and stored in the PC is retrieved by the printer either as whole or part by part and interprets using its own software and the hardware prints each pixel element of the original design as a number of colour drops in 4x4 or 8x8 matrix The shade of a particular area will depend on the proportion of each primary colour (CMYK) present in that area. The number of possible shades produced thus depends on the size of the matrix. If the size of the matrix is increased from 4x4 to 6x6 and to 8x8 the possible shades which can be produced increases by 50 and 100 times respectively. The shades which can be produced becomes infinite by further increasing the number of primary colours used [instead of CMY (3 colours + Black) and 6 colours + Black or more colour system can be used) and also by modulating the size of the droplets. Thus whilst a typical DOD printer using CMY inks and an 8 × 8 pixel matrix can reproduce about 275 000 shades, a multi-level charge-drop printer can attain a theoretical gamut of 16.7 million (in a 24-bit (8 bits per colour) RGB image, there are 256 possible values of Red, 256 of Green, and 256 of Blue, for a grand total of—are you ready?...16,777,216 possible values, tones, or colours for each pixel (see figure below). This may appear a complex process but much additional computation has already been imposed on the original design data by the colour management systems in the host computer.

It is the number of pixels per inch (ppi) that determines design definition. Basically ppi is same as dpi [If you have a scanned image and can count 100 pixels across (or down) one inch of the image, then the resolution is 100 pixels per inch or 100 ppi. Technically, it's pixels per inch (ppi) when you're talking about image files, monitors, and cameras. But it's dots per inch (dpi) when it comes out of a printer because, if it's an inkjet, the printer's software translates the pixels into tiny little marks or dots on the paper].However, there's a problem to more pixels. The higher the ppi and/or the greater the bit depth, the more space the files take up, the slower they are to edit and work with, and the harder they are to print since extra pixels are simply discarded by the printer or can cause it to choke, stall, or even crash. Mostly textile printers use a definition of either 300/360 or 600/720 dpi. Higher definition printers, giving 1200 dpi or greater and often capable of printing with very small-volume drops (5–10 pl) but this is rarely used since on fabric it is difficult to notice the difference of increased dpi.

The printer software also dither the placement of the individual droplets from pixel to pixel to avoid the undesirable moiré, mottled or chevron patterning effects. In most cases the colour increments represented by single drops are too large to give the correct mixed colour. For that reason a larger area of fabric is chosen and the ink drops are applied over that area to give "average colour" which best matches the correct colour. In screen printing with 10–12 dyes (primaries) mixed in any proportion, give the precise colour desired. Since they are premixed they do not suffer from variations due to order of printing the primaries. This gives screen printing the advantages of a closer match in colour, less variation in colour in solid areas and a "cleaner", brighter colour with less grey component to the mixture. Process colour involves subtractive colour. Each primary must be transparent so that light passing one colour ink will not absorb or scatter the light from another colour ink beneath it. Pigments must be chosen which are inherently transparent and they must reduce particle size to less than half the wavelength of light. Larger particles will scatter light and cause "greying", making it appear dull and dirty looking. This is a much smaller particle size than is usually found in commercial spot colour inks used in screen printing. This attribute is also important when one wants to print variable information on coloured materials, like logos for example. Process colour inks will not hide the underlying colour since they are transparent.

8.13.4   Transfer printing

This method uses the digital printing on a transfer medium, such as transfer paper, which is easily suited for the process and then transferring the image only to the fabric by applying heat. In this approach no pre-treatment of the fabric is needed prior to printing, the dye/pigment is not directly bound to the fabric but to the intermediate layer of plastic or adhesive which acts as binder in between the fabric and pattern. The prints are not as durable as ones which are directly printed.

Transfer papers require the artwork to be printed in reverse. This is due to the mirroring effect when you lay the transfer paper down with the printed side facing the shirt. If you were to print it normally and place the printed side down onto the shirt the design would be backwards. Most heat transfer papers will be made for laser or ink jet printers and will fall under two categories regardless if they are professional grade or not. They will either be a cold or hot peel. This simply means that the backing will be peeled either while hot or cold. For heat transferring normally heat press is used with high heat settings of 350 to 3750F. Standard inkjet heat transfers are much different in that the inkjet ink is transferred with a polymer adhesive layer that encapsulates the ink and fixes on the medium. Inkjet heat transfers are made to work with most any inkjet printer and ink. It is the polymer adhesive layer that does all of the actual transfer work. The ink is printed onto the adhesive layer and it soaks in a bit. After the ink sets into the adhesive layer, the image is ready to be transferred. The heat press causes the adhesive layer with the image to release from the paper and adhere to the material.

Pre-pressing your medium before pressing the transfer to them will help improve the application by eliminating any moisture in the fabric. Any water in the material during pressing can vaporize and affect the quality of the transfer process which will in turn affect the longevity and durability of the transferred image. If you live in a high humidity area such as coastal regions or even very wet inland environments, this process is more advantageous.

Sublimation transfers are little different than heat transfers. Actually, both are heat transfers but in sublimation transfers there is not adhesive or cross linkers. Dye sublimation uses a gaseous process to transfer the image to the substrate. This process is triggered in the heat press. The substrate is normally synthetic fabric, but can be done on other material with a polymer coating. It is better to use a dedicated sublimation printer for this process. The art is created in the computer with any graphic software and then sent to the printer to be printed on the sublimation transfer paper. Normally, sublimation transfer is done on polyester, Nylon, acetates, polyester/cotton with polyester 50% or higher.

However the synthetic material can be printed by this method, using disperse dyes which can give results equal to direct printing, taking advantage of the sublimation characteristics of the dyes and fixing them by applying heat, which directly fixes the dyes without any intermediate binding agent. The transfer-printing process (sometimes referred to as disperse dye-sub printing) is the easiest and most ecological printing process for polyester substrates. The textile design is printed (in mirror image format) on to special paper, dried and then thermally transferred, using a heated calendar press (under typical conditions of 2100C for 30-45 seconds). This is a transfer printing method which is practiced in bulk and explained in this book. After the transfer, the polyester requires no further washing treatments, so this is an ideal route for printers with limited textile-processing equipment. The disperse dyes used in this process are well known and are selected from the 'lower energy' class of disperse dyes, which have been well used in the analogue disperse-dye transfer-printing industry. Typical disperse dyes used in disperse-dye inkjet ink sets include: C.I. Disperse Yellow 54, C.I. Disperse Red 60, C.I. Disperse Red 11, C.I. Disperse Blue 359, C.I Disperse Blue 72

8.13.5   Direct printing

Direct printing can be done either with Pigment or Dyes. Since the post process is straight forward in case of pigment – Print, Dry and Bake – it is simpler than printing with dyes which will have longer wet process for fixation which is mainly – Print, Dry, Steam, Wash off, Dry .In most cases, dyes printing, especially Reactive dyes, are more in demand, because of the wide ranges of shades and brightness achievable and least effect on the handle of the fabric.

We shall go into details if fabric preparation and printing with each class of dyes

8.13.5.1 Preparation of the fabric

All fabrics meant for digital printing should be made ready for printing. Ready for printing fabrics need a pre-treatment (padding) before being digitally printed. The pre-treatment has chemicals which inhibit the dye migration. It also has chemicals aid in the fixation of dyestuff and controlling pH. Maximises thedepth of colour and sharpness on the surface of the fabrics.

For a professional textile processor preparation of the fabric may not be a big task. Normal fabric prepared for Rotary or any other printing may be sufficient as a base fabric. But the Digital printing is practiced by non-professionals also and hence they should have basic knowledge in preparation of the fabric.

Pre-treatment of fabric in digital printing terms is little different than what it means for a textile processor. Here it means only application of chemicals on the fabric before digital printing. It differs for each class of dye. Except in case of pigment, the thickeners and other chemicals are separated from dyes and applied on the fabric is different process called pre-treatment. The main reasons for this are

1. Unlike in normal textile printing where the mixing of dyes and chemicals are done at shopfloor, in case of digital printing dyes solution (or inks) is prepared by different manufacturers and it has to pass certain criteria to be able to be used in digital printers 'all-in' inks are not popular. Moreover, all-in dyes are less stable and have lower storage stability, e.g. reactive dyes are more likely to hydrolyse when alkali is present in the ink.

2. Some chemicals in the ink can cause corrosion of jet nozzles, e.g., the deleterious effect of sodium chloride on steel surfaces is well known, for instance; inks for use in 'charged drop' continuous printers should have low electrical conductivity.

3. Thickeners in the ink often do not have the desired rheological properties.

4. Some chemicals can be utilised in pre-treated fabric but would cause stability problems in the ink; e.g. sodium carbonate as alkali for reactive dye fixation is acceptable on the fabric but not in the ink.

5. The presence of large amounts of salts in aqueous inks reduces the solubility of the dyes; concentrated inks are required in jet printing due to the small droplet size.

Because of the above reasons it is better to use dyes separately which is made suitable for the digital printing and other chemicals used for the fixation of dyes and image improving chemicals are applied on the fabric ('Pre-treatment') and used. In this, the application of thickener is basically helps in the improvement of the printed image. It helps in controlling the wettability of the fabric so that the lateral spreading of the ink when the droplet impinges the surface of the substrate due to the capillary forces that are present in the narrow interstices between fibres and yarns.

8.13.5.2   Ingredients of pre-treatment

Thickeners: Generally neutral thickeners, which doesn't react with the dyes at does not colour the cloth by themselves, even after drying are used. But at the same time it should hold the dye within a certain area with enough time and some moisture and allow to penetrate through the fibers of the fabric. As in the case of normal printing (say, rotary or screen printing) the selection of thickener depends on the chemical composition of the colorant to be used, including its requirements for curing or the fixing of the print in post treatment. Common examples include sodium alginate, guar gum etc.

Alkali: In case of reactive dyes, one needs an alkali for the fixation of the dye to the fibre.

An alkali has a relatively low concentration of hydrogen ions and a pH of more than seven, as opposed to an acid. In conventional dyeing, this can result in significant effluent. Sodium carbonate or "soda ash" is a common alkali.

Urea: In digital pre-treatments it functions primarily as dissolving assistant for the dyes in case where stronger dye solutions are employed. It also acts as a 'hygroscopic agent' which supports in steaming and as "humectant," aiding moistening or wetting, and this moisture allows the dye to more completely travel into the fibers and reaction with substrate. Since it acts in many ways, generally there is no substitute for urea.

Cationic agents: Cationisation of many substrate is a way to increase the fixation rate of a dyestuff, and thus reduce the need for additional pre-treatment chemicals, and reduce dye lost as effluent due to wash-off. Cationically treated, positively charged fiber will attract anionic or negatively charged dye molecules that can then join to form a strong "ionic" or "covalent" bond, that is, joined at the molecular level. Cationisation also helps in reducing the ink consumption and faster fixation of the dye. Care should be taken regarding the compatibility when cationic components are used in pre-treatment mixture.

Wetting agents: These products function the same way as in normal printing recipes helping in wetting out and dye absorption to the substrate. Cationic, anionic or non-ionic wetting agents are available. It has to be used checking the compatibility with the mixture.

Handle modifiers and Softeners: Some of the digital printed materials, especially printed with pigments etc., will affect the handle of the substrate. When the fabric as to undergo wet post treatment and washing the substrate can be treated with softeners as a final finishing treatment, which may not be the case always. In standalone digital printing units separate finishing treatment may not be possible. In such cases, in the pre-treatment softeners can be added which does not affect the fixation and fastness of the ink (dye). One such chemical is silicone softeners.

Pre-treatment recipes given below are guideline purpose only, actually recipes are influenced by design, substrate type and construction, and fixation conditions.

8.13.5.3   Pre-treatment process for different class of dyes and substrates

Reactive dyes

During the production and standardisation of commercial dyes, electrolytes like sodium sulphate or sodium chloride is added to the dye. The presence of these chemicals may corrode the jet nozzles and hence the commercial dyes which is normally used in textile processing are not suitable for inkjet dyeing. Another problem is the low solubility of sodium salt of sulphonates of dyes (Normal reactive dyes are sodium sulphonates). For manufacturing digital inks, in manufacturing the sodium is replaced with lithium wherever necessary and the salt is removed by reverse osmosis process. Reactive inks also contain hygroscopic (or hydrotropic) agents such as diethylene glycol, propylene glycol, and diethyleneglycol monobutyl ether, to avoid drying out of the ink in the nozzles, surfactants and phosphoric acid-based buffers.

Pre-treatment recipe for reactive printing on cotton

The following recipe is applied RFP fabric at 70-80% pick up

Recipe 1	Recipe 2	Unit	Additions
x	x	ml	Water
130	100	g	Urea
150	150	g	Sodium Alginate (6%)*
50	50	g	Wetting agent
30-40	30	g	Sodium Bicarbonate
1000	1000	g	

* Suitable synthetic thickeners can be used alone or in mixture with natural thickeners

Alternate recipe with synthetic thickener

Cotton (Poplin 40 s)	Unit	Additions
5	g/l	Mild oxidising agent (Resist salt)
50	g/l	Synthetic thickener (e.g. Lyoprint RD HT)
10	g/l	PH Adjuster (e.g. Lyoprint AP)
50	g/l	Sodium Carbonate
100	g/l	Sodium Chloride
		Pick up 70%, dry at 100 0C

Recipe for reactive ink on viscose

Viscose	Unit	Additions
25	g/l	Mild oxidising agent (Resist salt)
40	g/l	Synthetic thickener (e.g. Lyoprint RD HT)
10	g/l	PH Adjuster (e.g. Lyoprint AP)
200	g/l	Urea
40	g/l	Sodium Carbonate
		Pick up 80-90 %, dry at 100 0C

Recipe for Silk (Satin)

Silk	Unit	Additions
20	g/l	Mild oxidising agent (Resist salt)
30	g/l	Synthetic thickener (e.g. Lyoprint RD HT)
10	g/l	PH Adjuster (e.g. Lyoprint AP)
70	g/l	Urea
40	g/l	Sodium Carbonate
		Pick up 80-90 %, dry at 100 0C

When bulk processing facility is available, one may pad with 70-90% pick up and dried on stenter and batched into roll size which can be loaded on the printer. But this is not the case with standalone digital printers. For such cases other pre-treatment method can be used. (1) Spraying on to the fabric using a spray bottle or a spray gun and drying is one possibility. (2) Another method is application using screens which is more suitable for chest printing where such screen printing machines are available which, probably, may not be used due to the advent of digital printing.

In any case, after pre-treatment and drying the material may be conditioned equally through out to the same conditions as the digital printing atmosphere.

Printing

Print using digital printer, dry and post treatment as follows:

Steam for 8 min at 102 °C with saturated steam. Rinse for 3 min in cold soft water, and soap at 700C for 30 min with 3 g/l suitable soaping agent, Rinse cold. If necessary soften dry at 125°C

Acid dyes

Acid dyes are water soluble dyes for protein fibres like Silk, wool, polyamide etc., which is fixed in the acid medium. They do not react with the fibre to form covalent bonds, but instead are attracted to positively charged dye sites on the fibre. These dyes are also purified almost the same way as the reactive dyes as explained earlier – RO process to remove the salt content in the dye. Ink formulations contain hygroscopic agents, such as glycerine, diethylene glycol and triethylene glycol monobutyl ether, surfactants, etc.

Pre-treatment recipe is given below. The recipe usually includes an acid donor or non-volatile organic acid (depending also on class of acid dyes used), besides wetting agent and thickener as shown below. Natural or synthetic can be employed.

Digital print, dry and fix by atmospheric steaming at 101-1030C for 30-40 minutes, followed by washing and drying.

Polyamide	Silk	Unit	Additions
x	x	ml	Water
70	70	g	Urea
150	150	g	Bean Gum ether (6%)*
50	50	g	Wetting agent
30-40	30-40	g	Ammonium Sulphate (1:2)
1000	1000	g	

* Suitable synthetic thickeners can be used alone or in mixture with natural thickeners

Alternate recipe with synthetic thickener (Silk satin)

Silk	Unit	Additions
15	g/l	Mild oxidising agent (Resist salt)
30	g/l	Synthetic thickener (e.g. Lyoprint RD HT)
5	g/l	PH Adjuster (e.g. Lyoprint AP)
50	g/l	Urea
30	g/l	Citric acid
		Pick up 70-80 %, dry at 100 0C

PA/Wool	Unit	Additions
100-150	g/l	Urea
200-300	g/l	Synthetic thickener (e.g. Lyoprint RD HT)
10-20	g/l	Ammonium Tartrate

Pre-treatment of woollen fabrics

The fabric has to be prepared initially the same way as it is prepared for normal printing and the additional pre-treatment has to be done on such prepared material for inkjet printing. Pre-treatment process for inkjet printing is necessary for any fabric because (1) all the chemicals required for the fixation, exhaustion, diffusion, print clarity, rheology etc. can affect the stability of the ink and decrease storage stability. (2) Chemicals and salts can corrode the printhead nozzles.(3) Viscosity modifiers can unduly affect the rheological properties of the ink and make it unjettable. (4) As the dyes solution is of high concentration to be able to print all depth of shades large salt concentrations can have a detrimental effect on the solubility of the dye.

SA typical pre-treatment solution applied on the wool fabric for the printing of acid or metal-complex acid dye-based inks is given below:

Quantity	Unit	Additions
150	g	Guar gum or Locust bean gum
100	g	Urea
50	g	Ammonium tartrate (2:1)
x	g	Water to bulk to

When applying reactive inks the following pre-treatment padding is done:

Quantity	Unit	Additions
100	g	Sodium Alginate of medium viscosity
100	g	Urea
x	g	Water to bulk to
1000	g	

During the print fixation by steaming, it is usual that the chlorinated woollen material to turn yellowish which in most cases mask the brightness of the shades. To prevent this an untreated material can be padded with the following solution before ink-jet printing. Such padded material after printing and steaming (104°C for 10 min.) has shown highest colour yields and almost-complete reactive dye fixation and no yellowing of the unprinted white areas.

Quantity	Unit	Additions
20	g	Sodium carboxy methyl cellulose
300	g	Urea (or 100g Formamide)
10	g	Sodium Bisulphite
5-10	g	Wetting agent (e.g., Alcopol O60 - Huntsman)
x	g	Water to bulk to
1000		

Disperse dyes

Disperse dyes are used for synthetic-fibres textiles as finely dispersed particles as they have very limited water-solubility. Dyes are not water soluble and commercially supplied as dispersions. From dispersions the dyes are dissolved in to the fibres at high temperature. These dye dispersions has to be nano-scale particles for excellent runnability on digital printers.

As explained earlier these dyes are directly printed or transfer printed on to the substrate. As per the applicable temperature for fixation disperse inks are supplied as two sets of dyes. One set of dyes of low energy which can be transfer printed using the sublimation fixation method. These dyes are low molecular, more volatile (Sublimation) dyes than the second sets of dyes and naturally the fastness of these dyes are lower. These dyes also can be directly printed on polyester and can be fixed at lower temperature and for less time. Since high definition and tonal prints achieved by transfer printing can be matched by new printers or by printing with higher resolution there is tendency to print these dyes directly and the extra expense of the paper and a second printing operation is avoided. The second set of colours of higher molecular weights can be printed directly but it has to be fixed by dry heat at 2100C for 30s as against the former set dyes which can be fixed at say 150-180°C.

Pre-treatment recipe

Polyester	Unit	Additions
x	ml	Water
50	g	Sodium Alginate (6%)
100	g	Bean Gum ether (6%)*
5	g	Monosodium Phosphate
10	g	Fixation accelerator (Optional)
1000	g	

Alternate recipe

Polyester	Unit	Additions
100-180	g/l	Colour yield improver (e.g., Thermacol MP*)
200-300	g/l	De-areating agent (Lyoprint AIR*)

*Auxiliary by Huntsman

Print (The optimal printing conditions are: 20–22°C and relative humidity ≥60%) dry and fix by dry heat as mentioned above or by high temperature (HT) steaming at 170-180°C.

Post treatment

Rinse 5 min with cold water; Soap 5 min at 40°C with 1 g/l soaping agent and Reduction clearing at 40 - 70°C with

1 - 2 g/l Hydrosulfite

2 g/l Caustic soda

2 g/l Soaping agent

Note: For transfer printing on paper no preparation is needed. Transfer printing from paper to polyester: Transfer/fixation with hot presses or calenders:

30s at 200–210°C. Transfer technique does not require subsequent washing off.

Pigments

Because of the easiness of process, pigment is most used inks in digital printing whether it textiles or other media. The advantage of the pigments is that it can be applied almost all the fabrics. We don't have to explain the method of fixation of the pigments and the function of binder etc. here as it is explained in this book elsewhere.

Pigments are insoluble colouring substances which is supplied as finely divided dispersions, in a similar manner to disperse dyes. The particle size of dispersions supplied for normal printing (rotary) may not be suitable for digital printing. Nowadays, micron size particles size dispersions are available. The pigment (also disperse) inks must be prepared with high degree of expertise so that the particles will not settle or agglomerate (flocculate). The particle size must have an average of 0.5 micrometer and the particle size distribution must be very narrow with more than 99% of the particles smaller than 1 micrometer in order to avoid clogging of the nozzles.

Since pigment printing accounts for over 50% of all conventional textile printing, it is main business for ink manufacturers. Several of the major jet ink producers have recently launched new pigment systems. Although still prone to some problems of handle and rub fastness, they offer excellent wash and light fastness and have the great advantage of universal application to almost all fibres and substrates. The main problem faced is the incorporation of binder which the main ingredient in fixation of pigment particles on substrate. Incorporation binder can cause polymerisation of the binder inside the print nozzles causing blockage. Several different approaches, from spraying resin through a separate jet head to screen printing binder over an inkjet printed colour have been suggested.

With pigment-based inks there are other considerations, because the pigment ink composition is dependent upon the print head technology used. In particular, the pigment must be bound to the textile fabric surface using a binder. In screen printing the binder is put into the print paste but in ink jet printing the binder could be supplied:

in the ink dispersion formulation;

by a separate nozzle (firing channel) system; or

by application of a textile binder after ink jet printing.

Two types of pigment inks are available in the market. One contains pigment only and no binder and one contains pigment and binder. The ink is selected depending on the print head – The pigment ink which contains binder should not be used in Thermal print head (Bubble-jet) (see Table below). Both inks can be directly printed on to the fabric. The first type ink (e.g., Helizarin P – BASF) after printing the material has to be padded with binder and cured. The second type ink, which is more common now (e.g., Irgaphor TBI HC-Huntsman) can be directly printed on the fabric and cured for fixation.

Printing Ink	Low Viscosity Piezo Printhead	High Viscosity Piezo Printhead	Thermal Printhead (Bubble-jet)
With no Binder	Yes	Yes	Yes
With Binder	No	Yes	No
With Special Binder	Possible	Yes	No

Primary requirement of Digital printing of textile is to produce an image comparable, equal or better to the quality of the printed images that are produced by means of Rotary Printing.

8.13.5.4 Printing process

The print quality of digital printing textile depends on

1. Ink and Ink system (Dyes, binders, monomers, or oligomers used in liquid dyes preparation.

2. Print head (Resolution, Jet straightness, nozzle reliability, etc.

3. Substrate: as mentioned earlier each substrate may need different adjustments of heads and approach. Surface properties of the substrate affect penetration of dyes, surface, etc.

4. Pre-treatment and post treatment of the fabric help in the image quality.

5. Hardware Fabric transport and Fixing while printing – Mechanisms supporting these are very important. Motion of the fabric affects registration between dots and cause artifacts like bands or stitching lines.

6. Software: Image Processing

Print heads, ink systems, substrate and pre-treatment have been explained earlier.

Digital printing machine

Hardware

Digital printing system essentially reproduces design patterns by means of a technique similar to digital printing (dietering): by using four (four-colour printing with 4 printing heads) or seven colours (seven-colour printing with 7 printing heads) and by mixing and/or spreading on the fabric dye drops of different colours (this step is controlled completely by special software) different shades can be accurately reproduced. The intensity can be adjusted:

- VDS Process (Variable Dot Size by analogue modulation): the drop size determines the intensity of the colour shade.
- FDS Process (Fixed Dot Size by digital modulation): the number of drops per surface unit determines the intensity of the colour shade.

This system allows (by using 4 or 7 standard colours and the same number of plates with nozzles) a quite easy reproduction of a wide coloured area, but creates problems in reproducing patterns with backgrounds in pastel shades (marked dotting, poorly uniform backgrounds), and very unsatisfactory penetration of pastel colours (poorer quantity of liquid ink). Some problem could arise also for the combination of shades and light colours. It is worth considering that the manufacturers of these printing systems are always more oriented towards nozzles granting more and more sharply defined printing (1200-2000 dots per inch), with always smaller droplets: this is an excellent technique for paper printing but not for textile printing, since very high definitions are not necessary and cause reliability problems (nozzles with very small holes which often clog), poor penetration and slow printing.

In the continuous' ink jet print head technology a charging plate is used to induce an electrical charge on electrically-conductive ink drops. This electrical charge is of opposite charge to that on the deflecting plate. Thus, a continuous stream of droplets is ejected from the nozzles and the charged droplets are deflected by the deflecting plate, collected in a catcher, and then recycled for subsequent printing. The undeflected droplets carry on to the fabric surface to print the design image. In another variant of continuous ink jet printing, a multi-deflection system can be used to give variable charges on the ink drops, which allows a single jet to print multiple positions on the fabric.

As explained earlier, the major type of print head technology currently being used is the DOD piezo-based ink jet system. Basically, this system is an open push-pull system. A pressure wave is created by the piezo crystal that overcomes the pressure lost at the nozzle and the surface tension forces at the meniscus, forming a drop at the nozzle. The pressure drop is also large enough to eject the formed liquid droplet on to the fabric. However, the pressure drop could equally well also force the ink back into the supply pipe system. The force which prevents this happening is the viscosity of the ink. Thus, if the ink viscosity is too low, the ink will be prevented from being ejected from the nozzle, while if it is too high, the suction force of the piezo crystal is too low to pump the ink into the chamber and printing is again prevented. An optimised ink viscosity is therefore required to operate satisfactorily in between these two extreme conditions.

Another problem can arise during the suction phase of the piezo crystal when the pressure in the ink falls. Any dissolved gasses may then form bubbles, interrupting the ink flow and, blocking the nozzles. For this reason, it is necessary to deareate the inks, usually by ultrasonic treatment before filling the ink tanks.

Irrespective of the type of DODink jet printing system, the genesis of the individual droplets at the nozzle outlet follows a similar sequence, forming approximately circular spots when they land on the fabric. As the jetted liquid leaves the nozzle (or jet orifice) the liquid forms a 'tail' which under normal circumstances will eventually collapse into the head of the drop, which becomes spherical before hitting the fabric surface. In some instances, particularly with thermal pulse (bubble jet) systems there can be a tendency for small satellite droplets to be generated from the long tail of the drop, but this does not necessarily affect the printed design. An alternative method of digital ink jet printing is the use of valve-jet technology which can be used on many forms of textiles, including rugs and carpets. The valve jet system differs from the other digital ink jet printing technologies in that it utilises a closed system in which the ink is maintained under pressure in a chamber. The valve is lifted momentarily off the nozzle by an electromagnetic coil which allows an ink droplet to be projected through the nozzle. The nozzle is then resealed by the return of the valve to its original position. The drop size is controllable in this system and the coverage can be varied widely. It has been considered that a 1600 valve jet continuous fabric printing system should be capable of printing fabric at 21 m/min, providing a potential printing rate of around 2772 m²/h.

Ink feeding

It is needless to say, only liquid inks are used. Inkjet printers are made for sampling, small batches and for bulk production yardages. Hence the ink feeding is also as per these requirements of machines. Ink feeding should be in such a way as the flow regulation should be in such a way as it should not choke the print heads and reduce down times due to any reasons related to ink feeding.

Some of the ink feeding systems is shown below:

Refillable catridges   Cans   Bottles

Elec. system   Bulk feed system

Print resolution technique

(A) Different printing heads offer wide range of resolution. _ Industrial digital textile printers usually print at 300 720 DPI.

Lower resolution permit higher printing speed.

300 DPI is applicable in most designs, only intricate designs need higher DPI.

(B) By changing the drop size emerging from print head resolution can be changed effectively. This way resolution can be extended from 360 DPI to 720 or 1440 DPI.

5- level grey scale with three drop sizes.

(C) Print head configuration can change printresolution.

For 90 dpi, if two print heads are aligned exactly, resultant resolution is 90 dpi

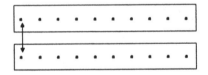

Case 1 : 90 dpi

By offsetting the heads by 1/180 inch, the effective resolution is 180 dpi.

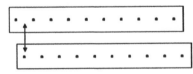

Case 2 : 180 dpi

Textile effects

DPI (or PPI) Dots/Pixels per Inch

If the size of an image is 5 inches horizontally and 4 inches vertically then, What is the size of the image you must create in pixels? (For 300 dpi)

Horizontally : 5 × 300 = 1500 pixels

Vertically : 4 × 300 = 1200 pixels

Changing DPI after creating the design might worse the quality. So the design should be started at the correct resolution. Continuous efforts is being made by manufacturers to improve production on the inkjet printer to make it into a fully bulk production machine.

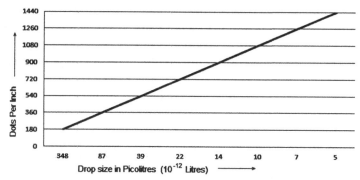

Resolution versus Drop size (for 17.5 ml/sq. metre coverage

An inkjet printer for the professional textile printer should have the technology based on following points ideally

1. Production capacity in the same range of rotary screen printing
2. Image quality equal or better than rotary screen printing
3. Using the same dyes and chemical technology as screen printing
4. Fitting within the existing production process of textile printing
5. Economical cost price.

Continuous flow inkjet technology

The inkjet technology may not be necessarily same as the smaller machines. One of the new technologies developed for bulk printing machine is the continuous flow inkjet head with multi-level deflection (originally developed by Markem. Imaje)

The technology can be illustrated as follows:

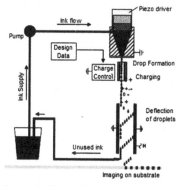

Schematic digram of a continuous flow jet

Software

Software: Normally is CAD-CAM software which can be used for Creation/ Scanning of a design, Colour Calibration, Colour Separation, Control of the Inkjet Printer. (e.g., earlier (RIP Software, Ergosoft, Aleph, Inedit, Hightex Software, etc.)

In a traditional screen printing you have one screen for each colour including the background colour (if not dyed).

Screen printing

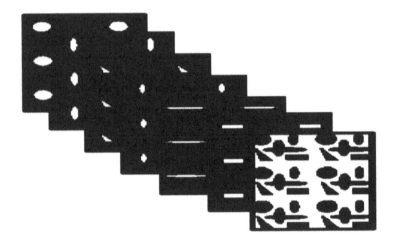

A seven colour design and seven screens

In a digital printing no of colour is no limitation. The number of colours are made by mixing colours in printing. But the colours are generally separated in to basic 4 colours (CYMK). In an inkjet printer different colours are made by mixing of colour dots which can be illustrated as follows:

In the following illustration let the shade at the top is Red and the shade at the bottom Yellow. If the rectangles are divided into a number of dots, the top rectangle all the dots will be red and the bottom rectangle all the dots will be yellow. In between rectangles will have a mix of red and yellow – from top to bottom red in the decreasing order and yellow in the increasing order thus making all the colours between red and yellow, where as in normal screen printing you need seven colours (for the shades shown in the fig.). In inkjet, one can see that the seven colours are made by mixing two colours only (red and yellow) Thus mixing 4 or more colours and making different colours in inkjet printer is infinite.

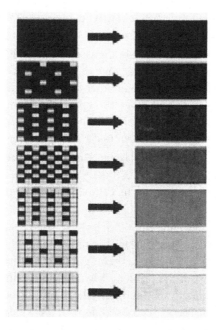

Inkjet Printing (2 colours)   Normal Printing (7 screens)

Colour reproduction in inkjet printing can be represented as follows:

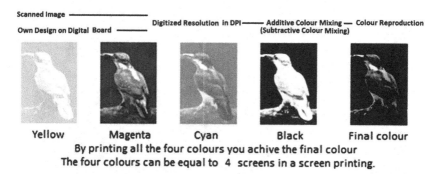

Scanned Image ──────────────
Own Design on Digital Board ────────
Digitized Resolution in DPI ──── Additive Colour Mixing ── Colour Reproduction
(Subtractive Colour Mixing)

Yellow   Magenta   Cyan   Black   Final colour
By printing all the four colours you achive the final colour
The four colours can be equal to 4 screens in a screen printing.

Fabric feeding

In digital printing machines, fabric feeding is of utmost importance, both for printing a perfect image and considering the safety of the print head. The ink should be falling perfectly perpendicular to the fabric surface. This is accomplished by different fabric movement and fabric path in different machines. It is more important in wide format printer and also when printing stretchable fabrics. The fabric batch has to be made as per the specification of the printer. Bulk printing machines there are provisions to feed from A-frames.

Once the printing is over the fabric passes through a heater, which dries the ink on the surface, or ink that has gone through the material, so that it will not continue to spread, drip, or stain another layer of the fabric as the material is wound up on the take-up reel. Even though heating arrangement can reach higher temperature it is safe to keep it   limited to a maximum of 60 degrees C. Most manufacturers give IR heater. (say, 3 Kw)

Fabric can be fed flat, which is held in place sometimes by a gentle vacuum from beneath the bed. This mostly used in case of T-shirt either as cut pieces or as T-shirt itself. Patterns can be printed along the length or as a "placement" print in a specific spot.

When a printer which can print fabric rolls it is fed from rolls through side guide and tension adjustments so that it is presented to the print heads clean without creases, and as per the required height from the print head which also can be adjusted in most of the printers. Most modern printers incorporate an "endless" belt or "blanket" that supports the fabric throughout the process as in case of analog printers. Such belts are often coated with adhesive like thermoplast which makes the belt little sticky to help the fabric temporarily hold but easily release fabrics. This is more useful for stabilising knit or stretch fabrics as well as helping to prevent expansion and shrinkage caused by the wet ink soaking into the fibres.

Fabric Roll   Fabric Tension Adjuster

Fabric feeding in a Digital Printer

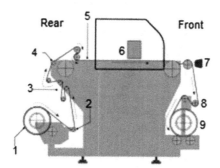

1. Feeding roll, 2. Tension roll, 3. Tensioner, 4. Guide roll, 5. Fabric laid on the transport belt, 6. Printhead, 7. IR drier, 8. Tensioner, 9. Printed roll.

Fabric Threading

Following care has to be taken so that the print head is taken care off and printer prints the image properly:

1. The rolls taken for printing should have no wrinkles or puckering and creases.
2. If the fabric is not flat over the print table the resulting printer places uneven prints at trough or ridges of the fabric. Padding with a suitable thickener helps in this matter (see padding recipes).
3. Very stiff fabric may not lie flat in the belt, and hence care has to be taken that the thickener percentage in padding recipe is not too high.
4. There should not be bowing or skewing in the fabric and the grain of the fabric should be straight so that the print is also placed straight in the fabric so that it does not pose a problem in cutting and sewing later.

Production printers or bulk printing units

Digital printers have come of age, is now is used as production machines which can almost meet the productivity of bulk rotary machines. Even though, the printers, when it was introduced, it was used for strike offs and photo shoot samples, it has soon emerged as production machines with many advantages, thanks to the technological advances in digital technology. Today the customer requirement per design/colourway has come down drastically come down and cost per meter is increased. In such a scenario the digital printing has come more relevant. Thus bulk production machines are becoming more and more popular. In such machines the basic technology remains same but there are number of print head has been increased for higher productivity

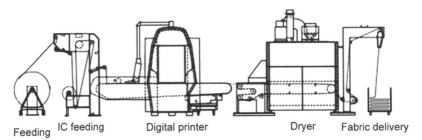

Feeding   IC feeding   Digital printer   Dryer   Fabric delivery

Fabric path in bulk digital printer

Given below the specification for a bulk production machines

Ink setup for CMYK, CMYK+2 or CMYK+4 or with 2 × CMYK for double speed printing

  Print head – Addressable jets -1024, Drop size: 8-170pl

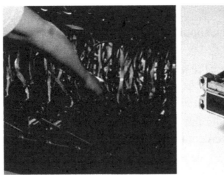

Print head arrays in Colaris (Zimmer)   Single Printhead

In textile printing most important is to have the right ink amount to get a good penetration and vibrant colours. This amount is normally specified by test prints. The ink depositionin a particular point is specified by the following parameters:

  Resolution in X and Y axis

  Dropsize in picoliters

  Number of passes (number of passes is used to increase the x-dpi and therefore to increase ink amount)

  The ink deposition on a single colour can be calculated as follows:

  Ink lay-down (ml/m2) = X-dpi*1000mm/25.4mm * Y-dpi*1000 mm/25.4mm * dropsize (pl) * number of passes /10-8

Simplified: Lay-down (ml/m2) = x-dpi * y-dpi * dropsize / 645 000

Example

2 pass with smallest drop size: 400 dpi * 800 dpi * 8pl / 645 000= 4 ml/m2

3 passes with biggest drop size: 400 dpi * 1200 dpi * 150pl / 645 000 = 112 ml/m²

The actual ink deposition of a design is in most cases below the single colour max. ink deposition

The print speed depends on the following parameters:

• Required ink deposition
• Number of passes
• Print coverage
• Resolution

Below table gives the production in m2/h for the same design on various print specification to give an idea about the relation between these:

Print Specifications	Production m²/h (depend on coverage)	Passes	Resolution (dpi) x - axis * y - axis	Maximum ink Deposition (g/m²)
Regular Uni-directional	185-230	single	1200 x 1200	16
Regular Bi-directional	270-330	single	1200 x 1200	16
Dual Uni-directional	95-115	Double	1200 x 1200	32
Dual Bi-directional	135-150	Double	1200 x 1200	32
Fast uni-directional	260-325	Single	600 x 1200	8
Fast Bi-directional	455-540	Single	600 x 1200	8
Dual Fast Uni-directional	130-160	Double	600 x 1200	16
Dual Fast Bi-directional	235 -295	Double	600 x 1200	16

One has to play between the parameters to get maximum output for a given design.

When you compare the performance of inkjet printers you should always use a printed sample and compare - print quality, penetration and the speed achieved on this sample.

The theoretical print speed using a single pass is in many cases not sufficient to reach a good quality.

The fabric is fed to the printer on an endless blanket the fabric is flat and faultless when it approaches the print heads. This is more important since the printing is done in an array of print heads. The blanket will be run by drive rollers after leaving the fabric to the drier goes down below the printing plain and passes through a washing and drying unit. As the blanket comes to the top the fabric is pressed on to the blanket by a press roller to allow it to lay flat on the blanket surface. This supported by thermoplast gluing device with IR heater incl. blanket surface temperature controller.

Different special pre-treatment, press station, fabric drying, fixation units etc., can be combined in line to get the desired print results. The production printers available with 4, 6 or 8 colour set-up, with up to 16 print heads or even more per colour for higher productivity.

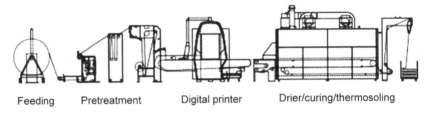

Feeding   Pretreatment   Digital printer   Drier/curing/thermosoling

Production printer with inline pretreatment

Fabrics – Post treatment
Every printed fabric needs a post treatment to complete the process. The fixation process is done as per the applied class of colours.

Reactive and acid dyes
Steaming allows the fibers within the fabric to open up and allow the dyes to be fixed. Steaming for a short time (8-10 min at 102°C) used for producing prints with weak colours. For stronger, more vibrant colours, more steaming time is necessary.

It can be also fixed by baking, making necessary changes in the pre-treatment padding recipe if a steamer is not available.

Disperse dyes
Disperse inks can be fixed by dry heat (thermosoling) on a stenter at 200-210 0C, in a universal ager with high temperature steam, or by pressure steaming (explained under printing with disperse dyes in this book). When printed with

transfer printing method fixation can be done by contact heat of about 160-1800C. If low energy disperse dyes are used it may be passed through dry heat at 160-1700C

Pigments

Pigments are printed and padded with binder, if not printed without binder in the ink, and may be fixed by curing at 160-165 0C. Curing condition is employed for the polymerisation of the binder, which will fix the pigment particles on the surface. The polymerisation also can be initiated by controlled exposure to UV light using UV curable inks, which is more used in the digital garment printing. The process is more environmental friendly since this method uses minimal heat and also generates minimal solvent evaporation. UV curing systems are more economic as it require less space than conventional fixing systems, and the production of the UV light by light-emitting diode lamps requiring less energy.

Washing off

Reactive and acid dyes needs a washing of process to remove the unfixed dyes. In case of disperse dyes need a reduction clear and soaping but the pigment printing do not need a washing off process and hence this method is more popular in digital printing especially garment printing.

Printed in the United States
by Baker & Taylor Publisher Services